BCC/UCF LIBRARY, COCOA, FL 32922-6598

Flow Measurement

Flow
Measurement

Béla G. Lipták

EDITOR-IN-CHIEF

CHILTON BOOK COMPANY | RADNOR, PENNSYLVANIA

To my colleagues, the process control and instrument engineers,
with the hope that they will make the world a better,
safer and happier place by applying the knowledge
found on these pages

Copyright © 1993 by Béla G. Lipták
All Rights Reserved

Published in Radnor, Pennsylvania 19089, by Chilton Book Company

No part of this book may be reproduced, transmitted or stored
in any form or by any means, electronic or mechanical,
without prior written permission from the publisher

Designed by Arlene Putterman
Manufactured in the United States of America

Library of Congress Cataloging-in-Publication Data
Flow measurement / Béla G. Lipták, editor-in-chief.
    p. cm.
   "An offprint book from the third edition of Instrument engineers'
handbook"—CIP data sheet.
   Includes index.
   ISBN 0-8019-8386-X (pbk.)
    1. Process control—Equipment and supplies. 2. Flow meters.
I. Lipták, Béla G. II. Title: Flow Measurement
TS156.8.F646   1993
681'.2—dc20   92-56585
                                                                          CIP

1 2 3 4 5 6 7 8 9 0   2 1 0 9 8 7 6 5 4 3

CONTENTS

CONTRIBUTORS ix

INTRODUCTION xi

ORIENTATION TABLES 2

1
APPLICATION AND SELECTION 5
Meter Selection 5
Differential Pressure 5
    Energy Costs 5
    Orifice Plates 6
    Venturi Tubes and Nozzles 6
    Pitot Tubes 6
    Elbow Taps 6
    Target (or Impact) Meters 7
Electromagnetic Flowmeters 7
Turbine Meters 7
Vortex Meters 7
Variable-Area Meters 7
Positive Displacement Meters 8
Ultrasonic Meters 8
Metering Pumps 8
Mass Flowmeters 8
Cross-Correlation 9
Sonic Venturi 9
Specialized Flowmeters 9
Flowmeter Selection Procedure 9
    Inaccuracy 14
    Reynolds Numbers 14
    Safety 15
    Installation 15
    Cost 16

2
BTU FLOWMETERS FOR HEAT EXCHANGERS 17
Mechanical BTU Meters 18
Electronic BTU Meters 18

3
BTU FLOWMETERS FOR GASEOUS FUELS 19
Measuring Heat Flow 19
The BTU Flowmeter Loop 19
Applications 20
Conclusions 21

4
CROSS-CORRELATION FLOWMETERING 22

5
ELBOW TAPS 24

6
FLOW SWITCHES 26
Design Variations 27
Solids Flow Swtiches 28

7
JET DEFLECTION FLOW DETECTORS 30
Hop-Tapping 31
Conclusion 32

8
LAMINAR FLOWMETERS 33
Theory 34
    Hagen-Poiseuille Law 34
    Design Parameters 34
    Design Calculations for Liquid Service 35
Errors 35
Range Extension Techniques 36
Commercially Available Units 38
Conclusions 38

9
MAGNETIC FLOWMETERS 39
Theory 39
AC and DC Excitation 40
    Dual-Frequency Excitation 41

Construction 41
   Ceramic Liners 43
   Probe-Type Units 43
   Magmeter Electronics and Intelligence 45
Capacity and Range 45
Applications 45
Installation 47
Advantages 48
Limitations 49

10
MASS FLOWMETERS—CORIOLIS 50

Theory 50
Construction 53
   Sensor 53
   Electronics 56
Calibration 56
   Zero Offset or Zero Stability 59
   Inaccuracy 59
Fluid Properties 59
   Temperature 59
   Density 59
   Pressure 59
   Viscosity 60
   Pressure, Plugging, Erosion, Corrosion 60
Environmental Conditions 61
   Temperature 61
   Humidity 61
   Vibration 61
   Pipe Stress 61
Installation 61
Applications 62
Advantages 62
Limitations 63

11
MASS FLOWMETERS—MISCELLANEOUS 64

Radiation-Type Mass Flowmeters 64
Angular-Momentum-Type Mass
   Flowmeters 65
   Impeller-Turbine 65
   Constant-Torque-Hysteresis Clutch 66
   Twin-Turbine 66
   Coriolis 66
   Gyroscopic 67
Linear 67
Conclusions 67

12
MASS FLOWMETERS—THERMAL 68

Heat Transfer Flowmeters 68
   Bypass-Type Designs 70
Hot Wire Probes 71

13
METERING PUMPS 73

Peristaltic Pumps 74
Piston Pumps 74
Diaphragm Pumps 75
   Hydraulic-Actuated Metering Pumps 76
   Pulsator-Head Pumps 77
Proportioning Pumps 77
Conclusions 78

14
ORIFICES 79

Head Meters 79
   Theory of Head Meters 80
   Head Meter Characteristics 81
   Reynolds Number 81
   Compressible Fluid Flow 82
   Choice of Differential Pressure Range 82
   Pulsating Flow and Flow "Noise" 82
The Orifice Meter 83
   Flow Through the Orifice Plate 83
   Location of Pressure Taps 84
Eccentric and Segmental Orifice Plates 85
Quadrant Edge and Conical Entrance Orifice
   Plates 86
The Integral Orifice 87
   Installation 88
   Limitations 90
Orifice Bore Calculations 91
Orifice Metering Accuracy 95

15
PITOT TUBES, AVERAGING AND DUCT SECTION UNITS 97

Theory 97
Static Pressure Measurement 98
Single-Ported Pitot Tube 99
   Calibration of Pitot Tubes 100
Multiple-Opening Pitot Tubes 100
   Pressure Differential Produced 103
Area-Averaging Pitot Stations 104
Special Pitot Tubes for Pulsating Flow 105

16
POSITIVE DISPLACEMENT GAS FLOWMETERS 107

The Diaphragm Meter 107
The Lobed Impeller 108
High-Precision Gas Flowmeter 109
Application Notes 109
Testing and Calibration 110

17
POSITIVE DISPLACEMENT LIQUID METERS AND PROVERS 111

Nutating Disk 111
Rotating Vane 112
Oscillating Piston 112
Reciprocating Piston 113
Rotating Lobe and Impeller 113
Oval-Gear Flowmeters 114
Viscous Helix 114
High-Precision Displacement Flowmeter 115
Accessories and Intelligent Electronics 115
Provers 115
Positive Desplacement Flowmeter Summary 116

18
PURGE FLOW REGULATORS 118

Purge Rotameters 118

19
SEGMENTAL WEDGE FLOWMETER 121

20
SIGHT FLOW INDICATORS 124

21
SOLIDS FLOWMETERS AND FEEDERS 127

Hoppers and Their Accessories 128
Throttling the Flow to the Feeders 129
    Vertical Gate 129
    Rotary Vane 130
    Screw Feeder 130
    Vibratory Feeder 131
    Shaker Feeder 131
    Roll Feeder 131
    Revolving Plate Feeder 132
Belt Feeders 132
    Feed Rate Control 133
    Belt Speed Control 134
    Digital Control 135
Vertical Gravimetric Feeder 136
Nuclear Belt Scales 137
Loss-in-Weight Flowmeters 137
    Equipment 138
    System Sizing 138
Dual-Chamber Gravimetric Feeder 139
Dynamic Solids Flowmeters 139
    Impulse-type Solids Flowmeters 139
    Accelerator-type Flowmeter 140
    Volumetric Flowmeters 141

22
TARGET METERS 142

23
TURBINE AND OTHER ROTARY ELEMENT FLOWMETERS 144

Liquid Turbine Meters 145
    Electronic Display Units 145
    Linearity and Repeatability 146
    Viscosity and Density Effects 147
    Meter Sizing 147
    Pelton Wheel Meters 149
    Meter Characteristics and Features 149
    Mechanical Installation 150
    Electrical Installation 151
Gas Turbine Meters 151
Impeller and Shunt Flowmeters 152
Insertion-type Flowmeters 153
    Optical Flow Sensors 154
    Paddlewheel Flowmeters 154

24
ULTRASONIC FLOWMETERS 155

Transit-Time Flowmeters 156
    Frequency Difference-type 156
    Flowmeter Construction 157
    Application and Performance 158
Doppler Flowmeters 158
    Application and Performance 159
Displays, Receivers, and Intelligent Units 159

25
VARIABLE-AREA, GAP, AND VANE FLOWMETERS 162

Rotameters 163
    Sizing 164
    Rotameter Characteristics 164
    Rotameter Types 165
    Bypass and Pitot Rotameters 166
Tapered Plug and Piston Meters 167
Gates and Vanes 167

26
V-CONE FLOWMETER 170

27
VENTURI TUBES, FLOW TUBES, AND FLOW NOZZLES 172

The Classic Venturi 173
Short-Form Venturis 173
    Installation 174
    Flow Calculations 175
Flow Tubes 175
Flow Nozzles 176
    Application Considerations 178
    Critical-Velocity Venturi Nozzles 178
Accuracy 178

Differential Pressure Measurement   179
Conclusions   179

28
VORTEX AND FLUIDIC FLOWMETERS   181

The Vortex Shedding Phenomenon   182
The Detector   182
   Features   184
   Selection and Sizing   184
   Installation Requirements   188
Vortex Precession (Swirl) Meters   188
Fluidic (Coanda Effect) Meter   189
   Characteristics   189
Oscillating Vane in Orifice Bypass   190
Conclusions   191

29
WEIRS AND FLUMES   192

Weirs   192
The Parshall Flume   194
The Palmer Bowlus Flume   195
The Kennison Nozzle, Parabolic Flume, and
   Leopold Lagco Flume   195
Detectors for Open-Channel Sensors   196

A.1 INTERNATIONAL SYSTEM OF UNITS   197

A.2 ENGINEERING CONVERSION FACTORS   198

INDEX   208

CONTRIBUTORS

CATHY APPLE	BSChE, Project Engineer, Micro Motion Inc.
JAMES B. ARANT	BSChE, Senior Consultant, E.I. du Pont de Nemours Co.
GILES M. CRABTREE	BSEE, PE, Principal Engineer, GIMACA Engineering
LOUIS D. DINAPOLI	BSEE, MSEE, Director, Flowmeter Marketing and Technology, BIF Products of Leeds & Northrup Co.
ALBERT P. FOUNDOS	BSChE, MBA, President, Fluid Data Inc.
JOEL O. HOUGHEN	PhDChE, PE, Consultant, Professor Emeritus, University of Texas
WILFRED H. HOWE	BSEE, MBA, PE, Chief Engineer, The Foxboro Co.
DAVID S. KAYSER	BSEE, Senior Instrument Engineer, Texas City Refining Inc.
JOHN G. KOPP	BSME, PE, Senior Product Marketing Manager, Fischer & Porter Co.
BÉLA G. LIPTÁK	ME, MME, PE, President, Lipták Associates P.C.
DAVID J. LOMAS	Marketing Support Executive, Kent Process Control Ltd.
DAVID C. MAIR	BCE, PE, Manager, Sales Services, Wallace & Tiernan Div. of Pennwalt Corp.
KURT O. PLACHE	BSChE, PE, Vice-President, Marketing, Micro-Motion Inc.
MICHAL PTACNIK	PhD, Aeronautical Research and Test Institute, Czechoslovakia
STEPHAN RUDBÄCH	MSc, President, Matematica AB, SWEDEN
ROBERT SIEV	BSChE, MBA, CE, Engineering Specialist, Bechtel Corp.
JOAN B. STODDARD	PhD, President, Stoddard Productivity Systems Inc.
EUGENE L. SZONNTAGH	MSChE, PhD, PE, Consultant

INTRODUCTION

This is one of three "spinoff" books from the *Process Measurement and Analysis* volume of the *Instrument Engineers' Handbook*. Its purpose is to provide timely and cost-effective information on selected subject matters. The *Instrument Engineers' Handbook* was first published in 1968–69, with a supplemental volume issued in 1972; then it was completely revised in 1982–85. The third complete revision is scheduled for 1994–95. While a complete handbook cannot be updated more often than once a decade, some of its chapters can. The chapters covering topics that are fast changing, and should therefore be updated more often, were selected as "spinoff" volumes.

Another consideration in selecting chapters for separate publication was economic: If a reader works with only one category of instruments on a daily basis, it is probably efficient to visit the company library when information is needed on other categories of devices. But the chapter dealing with the area of his or her specific interest, should be permanently on that reader's desk. Publishing "spinoff" volumes provides readers with this flexibility at low cost.

In selecting subjects for separate publication as "spinoff" volumes, there was little question that flow measurment should be one of them. This is partly because flow is the most often used sensor partly because it provides the most critical measurements for recipe formulation and material balance control, but, most importantly, because it is a fast-changing technology. The addition of the microprocessor has changed—and is still changing—the capabilities of flow transmitters. When an international fieldbus standard results in the same kind of standard digital signal as the 4–20 mA signal in the analog age, each smart flow transmitter will be able to be directly connected to any DCS system. The smart transmitters will improve accuracy not only because they eliminate the need for A/D conversion, but also because they will be able to self-calibrate their own spans and will be able to match the selected span to the actual flow rates occuring in the process. The end result will be a combination of reduced measurement error and higher rangeability. The time is near when a smart d/p cell will not only reduce its own span when the flow is low, but will also offer built-in PID algorithms, inaccuracies of 0.1% of actual reading, and rangeabilities of 40:1.

It is also time for the "smart" pitot tube and ultrasonic flowmeter to be marketed. I expect that, in the next few years, microprocessor-based "smartbars" will be introduced, having the capability to measure two-point velocities, or a point velocity and the Reynolds Number: Using these measurements, the microprocessor will accurately calculate the prevailing velocity profile and, therefore, flow. I also expect that the transit-time and the Doppler-type ultrasonic flowmeters will merge into a "smartsonic" meter which switches its own mode of operation between transit and Doppler, depending on the characteristics of the process fluid.

Other changes that are expected in the field of flow measurement include more extensive use of fiber optics; further improvements in the dual-crystal sensors of vortex shedding sensors; Coriolis mass flowmeters with less sensitivity to temperature and vibration; orifice d/p cells which automatically match their range to the actual process flows; multi-opening pitot tubes which use a microprocessor to calculate the velocity profile across the pipe; cross-correlation flowmeters for hard-to-handle process materials; and flow sensors for multi-phase process streams, separately detecting solid, liquid, and gas phases. I hope that we will be able to report on these developments, and also on the changes in the international marketplace, more frequently in these spinoff volumes.

Béla G. Lipták
Stamford, Connecticut
June, 1993

Flow Measurement

TABLE A
Orientation Table for Flow Sensors

Type of Design	Applicable to Detect the Flow of					Direct Mass—Flow Sensor	Volumetric Flow Detector	Flow Rate Sensor	Inherent Totalizer	Direct Indicator	Transmitter Available	Linear Output	Rangeability	Pressure Loss Thru Sensor	①Approx. Straight Pipe-Run Requirement (Upstream Diam./Downstream Diam.)	Accuracy * ± % Full Scale / ** ± % Rate / *** ± % Registration	FLOW RANGE
	Clean Liquids	Viscous Liquids	Slurry	Gas	Solids												
Elbow Taps	√②	L	L				√	√			√	SR	3:1②	N	②25/10	5–10*	gpm —— m³/hr
Jet Deflection	√			√			√	√				√	25:1	M	②20/5	2*	———— SCFM—Sm³/hr
Laminar Flowmeters	√	√		√			√	√			√	√	10:1	H	15/5	½–5*⑩	gpm—m³/hr / SCFM—Sm³/hr
Magnetic Flowmeters	√⑦	√⑦	√⑧				√	√			√	√	10:1⑤	N	5/3	½**–2*	gpm—m³/hr
Mass Flowmeters, Misc. Coriolis	√ √	√ √	√ L	√ √	SD	√ √		√ √	SD SD	SD SD	√	√ √	100:1 / 20:1	A H	N N	½** / 0.15–½**	lbm/hr–kgm/hr / SCFM—Sm³/hr
Metering Pumps	√	L					√	√	√		SD	√	20:1	—	N	¹⁄₁₀–1*	gpm—m³/hr
Orifice (Plate or Integral Cell)	√	L	L	√			√	√			√	SR	3:1②	H	②20/5	½**–2*	gpm—m³/hr / SCFM—Sm³/hr
Pitot Tubes	√		L	√			√	√			√	SR	3:1②	M	②30/5	0.5–5*	gpm—m³/hr / SCFM—Sm³/hr
Positive Displacement Gas Meters				√			√	√	√	√	SD	√	10:1 to 200:1	M	N	½–1***	———— SCFM—Sm³/hr

Flow Meter Type										Ratio		N	%	Units
Positive Displacement Liquid Meters	✓									10:1①	H	N	0.1–2**	gpm—m³/hr
Segmental Wedge	✓			✓			✓			3:1	M	15/5	3**	gpm—m³/hr
Solids Flowmeters		SD	SD	✓	✓	SD				20:1	—	5/3	½**–4*	lbm/hr—kgm/hr
Target Meters	✓	L	✓	✓		SD	SR			4:1	H	20/5	0.5*–5*	gpm—m³/hr / SCFM—Sm³/hr
Thermal Meters (Mass Flow)	✓	L	✓	✓			L			20:1②	A	5/3	1–2*	gpm—m³/hr / SCFM—Sm³/hr
Turbine Flowmeters	✓	L	SD	✓			✓			10:1⑥	H	15/5⑦	¼**	gpm—m³/hr / SCFM—Sm³/hr
V-Cone Flowmeter	✓	L	✓	✓			SR			3:1②	M	2/5	½*–2**	gpm—m³/hr / ACFM—Sm³/hr
Ultrasonic Flowmeters — Transit	✓	L	L	✓			✓			20:1	N	⑦15/5	1**–2*	gpm—m³/hr / SCFM—Sm³/hr
Ultrasonic Flowmeters — Doppler		L	L	✓			✓			10:1	N	⑦15/5	2–3*	
Variable-Area Flowmeters	✓	L	✓	✓		✓	✓			5:1	A	N	½*–10**	gpm—m³/hr
Venturi Tubes	✓	L	✓	✓			SR			3:1②	M	⑦15/5	½**–1*	gpm—m³/hr / SCFM—Sm³/hr
Flow Nozzles	✓	L	✓	✓			SR			3:1②	H	⑦20/5	1**–2*	
Vortex Shedding	✓		✓	✓			✓			10:1⑧	H	20/5	0.5–1.5**	gpm—m³/hr / ACFM—Sm³/hr
Fluidic	✓		✓	✓			✓			20:1⑧	H	20/5	1–2*	
Oscillating	✓		✓	✓			✓			10:1⑧	H	20/5	0.5*	
Weirs, Flumes	✓	L		✓		✓	SD			100:1	M	See Text	2–5*	gpm—m³/hr

---- = Non Standard Range
L = Limited
SD = Some Designs
H = High
A = Average
M = Minimal
N = None
SR = Square Root

① = The data in this column is for general guidance only.
② = Inherent rangeability of primary device is substantially greater than shown. Value used reflects limitation of differential pressure sensing device, when 1% of actual flow accuracy is desired. With multiple-range intelligent transmitters the rangeability can reach 10:1.
③ = Pipe size establishes the upper limit.
④ = Practically unlimited with the probe type design.
⑤ = Must be conductive.
⑥ = Can be re-ranged over 100:1.
⑦ = Varies with upstream disturbance.
⑧ = Can be more at high Re. No. services.
⑨ = Up to 100:1 with high precision design.
⑩ = Commercially available gas flow elements can be ±1% of rate.
⑪ = More for gas turbine meters.

TABLE B

Conversion of Volume or Flow Units

To Convert	Into	Multiply by	To Convert	Into	Multiply by
cubic feet	bushels (dry)	0.8036	gallons	cu. meters	3.785×10^{-3}
cubic feet	cu. cm	28,320.0	gallons	cu. yards	4.951×10^{-3}
cubic feet	cu. in.	1,728.0	gallons	liters	3.785
cubic feet	cu. meters	0.02832	gallons (liq. Br. Imp.)	gallons (U.S. liq.)	1.20095
cubic feet	cu. yards	0.03704	gallons (U.S.)	gallons (Imp.)	0.83267
cubic feet	gallons (U.S. liq.)	7.48052	gallons of water	pounds of water	8.3453
cubic feet	liters	28.32	gallons/min	cu. ft/sec	2.228×10^{-3}
cubic feet	pints (U.S. liq.)	59.84	gallons/min	liters/sec	0.06308
cubic feet	quarts (U.S. liq.)	29.92	gallons/min	cu. ft/hr	8.0208
cubic feet/min	cu. cm/sec	472.0	kilograms	dynes	980,665.
cubic feet/min	gallons/sec	0.1247	kilograms	grams	1,000.0
cubic feet/min	liters/sec	0.4720	kilograms	poundals	70.93
cubic feet/min	pounds of water/min	62.43	kilograms	pounds	2.205
cubic feet/sec	million gals/day	0.646317	kilograms	tons (long)	9.842×10^{-4}
cubic feet/sec	gallons/min	448.831	kilograms	tons (short)	1.102×10^{-3}
cubic meters	cu. ft	35.31	pounds	drams	256.
cubic meters	cu. in.	61,023.0	pounds	dynes	44.4823×10^{4}
cubic meters	cu. yards	1.308	pounds	grains	7,000.
cubic meters	gallons (U.S. liq.)	264.2	pounds	grams	453.5924
cubic meters	liters	1,000.0	pounds	kilograms	0.4536
cubic meters	pints (U.S. liq.)	2,113.0	pounds	ounces	16.0
cubic meters	quarts (U.S. liq.)	1,057.	pounds	ounces (troy)	14.5833
gallons	cu. cm	3,785.0	pounds	poundals	32.17
gallons	cu. ft	0.1337	pounds	pounds (troy)	1.21528
gallons	cu. in.	231.0	pounds	tons (short)	0.0005

1 Application and Selection

D. J. LOMAS (1982) **B. G. LIPTÁK** (1993)

METER SELECTION

The variety of choices facing an engineer confronted with a flow measurement application is vast. Table A lists many different categories of flowmeters. In nearly every case each category further subdivides into several distinctly different variants. For example, the positive displacement principle types include rotary piston, oval gear, sliding vane, and reciprocating piston. If these subvariants are included, the engineer faces a confusing list of well over 50 different meter types. Each type has advantages and limitations and no one type combines all the features and all the advantages.

A brief summary of the relevant advantages and limitation of each basic category is given below. Comparison tables on some of the key parameters, such as accuracy, flow range, and operating temperature, are also provided. It is hoped that this information, together with the basic selection philosophy outlined, will assist in finding a way through the maze of meter types and in selecting suitable meters for specific applications.

Having narrowed the list of choices, the engineer can then consult the other sections for detailed information on the meters concerned.

DIFFERENTIAL PRESSURE

This is undoubtedly the most widely used method of industrial flow measurement. A restriction is introduced into the pipe. The resulting pressure decrease is proportional to flow rate in accordance with the formula:

$$Q(\text{flow}) = K(\text{constant}) \sqrt{\frac{h(\text{differential head})}{d(\text{fluid density})}} \quad 1(1)$$

Differential pressure meters have the advantage that they are the most familiar of any meter type. They are suitable for gas and liquid, viscous and corrosive fluids. There are no significant pipe size or flow rate limitations and they have no moving components. All differential pressure meters exhibit a square law relationship between head and flow rate which severely limits the usable flow range (typically 4:1 maximum). Another disadvantage is that in addition to the sensor element, a separate transmitter is required. Furthermore, a manifold and shutoff valves are required to attach the transmitter to the sensor (causing potential leakage points), and the fluid density must be known or measured.

Energy Costs

The yearly energy cost of operating differential-pressure-type flowmeters can in some cases exceed the purchase price of the meter. The permanent pressure loss through a flowmeter is usually expressed in units of velocity heads. The velocity head is calculated as $V^2/2g$, where V is the flowing velocity and g is the gravitational acceleration (9.819 m/s² or 32.215 ft/s² at 60 degrees latitude).

Therefore, the velocity head at, say, 10 ft/s flowing velocity is calculated in the English units as $10^2/64.4 = 1.55$ ft of the flowing fluid. If the flowing velocity is 3 m/s, the velocity head is calculated in the metric units as $3^2/19.64 = 0.46$ meters of the flowing fluid. The velocity head is converted into pressure drop by multiplying it with the specific gravity of the flowing fluid. Different flowmeters require different pressure drops for their operation:

Flowmeter Type	Permanent Pressure Loss (in Velocity Heads)
Orifice plates	Over 4
Vortex shedding	Approximately 2
Positive displacement	1 to 1.5
Turbine flowmeter	0.5 to 1.5
Flow tubes	Under 0.5

One can calculate the yearly operating cost of any flowmetering installation by using the following equation:

$$\$/\text{yr} = (C)(\$/\text{KWH})(OT)(dP)(F)(SpG)/(\%) \quad 1(2)$$

where

- C = a correction factor for the units used (1.65 if flow is in GPM and pressure loss is in feet)
- $/KWH = the unit cost of electricity in the area
- OT = the operating time of the meter (1.0 if operated continuously)
- dP = the pressure loss; can be estimated on the basis of the velocity heads required by the particular meter (units are feet or meters)
- F = the flow rate (units are GPM or m³/s)
- SpG = specific gravity (water = 1.0)
- % = the efficiency of the pump (or compressor) expressed as fraction (70% = 0.7)

Example: An orifice sized for 100 in. H₂O pressure drop (dP = 8.34 ft) in 16 in. schedule 40 steel pipe is measuring the flow of 5000 GPM of water flow. The meter is operating continuously (OT = 1.0), the cost of electricity is \$0.1/KWH, and the pump efficiency is 60% (% = 0.6). What is the yearly cost of operation?

$$\$/yr = (1.65)(0.1)(1.0)(8.34)(5000)(1.0)/(0.6)$$
$$= \$11,467 \text{ per year} \quad \quad \textbf{1(3)}$$

If the cost of electricity is \$0.1/KWH and the pump efficiency is 60%, an estimate of the cost of any continuous pressure drop in any water pumping system can be calculated as:

$$\$/yr = (0.635)(GPM)(PSID) \quad \quad \textbf{1(4)}$$

When selecting a flowmeter, one should consider not only the purchase and installation cost, but also the operating cost during the life of the flowmeter. As was shown above, a major component of the operating cost of flowmeters is their pumping/compressor energy costs.

Orifice Plates

Orifice plates are the simplest and cheapest types of differential pressure flow elements. The system cost is relatively independent of pipe diameter because the differential pressure transmitter remains at a fixed price regardless of pipe size. Orifice installations are consequently relatively expensive in small pipe sizes but very economical in pipe sizes above 6 in. (150 mm). Availability in a variety of materials and types (concentric, segmental, eccentric) enables use on a wide range of applications. One advantage is that even when the orifice plate is badly worn or damaged, it will provide a reasonably repeatable output, albeit significantly inaccurate. Another convenient feature is the ability to work on or replace the transmitter, without removing the orifice or interrupting the process flow.

The limitations include high irrecoverable pressure loss (40 to 80% of generated head); deterioration in accuracy and long-term repeatability with edge wear or deposition; maintenance required due to potential manifold leakage, tapping blockage, etc.; and low flow range and accuracy. (See Figure 1a.) With multirange intelligent d/p cells, a 1% of actual flow accuracy over a 10:1 range can be reached.

Orifice-type flow measurement has been modified and new, special-purpose devices have been introduced to meet particular requirements. One such unique design is the annular orifice used, for example, to measure the hot and dirty gases in the steel industry. Here the process flow passes through an annular opening around a disk-shaped, concentrically located plate and the pressure difference is detected between the up- and downstream faces of that disk.

For slurry flow detection such as the measurement of paper pulp, the segmental and eccentric orifices, Venturi-Cones, and the segmental wedge elements have been developed. The Venturi-Cone is shaped as a restriction in the center of the flow path, forcing the flowing stream into an annular space between the cone and the pipe. The segmental wedge element is produced by forming an indentation in the top of the pipe and thereby restricting the flow passage. This

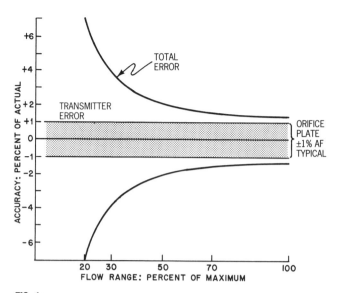

FIG. 1a
Total error of an orifice type flow measurement, using a ±½% full-scale d/p cell, is shown as a function of actual flow.

sensor is used on dirty fluids and fluids at higher temperatures.

Venturi Tubes and Nozzles

These tubes use specially designed and engineered shapes to give a low head loss. They are also more resistant to abrasion than an orifice plate and can be used on dirty fluids and slurries. They are, however, considerably larger, heavier, and more expensive than the orifice plate. Installation is also more difficult. These tubes are largely used to reduce capital expenditure on pumping equipment and save pumping energy costs.

Flow nozzles represent a transition between orifices and flow tubes, both in terms of cost and in terms of head loss.

Pitot Tubes

A pitot tube is a small tube inserted into the pipe. The differential between the total pressure on the impact port and the static pipeline pressure is measured. Pitot tubes offer a very low-cost measuring system (for large diameter pipes) with negligible pressure loss. They are also convenient for temporary measurements and pipe velocity traverses. Their principal limitation is that they measure a point velocity and, due to velocity profile changes, only provide very low-accuracy volumetric readings. They are also subject to contamination and fouling.

In order to reduce the effect of velocity profile changes, multiple-opening pitot tubes and area-averaging pitot traverse stations have also been developed. Their accuracy is somewhat better then that of the traditional pitot.

Elbow Taps

Elbow taps provide flow rate by measuring the differential pressure across an elbow. This results in an economical installation because the cost is that of one transmitter regardless of the pipe size. The system suffers from very poor accuracy; additionally, relatively high flow velocities and long upstream straight pipe lengths are required.

Target (or Impact) Meters

In this system, deflection of a plate supported in the fluid flow is measured and an electronic or pneumatic analog output provided. The target meter is more expensive than a comparable orifice installation but since there are no pressure taps to plug, the target meter is more suited for applications with suspended solids or "sticky" process fluids. There are no moving parts. Operating flow range is limited (3:1) but it can be re-ranged.

ELECTROMAGNETIC FLOWMETERS

In accordance with Faraday's Law, these meters measure the velocity of an electrically conductive liquid as it cuts the magnetic field produced across the metering tube. The principal advantages include no moving components; completely unobstructed bore and, therefore, no pressure loss and no wear and tear on components; chemical compatibility with virtually all liquids; indifference to viscosity, pressure, temperature, and density; linear analog output; suitability for bidirectional flow; extensive range of sizes; and ease and rapidity of re-ranging on site. The major limitation to the use of electromagnetic flowmeters is that the fluid must be electrically conductive. (This requirement eliminates their use on any and all gases.) The purchase price and the cost of maintaining magnetic field is high compared with alternative systems. A separate converter is required together with a power supply at the measurement point. Electromagnetic flowmeters are used for corrosive liquids, slurries, and low pressure loss applications.

The more recent advances include the various electrode cleaners, the magnetic flowmeter probe, and the on/off cycling of the magnetic field, which not only conserves electric energy but also allows for automatic rezeroing and therefore better accuracy. The use of ceramic flowtubes has reduced the overall cost while eliminating liner problems and electrode leakage, because the sintered electrodes cannot leak. Intelligent magmeters are capable of double range operation, wide turndown, empty pipe detection, and accuracy within 0.5% of actual flow over a range of 10:1.

TURBINE METERS

The speed of rotation of a bladed turbine, driven by the fluid, provides a digital output linear with flow rate. Turbine meters are available for liquids and gases, for very low flow rates, and as insertion designs. The liquid turbine meter is one of the most accurate meters available for low- to medium-viscosity products. Turbine meters offer good rangeability (10:1) and are suitable for virtually unlimited pressures and for extremes of high and low temperatures. They are easy to install and are small in size and weight relative to the pipeline diameter. The meter has a very fast speed of response and can be made hygienic. The principal limitations are incompatibility with high-viscosity or dirty liquids, possible damage due to overspeeding during a liquid/gas phase, and the necessity for secondary readout equipment. Filtration is also recommended. Turbine meters are widely used for high-accuracy product sale, blending, test rig duty, and general measurement.

Variations on the basic turbine flowmeter design include the non-electric (fiber-optic) turbine probe designs, the bearingless "hoverflow" units, and the various paddlewheel, impeller, and shuntflo designs. The impeller and paddleflow units cost less than the traditional turbine flowmeters, but their accuracy is also lower.

VORTEX METERS

The three types of vortex meters are vortex shedding, vortex precession, and fluidic oscillation. All three types utilize fluid oscillation to provide meters with no moving components, suitable for gas, steam, or liquid service. They generally have wide flow range capability, minimal maintenance, frequency or linear analog output, good accuracy, and long-term repeatability. The liquid vortex meters are unsuitable for use on viscous or dirty fluids.

The two main limitations are line size and Reynolds number. The frequency of fluid oscillation drops off as the line size increases. For this reason meters in excess of 8 to 12 in. (200 to 300 mm) diameter are not very practical. The other limitation is that vortices form only at Reynolds numbers exceeding 10,000, and therefore this meter is not usable under that limit. Vortex shedding meters are gaining increasing acceptance and are used as a general-purpose, low-cost alternative to the orifice plate and dp transmitter, but they are also used for many demanding applications in the chemical industry.

VARIABLE-AREA METERS

Variable-area meters are widely used for small flow rate measurement where local indication is required, in test rigs, and in general industry.

Variable-area meters are available in both glass and metal tube construction. In the glass tube design the visually observed position of the float inside the tapered tube is an indication of the flow rate. The advantages of this design include low cost, constant low pressure loss, direct flow indication without requiring any power supply, and ability to detect very low flow rates of both gases and liquids, including viscous fluids. Disadvantages include the limited availability of transmitters, larger sizes or higher pressure ratings, the relatively low accuracy, and the requirement for vertical installation.

The metallic tube units are readily available as transmitters, can be obtained in larger sizes with higher pressure ratings, and provide good rangeability and a linear output, but they are still limited to use on clean fluids and must be mounted vertically. A wide variety of spring-loaded (instead of gravity-operated) variable-area flowmeters are also available. In these units, an increase in flow causes the compression or deflection of a spring and the resulting motion operates the flow display. These units do not need to be mounted in a vertical position and are installed as flow-through pipeline devices.

POSITIVE DISPLACEMENT METERS

Positive displacement (PD) meters trap a known fixed volume of fluid and transfer it from the inlet to the outlet side of the meter. The number of "packages" of fluid is a measure of the flow. Designs include rotary piston, oval gear, sliding vane, and reciprocating piston. Liquid PD meters offer good accuracy and flow range and are particularly suited to high-viscosity products. The meters provide a local readout and do not require a power supply. They can provide a linear transmission signal. The PD meter relies on close meshing surfaces and is, therefore, only suitable for clean fluids. It requires regular recalibration and maintenance, particularly on nonlubricating liquids. Large sizes are bulky, heavy, and more difficult to install. PD meters are used for viscous product sales and general-purpose industrial metering.

ULTRASONIC METERS

Ultrasonic meters are available in two forms: Doppler and transit-time. With Doppler meters, an ultrasonic pulse is beamed into the pipe and reflected by inclusions, such as air or dirt. The Doppler meter is frequently used as a "clamp on" device which can be fitted to existing pipelines. It detects the velocity only in a small region of the pipe cross section and as such its accuracy is not good. It is not suitable for clean fluids or clean gas service. The meter cost is low and is independent of pipe size.

The single or multi-beam transit-time flowmeters project an ultrasonic beam right across the pipe at an acute angle, first with the flow and then in opposition to the flow direction. The difference in transit time is proportional to flow rate. This type of ultrasonic meter is considerably more expensive but offers better accuracy. Unlike the Doppler meter, it requires a relatively clean fluid. There is no obstruction to flow and hence little pressure loss. It is ideally suited to measure the flow of very corrosive liquids. The approach pipework conditions are fairly critical. This type of meter is gaining limited acceptance in the water and chemical industries.

METERING PUMPS

These units pump and meter the fluid at the same time. The two basic types are the plunger pump, which provides reasonable accuracy, and the diaphragm-type, used for dangerous or contaminated fluids. The advantages of a metering pump include that it is self-contained, and is simple to install. It is reasonably accurate, but subject to errors caused by variations in pressure and viscosity. The limitations are that their cost is fairly high, the process fluid must be clean, there is a need for periodic recalibration, and pump operation is no guarantee of product discharge.

MASS FLOWMETERS

The mass flow of homogeneous gases is most frequently measured by thermal flowmeters. The main advantage of these detectors is their good accuracy and very high rangeability. Their main disadvantage is that when the specific heat of the process fluid changes, due to composition or temperature changes, this will register as a change in mass flow. The volumetric flow rate or the flowing velocity of process streams can also be detected by thermal devices, such as the hot wire anemometers or thermal flow switches.

The mass flow of liquids and gases has been detected by angular momentum devices and indirectly through the measurement of volumetric flow and density. These approaches have been overshadowed by the Coriolis-based mass flowmeters. These units detect the twisting of an oscillating, usually stainless steel flow tube. The amount of the twist is a function of the mass flow through the tube. The meters can handle flow velocities from 0.2 to 20 ft/s (0.061 to 6.1 m/s), which give them a rangeability of 100:1. Their accuracy is high (0.2% of actual flow), their pressure and temperature ratings are acceptable, and they can be provided with additional outputs signaling alarm conditions or detecting the process fluid's density in addition to its mass flow. Some of their limitations include vibration sensitivity, inability to handle high temperatures (over 400°F [205°C]), and not being available in sizes over 6 in. (150 mm). The Coriolis-based mass flowmeters are very popular in the measurement of fuel flows, reactor feed flows, and other measurements where the mass and not the volume of the process fluid is of interest.

At low flow rates the Wheatstone-type mass flowmeter can provide a ±0.5% of actual flow performance over a 100:1 range.

The mass flow of solids in gravity flow can be detected by impact flowmeters, which are relatively low-accuracy devices. Better accuracy and rangeability is provided by belt-type gravimetric feeders, which measure both the speed and the loading on the belt. In addition to the belt-type gravimetric feeders, the loss in weight-type systems can also measure the mass flow of liquids or solids. These devices operate by differentiating the load cell signal from tank weighing systems. The rate at which that signal is dropping is the mass flow out of the tank. These systems are recommended for the measurement of hard-to-handle process flows, as they do not make physical contact with the process stream.

(Cross-correlation flowmeters are also available for the measurement of the mass flow of solids in pneumatic conveying systems.)

CROSS-CORRELATION

The cross-correlation flowmeter uses statistical means to average the time it takes for particles in fluid to travel a known distance. The meter can be non-invasive and is suitable for solids flow detection, two phase flows, including heavy slurries and very corrosive and difficult applications. Limitations are a high selling price, fairly high minimum Reynolds number requirement, and reasonable to poor accuracy.

SONIC VENTURI

This design in illustrated in Figure 1b. The meter body is that of a multiport digital control valve, where the area of each port is twice the size of the next smaller one. The on/off ports are opened through binary manipulation and, therefore, the meter rangeability is a function of the number of ports used. With 8 ports the rangeability is 255:1; with 10 it is 1023:1; with 12 it is 4095:1; and so on. The digital control valve is converted into a flowmeter through the insertion of a sonic velocity venturi into each of the ports. A sonic velocity venturi element passes a predictable, known, and constant flow rate when the flow velocity reaches sonic at its throat. Therefore, this flowmeter requires the continuous presence of sonic velocity of its throat, which in turn requires that the meter ΔP be greater than 40% of the upstream absolute pressure. Because of the inherent requirement for high pressure drops, this meter is ideal for applications where a combination device is required to both meter the flow and control the downstream pressure.

The meter accuracy of ½ to 1% of actual flow is the same at high or low flow rates throughout the meter range. With the addition of inlet gas pressure, temperature and/or density sensors, the meter can be converted for mass flow measurement. It can also meter liquids. This flowmeter is available in sizes from 1 to 8 in. (25 to 200 mm). Units have been built for up to 10,000 PSIG (69 MPa) pressure services and for temperatures from cryogenic to 1200°F (650°C).

SPECIALIZED FLOWMETERS (LOW FLOW)

In addition to the flowmeters discussed in the previous paragraphs there are a number of other, more specialized designs that are also covered in the Orientation Table (Table A) and the sections of this chapter but need not be discussed here. One exception to this approach is treatment of the low flow rate detectors, which are not covered in a separate section but are discussed here.

The measurement and control of low flow rates is a requirement in purging applications, in bioreactors, in leak testing, and in controlling the reference gas flow in chromatographs and in other analyzers such as plasma emission spectrometers.

The most traditional and least expensive low flow sensor is the variable area flowmeter, frequently made out of a transparent acrylic material. It requires little pressure drop and has an acceptable rangeability of 10:1. Due to its relatively low accuracy, it is most often used on purge and leak-detection applications.

A much more accurate low flow detector for gas metering applications is the sonic flow nozzle. It accurately maintains the flow through the nozzle as long as sonic flow is maintained, which is guaranteed by keeping the inlet pressure about 1.5 times the outlet. Disadvantages include the low rangeability and therefore the difficulty in modulating flow; the high pressure drop; and cost.

Laminar flow elements guarantee that pressure drop and flow will be in linear relationship. The laminar flow element can be used in combination with either differential pressure or thermal detectors. Their costs are similar to the sonic nozzle, but their rangeabilities are better. If mass flow control is desired then the thermal sensor design has the edge, because it is an inherent mass flow sensor, it has a 100:1 rangeability, and control capability is readily available.

Thermal flowmeters can also detect low mass flows without laminar elements. These units are installed directly in the pipeline as either thermal flowmeters or anemometers.

FLOWMETER SELECTION PROCEDURE

Meter selection should take place in two steps. First, identify the meters which are technically capable of performing the

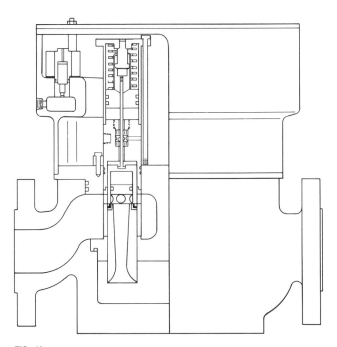

FIG. 1b
Sonic venturi digital flowmeter featuring extremely wide rangeability.

required measurement and are available in acceptable materials of construction; then, select the best choice from those available. A list should be made of the key features which the meter must have. By comparing these requirements with the information given in Table A, 1c, 1d, and 1e, a first-pass elimination of technically unsuitable meters can be made. The list can then be further refined by a more detailed consideration of the applicational requirements against the "Features Summary" at the start of each appropriate section in this volume.

In order to cover special features such as reverse flow, pulsating flow, response time, and so on, it is necessary to study the individual meter specifications in detail and/or obtain the manufacturer's comments and advice.

Although the above steps will eliminate technically unsuitable meters, it does not necessarily follow that there will be available meters which are technically suitable for the application. A meter may possess some of the individual features required, but it may not be possible to find a combination of all of the desired features in one meter. Electromagnetic flowmeters, for example, are readily available for operating pressures of 1500 PSIG (10.3×10^6 N/m^2). They are also readily available for flow rates of 500,000 GPM (31.5 m^3/s) but the availability of an electromagnetic flowmeter for use at 1500 PSIG *and* a flow rate of 500,000 GPM is an entirely different proposition. Although such an applicational requirement may not exist, the example does illustrate the danger of assuming that all of a meter's features can automatically be incorporated in one individual meter.

The length of the list of technically suitable meters will depend on the complexity of the application. On an extreme application such as a highly corrosive, nonconductive liquid with large solid content, the list will probably consist of a single meter (cross-correlation meter). On a straightforward clean water application, the list will consist of nearly all the flowmeters listed in the Orientation Table (Table A).

In order to narrow down that choice, the engineer should concentrate on the reasons for measuring the flow. The key requirements should be identified and prioritized. For example, is high accuracy the most important requirement, or is

TABLE 1c
Fluid Duty Selection Table

Fluid Details	Correlation	Elbow Taps	Laminar	Electro-Magnetic	Angular Momentum	Metering Pumps	Orifice	Pitot	Gas Displacement	Liquid Displacement	Solids Flowmeter	Target	Thermal	Liquid Turbine	Gas Turbine	Doppler U-Sonic	Transit U-Sonic	V.A.	Venturi	Vortex Shedding	Vortex Precession	Fluidic Oscillation
Liquid — Clean	X	√	√	*√	√	√	√	√	X	√	X	√	√	√	X	X	√	√	√	√	X	√
Dirty	√	?	√	*√	√	√	?	?	X	X	?	√	√	?	X	√	?	√	√	?	X	?
Slurries	√	X	?	*√	?	√	X	X	X	X	SD	?	?	X	X	X	X	X	?	X	X	X
Low Viscosity	√	√	√	*√	√	√	√	√	X	?	X	√	√	√	X	√	√	√	√	√	X	√
High Viscosity	√	?	?	*√	?	√	?	X	X	√	SD	?	?	X	X	?	?	?	?	?	X	X
Corrosive	√	√	?	*√	√	?	√	√	X	?	X	?	?	?	X	√	√	√	?	?	X	?
Very Corrosive	√	?	X	*√	X	X	?	?	X	X	X	X	?	X	X	√	√	?	√	X	X	X
Gas — Low Pressure	X	√	√	X	√	√	X	√	√	X	X	√	√	X	√	X	X	√	√	√	√	X
High Pressure	X	√	√	X	√	√	√	√	√	X	X	√	√	X	√	X	X	X	√	√	√	X
Steam	X	X	?	X	X	X	√	X	X	X	X	√	X	SD	X	X	√	√	SD	X	X	
Reverse Flow	X	√	X	√	X	X	SD	X	X	X	X	X	SD	SD	√	√	X	X	X	X	X	X
Pulsating Flow	?	X	√	√	X	X	?	X	X	X	X	X	X	X	√	√	?	?	X	X	X	X

* = Must be electrically conductive
√ = Generally suitable
? = Worth consideration
X = Not suitable
SD = Some designs

TABLE 1d
Flowmeter Selection Table*

	Clean Liquids	Dirty Liquids	Corrosive Liquids	Viscous Liquids	Abrasive Slurries	Fibrous Slurries	Low Velocity Flows	Vapor or Gas	Hi Temp. Service	Cryogenic Service	Semi-Filled Pipes	Non-Newtonians	Open Channel
Differential Pressure													
Orifice	√	??	?	?	X	X	√	√	√	√	X	??	X
Venturi	√	?	??	??	??	??	??	√	??	??	X	??	X
Flow Nozzles and Tubes	√	??	??	??	??	??	??	√	??	??	X	??	X
Pitot Tubes	√	??	?	??	X	X	??	√	??	??	X	X	X
Elbow	√	?	?	??	?	X	X	√	??	??	X	??	X
Magnetic	√	√	√	?	√	√	?	X	??	X	??	?	??
Mass													
Coriolis	√	√	?	√	√	?	?	??	??	??	X	√	X
Thermal	??	??	??	??	??	??	?	√	??	X	X	??	X
Oscillatory													
Vortex Shedding	√	?	?	??	X	X	X	√	??	??	X	X	X
Fluidic	√	??	?	??	X	X	X	X	??	??	X	X	X
Vortex Precession	√	X	??	??	X	X	X	√	??	X	X	X	X
Positive Displacement	√	X	??	√	X	X	√	√	??	??	X	X	X
Target	√	?	?	?	??	X	??	√	??	??	X	??	X
Turbine	√	??	??	?	X	X	??	√	??	??	X	X	?
Ultrasonic													
Transit Time	√	??	??	??	X	X	??	??	X	??	X	X	?
Doppler	X	√	??	??	??	??	??	X	X	X	X	??	X
Variable Area	√	?	?	?	X	X	?	√	?	X	X	X	X
Weirs and Flumes	√	?	??	X	??	??		X	X	X	√	X	√

√ Designed for this service
?? Applicable for this service under certain conditions, consult manufacturer
? Normally applicable for this service
X Not applicable for this service

*Courtesy of Fischer & Porter.

TABLE 1e
*Flowmeter Selection Table**

Flowmeter	Pipe size, in (mm)	Gases (vapors) Clean	Gases (vapors) Dirty	Liquids Clean	Liquids Viscous	Liquids Dirty	Liquids Corrosive	Slurries Fibrous	Slurries Abrasive	Temperature, °F (°C)	Pressure, psig (kPa)	Accuracy, uncalibrated (including transmitter)	Reynolds number† or Viscosity
Orifice													
	SQUARE ROOT SCALE: MAXIMUM SINGLE RANGE 4:1												
Square-edged	>1.5 (40)	√	X	√	X	?	?	X	X	Process temperature to 1000°F (540°C); transmitter limited to −30–250°F (−30–120°C)	To 4000 psig (41,000 kPa)	±1–2% URV	$R_D > 2000$
Honed meter run	0.5–1.5 (12–40)	√	X	√	?	?	?	X	X			±1% URV	$R_D > 1000$
Integral	<0.5 (12)	√	X	√	√	X	?	X	X			±2–5% URV	$R_D > 100$
Quadrant/conic edge	>1.5 (40)	X	X	√	√	?	?	X	X			±2% URV	$R_D > 200$
Eccentric	>2 (50)	?	√	?	X	X	?	X	X			±2% URV	$R_D > 10,000$
Segmental	>4 (100)	?	√	?	X	√	?	X	X			±2% URV	$R_D > 10,000$
Annular	>4 (100)	?	√	?	X	√	?	X	X			±2% URV	$R_D > 10,000$
Target	0.5–4 (12–100)	√	√	√	√	√	?	X	X			±1.5–5% URV	$R_D > 100$
Venturi	>2 (50)	√	?	√	?	?	?	?	?			±1–±2% URV	$R_D > 75,000$
Flow nozzle	>2 (50)	√	?	√	?	?	?	X	X			±1–±2% URV	$R_D > 10,000$
Low loss	>3 (75)	√	X	√	X	X	√	X	X			±1.25% URV	$R_D > 12,800$
Pitot	>3 (75)	√	X	√	?	X	?	X	X			±5% URV	No limit
Annubar	>1 (25)	√	X	√	X	X	?	X	X			±1.25% URV	$R_D > 10,000$†
Elbow	>2 (50)	√	?	√	X	?	?	?	?			±4.25% URV	$R_D > 10,000$†

LINEAR SCALE TYPICAL RANGE 10:1

	Size range in. (mm)								Temperature range °F (°C)	Pressure range psig (bar)	Accuracy	Viscosity
Magnetic	0.1–72 (2.5–1800)	X	✓	✓	✓	✓	✓	✓	360 (180)	≤1500 (10,800)	±0.5% of rate to ±1% URV	No limit
Positive-displacement	<12 (300)	✓	X	✓	X	?	✓	X	Gases: 250 (120) Liquids: 600 (315)	≤1400 (10,000)	Gases: ±1% URV Liquids: ±0.5% of rate	≤8000 cS
Turbine	0.25–24 (6–600)	✓	X	✓	X	X	✓	X	−450–500 (−268–260)	≤3000 (21,000)	Gases: ±0.5% of rate Liquids: ±1% of rate	≤2–15 cS
Ultrasonic												
Time-of-flight	>0.5 (12)	X	X	?	✓	X	✓	X	−300–500 (−180–260)	Pipe rating	±1% of rate to ±5% URV	No limit
Doppler	>0.5 (12)	X	X	?	✓	✓	✓	✓	−300–250 (−180–120)	Pipe rating	±5% URV	No limit
Variable-area	≤3 (75)	✓	X	✓	✓	X	?	X	Glass: ≤400 (200) Metal: ≤1000 (540)	Glass: 350 (2400) Metal: 720 (5000)	±0.5% of rate to ±1% URV	<100 cS
Vortex	1.5–16 (40–400)	✓	?	X	?	✓	?	X	≤400 (200)	≤1500 (10,500)	±0.75–1.5% of rate	>10,000

cS = centi Stokes
URV = Upper range-value
✓ = Designed for this application
? = Normally applicable
X = Not applicable

*This material is reproduced by permission of McGraw-Hill, Inc., from R. W. Miller's *Flow Measurement Handbook*, 2nd edition, 1989.
†According to other sources, the minimum Reynolds Number should be much higher.

long-term repeatability, low installed cost, or easy maintenance?

It is essential that the requirements be objectively specified. Otherwise, a utopian but unavailable flowmeter can be specified. It should also be realized that certain flowmeters, such as those required for the measurement of two-phase flow, are still in the developmental stage and are not yet readily available.[1-4]

Inaccuracy

The one parameter which is specified more than any other is inaccuracy. Statements such as "best possible," "better than one-quarter percent," and the like are frequently made. If taken at face value, they will severely limit the meter choice and often result in an unnecessarily high cost. A realistic accuracy level should, therefore, always be set. In some instances—for example, repetitive batch dispensing—true accuracy is of no serious consequence, provided the meter gives a stable, repetitive, long-term output. On such applications accuracy should be given less emphasis and more emphasis should be placed on long-term repeatability. Where true accuracy is required, it is necessary to be very specific about the meter specification and also to establish the likely on-site accuracy, which may well be different from the manufacturer's performance figures.

When specifying accuracy, one should define the flow range over which the error should stay within the stated limits, and one should also specify if the allowable error is given as a percentage of full scale (%FS) or as a percentage of actual flow (%AF). With %FS sensors the absolute error increases as the flow drops, as shown in Figure 1f. Therefore, in a well-prepared specification, the accuracy requirement should be given in the form of: "1% AF between 10% and 100% flow" or as "0.5% FS over 5 to 100% flow." Linearity should not be confused with accuracy. Linearity is the closeness to which the curve of the output signal approximates a straight-line relationship to flow. The linearity of a flowmeter can be different on the factory calibration stand and under the installed conditions in the field.

The published meter performance specification is generally based upon stipulated installation and operating conditions. It is an indication of the performance which the meter can achieve but it is no guarantee that the same performance will be achieved on site under actual operating conditions. Substandard approach pipework, resulting in a high degree of swirl in the liquid, will cause a dramatic deterioration in the linearity tolerance and a shift in the nominal calibration constant in many meter types. Consequently, the manufacturer's installation recommendations should be adhered to, or, if this is not possible, the resulting deterioration in the meter's performance should be evaluated in advance.

Changes in the fluid characteristics can also alter the meter's performance. Figure 1g illustrates the effect of viscosity on two of the most accurate flowmeters, the turbine meter and the positive displacement meter. With the turbine meter, an increase in viscosity causes a deterioration in the performance, whereas with the PD meter, a reduction in viscosity causes a deterioration in the performance. Acceptability of the proposed operating conditions should always be verified for any application.

Wear, analog drift, calibration shift, etc., should also be investigated for their differing effects on the alternative meter types being considered. Figure 1h illustrates an inline ballistic prover that can be used to recalibrate flowmeters, without requiring the interruption of the process flow.

Reynolds Number

At constant flow the output signal of a head-type flowmeter will be constant if the Reynolds number does not change. Therefore it is necessary to calculate the Reynolds numbers at both maximum and minimum flow and check if the corresponding change in flow coefficients is within the acceptable error. If it is not, a different type sensor must be selected, such as the quadrant-edged orifice for low Reynolds numbers, or flowmeters, such as the magnetic type, that are not sensitive to Reynolds variation.

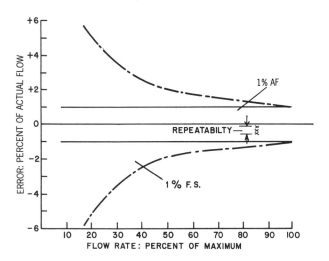

FIG. 1f
Comparison of 1% F.S. inaccuracy with 1 percent of flow inaccuracy.

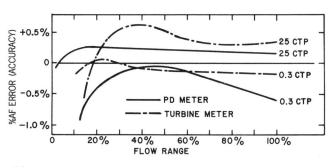

FIG. 1g
Differing effects of viscosity variation on a turbine meter and a positive displacement meter (CTP = centiPoises).

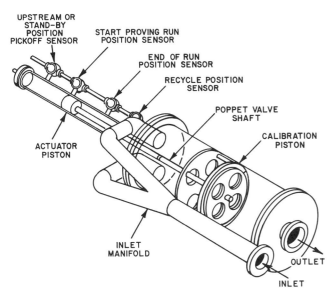

FIG. 1h
Inline ballistic flow prover. (Courtesy of Brooks Instrument Div. of Emerson Electric)

Figure 1i depicts the relationship between the pipeline Reynolds number and the discharge coefficient for various head-type flow elements. The Reynolds number is determined by the following equation:

$$\mathrm{Re} = \frac{3.160 G_f Q_f}{D\mu} \quad\quad 1(5)$$

where

G_f = process fluid specific gravity (at 60°F, or 15.5°C)
Q_f = liquid flow in GPM
D = pipe inside diameter (in inches)
μ = process fluid viscosity (in centipoises)

As shown by Figure 1i, the orifice plate discharge coefficient is constant within $\pm 0.5\%$ over a Reynolds number range of 2×10^4 to 10^6. This consistency of the discharge coefficient guarantees a corresponding limitation of inaccuracies based on *actual* flow, over a range that is wider than what the d/p flow transmitter can handle.

Safety

Safety is one of the most important considerations in the selection of any equipment. The meter and associated equipment must, obviously, be certified as being suitable for the electrical safety classification of the area in which it will be used. This may be achieved by use of purely mechanical or pneumatic equipment, or, more commonly, by use of certified intrinsically safe, flameproof or explosion-proof units.

Other safety aspects, which are more frequently overlooked, are the influence of the materials of construction on safety and the avoidance of leakage. Fluids such as oxygen or liquid chlorine can represent a most dangerous potential for causing explosions, because of their chemical reaction with certain materials. If the heat of the reaction cannot be removed and, especially, if pressure is confined, a violent explosion can result. Various organic and inorganic substances, including ordinary lubricants such as oil, grease, and wax, can cause explosions in the presence of oxygen or chlorine. It is, therefore, essential that any flowmeter operating in such services be thoroughly cleaned and degreased.

The choice of the materials of construction is also critical for oxygen or oxygen-enriched applications. Steels, for example, present an explosion hazard which increases as the oxygen velocity and pressure increases. The cleanliness and surface finish affect the limitations on safe velocity and pressure. Therefore steel can be used at higher pressures and flow rates if clean and provided with a high surface finish. This problem can be avoided by the use of alternative materials, such as phosphor bronze, gunmetal, brass, beryllium, copper, etc.

Where noxious or dangerous fluids are being metered, it is essential, from an operator safety standpoint, that leakage be kept to an absolute minimum. The probability of leakage is increased by the presence of manifolds, pressure taps, and fragile components, while the likelihood of leakage is reduced when the flowmeter does not penetrate the pipe.

Installation

Installation requirements vary dramatically among the various meter types. They can be the deciding factors in meter selection. An extreme example is a pipeline which cannot be shut down or the measurement point bypassed. In this case, the user's choice is limited to clamp-on meters, such as Doppler and cross-correlation, or hot-tap insertion meters, such as the various probe designs.

Even if it is not necessary to install the meter without a shutdown, the installation requirements are still important factors both with regard to cost and plant acceptability. It might be the case that, the requisite straight pipe lengths are not available and it is then necessary de-rate the perform-

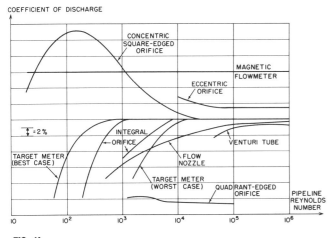

FIG. 1i
Discharge coefficients as a function of sensor type and Reynolds number. (Courtesy of The Foxboro Company)

ance or consider an alternative meter type, such as an electromagnetic one having a minimal requirement for straight upstream piping.

Specific application requirements affect different meters in different ways. Unavailability of an electricity supply at the measurement point eliminates the electromagnetic flowmeter from consideration. If a vertical pipe section cannot be found, the variable area meter cannot be used. A positive displacement meter requires a strainer, and so on. Even if the meter installation requirements can be met, their effect on the overall system cost should still be considered and quantified.

Cost

Cost is a critical factor in the selection of any equipment. To arrive at a reasoned decision one should not limit cost to purchase price. Other factors, such as operating cost, maintenance, spare parts inventory, the effect of downtime—these and many others should all be considered if a "reasoned" decision is to be reached. Hardware cost should always be balanced against the potential benefits of increased plant efficiency or product quality. These benefits are usually byproducts of increased sensor accuracy, repeatability and rangeability, which all tend to increase costs.

The initial purchase cost comparison should be based upon system cost and not merely the flowmeter price. If, for example, the meter requires a separate converter or transmitter to produce the required pneumatic output, digital or analog transmission signal, linear output, etc., then these costs should be included. Likewise, the cost of ancillary items, such as recommended upstream and downstream piping, flow conditioning elements, filters, power supplies, and so on, should all be considered. So too should the cost of the installation itself. Installation complexity varies significantly between meter types, and, with the current high labor rates, installation cost can be a significant factor.

Operating cost is a complex subject. The types of questions which should be asked about the various aspects of the operating cost include:

Routine Service and Maintenance—how much, if any, routine service is recommended? What level of personnel is required? Are special flow simulator units required or available?

Versatility—Can the secondary units be used on other meters? Can the meter be easily re-ranged? Can the meter be used on other applications?

Spare Parts Inventory—What level and value of inventory is required? Are the spares interchangeable with other sizes and models?

Reliability—Are there moving components? What is guaranteed and what is the estimated meter life? What are the likely failure modes?

The meter pressure loss is also part of the total operating cost. If the choice is between an orifice plate and a low loss flow tube, the orifice plate would obviously be far cheaper but its irrecoverable head loss would be far greater. As was discussed, pumping cost is a function of flow rate, electricity costs, running efficiency and pressure loss. Consequently, the higher pressure drop across the orifice plate will result in increased pumping costs throughout the life of the installation.

References

1. Linn, J.K., and Sample, D.G., "Mass Flow Measurement of Solids/Gas Stream Using Radiometric Techniques," *Report SAND-82-0228C,* Dept. of Energy (Washington DC), 1982.
2. Pursley, W.C., and Humphreys, J.S., "Two-Phase Flow Measurement at NEL," *Proceedings of the NEL Fluid Mechanics Silver Jubilee Conference,* National Engineering Lab (East Kilbride UK), 1979.
3. Hewitt, G.F., and Whalley, P.B., "Flow Measurement in Two-Phase (Gas-Liquid) Systems," *Proceedings of Interflow '80,* Institution of Chemical Engineers (Rugby UK), 1980.
4. John, H., and Riemann, J., "Test Facility for Tests and Calibration of Different Methods of Two-Phase Mass Flow Measurements," Institute Fuer Reaktorbauelemente (Karlsruhe Germany), February 1979.

Bibliography

Batur, C., "Measuring Flow with Machine Vision," *InTech,* May 1989.
De Boom, R.J., "Flow Meter Evaluation," 1991 ISA Conference, Paper #91-0509.
The Flowmeter Industry, 2nd edition, 1985–1990, Venture Development Corp., 1986.
Hall, J., "Flow Monitoring Application Guide," *Instruments and Control Systems,* February 1983.
Hardy, J.E., Hylton, J.O., and Moore, R.L., "Steam Flow Measurements," *InTech,* December 1985.
Hayward, A.J., "Choose the Flowmeter Right for the Job," *Processing Journal,* 1980.
Husain, Z.D., "Flowmeter Calibration and Performance Evaluation," 1991 ISA Conference, Paper #91-0508.
Krigman, A., "Flow Measurement: Some Recent Progress," *InTech,* April 1983.
Krigman, A., "Guide to Selecting Non-Intrusive Flowmeters," *InTech,* December 1982.
Laskaris, E.K., "The Measurement of Flow," *Automation,* 1980.
Lipták, B.G., "Applying Gas Flow Computers," *Chemical Engineering,* December 1970.
Lipták, B.G., "On-Line Instrumentation," *Chemical Engineering,* March 31, 1986.
Lomas, D.J., "Selecting the Right Flowmeter," *Instrumentation Technology,* 1977.
Mesch, F., "Speed and Flow Measurement by an Intelligent Correlation System," 1990 ISA Conference, Paper #90-0632.
Miller, R.W., *Flow Measurement Engineering Handbook,* New York: McGraw Hill, 1983.
O'Brien, C., "Flowmeter Terms, Types & Successful Selection," *InTech,* December 1989.
O'Brien, C., "Fueling Flowmeter Accuracy, Reliability," *InTech,* April 1989.
Robinson, C., "Obstructionless Flowmeters," *InTech,* December 1986.
Rusnak, J., "The Fundamentals of Flowmeter Selection," *InTech,* April 1989.
Spitzer, D.W., "Industrial Flow Measurement," ISA, 1984.
Watson, G.A., "Flowmeter Types and Their Usage," *Chartered Mechanical Engineer Journal,* 1978.
Welch, J.V., "Trends in Low Gas Flow Metering," *InTech,* February 1991.
Young, A.M., "Volumetric Flowrate Measurement with Coriolis Flowmeter," 1990 ISA Conference, Paper #90-0631.
Zapolin, R.E., "New Ways to Meet User Needs for SCADA Mass Flow Monitoring," 1990 ISA Conference, Paper #90-0633.

2 BTU Flowmeters for Heat Exchangers

B. G. LIPTÁK (1982, 1993)

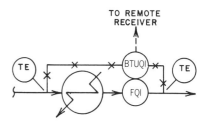

Flow Sheet Symbol

BTU Flowmeter Types:	A. All mechanical design B. Electronic BTU computer
Approximate Cost:	A. About 50% more than the cost of the positive displacement flowmeter. If, for example, a 6 in. (150 mm) propeller meter is the flow sensor, the total cost is about $3000 B. $2000 in addition to the cost of the flow and temperature transmitters
Inaccuracy:	A. ±2 to 5% of full scale B. ±0.5% of full scale
Minimum ΔT:	A. 5°F (2.8°C) B. 1°F (0.56°C)
Partial List of Suppliers:	American Sigma (B); Hersey Measurement Co. (A); ISTA Energy Systems Corp. (B); ITT Barton (B); Kent Meters Inc. (A); Onicon Inc. (B) Note: The type ''B'' units can be made by multiplying the output signals of any digital or analog flow and temperature difference transmitters.

The first step toward energy conservation and energy-efficient plant design is a reliable energy audit throughout the plant. An overall heat balance around the plant can only be prepared if the individual loads are accurately measured. This is illustrated in Figure 2a.

The efficiency of the boiler is measured by the ratio of integrated energy flows at points 1 and 2. This is usually done by totalizing the fuel and steam flows over some period of time. The total fuel consumed over that period is multiplied by its heating value to obtain the total energy input into the boiler. Multiplying the totalized steam flow by the difference between the enthalpy of the steam and of the feedwater gives the total useful energy obtained from the boiler. The ratio of the two is the boiler's efficiency.

Similarly, the coefficient of performance of the chiller is measured by the ratio of energy flows at points 10 and 11. The efficiency of individual heat exchangers is also detected by measuring the energy flows on the utility and on the process sides, such as points 3 and 4 or points 12 and 13 in Figure 2a.

The efficiency of the overall utility distribution system is determined by comparing the sum of the individual loads with the total supply at the source. The difference between the two represents the losses due to insufficient thermal insulation of the pipe lines, leaking steam traps, and other causes.

When various optimization techniques are being considered, one of the first steps in the cost-benefit analysis is to empirically measure the energy consumption prior to and after optimization. The measured saving is then used on the benefit side to calculate the payback period for the installation.

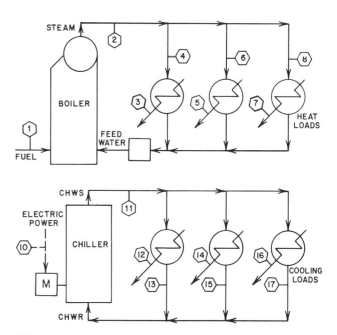

FIG. 2a
Plant-wide energy audit.

BTU flowmeters are required in many of the energy flow sensor locations noted in Figure 2a. These BTU computing units are available either in mechanical or electronic designs.

MECHANICAL BTU METERS

In the mechanical BTU meters, flow is detected by positive displacement or propeller-type sensors and is mechanically transmitted through gear trains. The temperature difference is sensed by filled thermal bulbs, which are connected to bourdon springs as illustrated in Figure 2b. Dual cam rollers are used in the computing mechanism which produces the digital displays of both total BTUs and total flow.

The advantages of this design include its simplicity, low cost, and the fact that it does not require any power supply. Transmitting attachments can also be provided where remote readouts are needed.

The limitations of this design include its relatively low accuracy: about ±2% of full scale error at temperature differences of 15°F (8.3°C) or higher. As the temperature

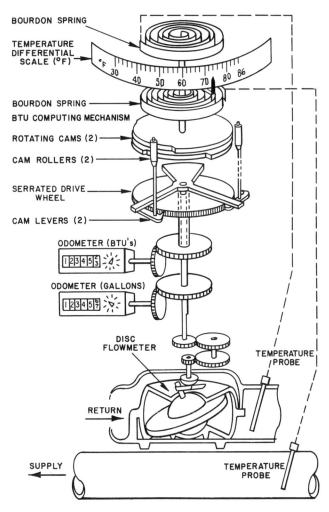

FIG. 2b
Mechanical BTU meter.

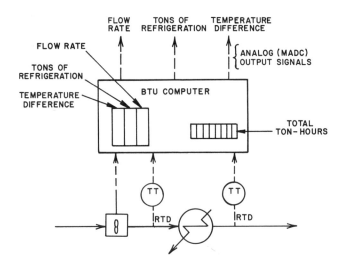

FIG. 2c
Electronic BTU meter.

difference decreases, the error tends to increase. These units are not recommended for temperature differences under 5°F (3°C).

The typical applications for the mechanical BTU meters are in the heating, ventilating, and air conditioning (HVAC) industry and in heat exchanger efficiency monitoring.

ELECTRONIC BTU METERS

In the electronic BTU computer packages, the flow sensor is usually a high-accuracy turbine flowmeter. The two temperatures are usually detected by resistance bulb-type temperature transmitters. Both the flow and the temperature sensors are accurate devices and they do provide high repeatability and turndown.

Therefore, the main advantage of electronic BTU computers is their superior accuracy. The error is held under ±0.5% of full scale. As illustrated in Figure 2c, the BTU computer digitally displays the accumulated total ton-hours. In addition, analog electronic retransmission signals are provided to facilitate the remote display of flow rate, BTU rate, and temperature difference.

While these units are more expensive than the mechanical BTU meters, the added cost can frequently be justified on larger or more critical installations, where accuracy is a prime concern.

Bibliography

"BTU Computer," Bulletin BTU-1, published by Barton-ITT.
Lipták, B.G., "On-Line Instrumentation," *Chemical Engineering,* March 31, 1986.
Reese, W.M., Jr., "Factor the Energy Costs of Flow Metering," *InTech,* July 1980.
Shinskey, F.G., *Energy Conservation Through Control,* New York: Academic Press, Inc., 1978.

3 BTU Flowmeters for Gaseous Fuels

A. P. FOUNDOS (1982)   **B. G. LIPTÁK** (1993)

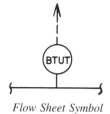

Flow Sheet Symbol

Design Basis:	Conventional head-type flowmeter and heat release (Wobble Index) measurements
Fluids:	Gaseous fuels
Applications:	Feedforward control loops for combustion optimization; BTU rate integration
BTU Flow Range:	From 100 BTU/min to very large flow rates limited only by pipe sizes
Inaccuracy:	±1.0 to ±2.0% of full scale based on accuracy of head meter
Costs:	The cost of a Wobble Index detector is $10,000 to $12,000 for general-purpose designs, and $16,000 to $18,000 for explosion-proof designs
Partial List of Suppliers:	ABB Kent-Taylor; Bailey Controls Company; Fischer & Porter Company; Fluid Data/Amscor; The Foxboro Company; Honeywell, Industrial Controls; Leeds & Northrup, Unit of General Signal

The heat flow rate of a fuel gas can be measured by detecting its mass flow rate and multiplying it by its heating value, detected by a calorimeter. The increase in energy costs necessitated a more careful utilization of waste gases and the burning of gases from a variety of sources (Table 3a) has made it necessary to measure their heat flow on-line and continuously.

Heretofore, standard procedure was to measure gas flow and use constants for the heating value and specific gravity of the gas, but this method is no longer satisfactory for efficient operation of burning processes.

In the past the heating value of gases could not be continuously measured in hazardous areas. Therefore, specific gravity measurements were used as an indirect means of estimating the heating value, in addition to using the specific gravity measurement to control the combustion process. Now that continuous and explosion-proof calorimeters are available one can measure the heating value of any fuel gas and compensate the combustion controls for variations in both specific gravity and heating value.

MEASURING HEAT FLOW

The heat flow rate (Q) of a gaseous fuel is the product of its volumetric flow rate at standard conditions (V_0) and of its calorific value (CV) or composition.

$$Q = V_0 \times CV = \text{SCF/hr} \times \text{BTU/SCF} = \text{BTU/hr} \qquad 3(1)$$

The composition effects both its heating value and the pressure drop through an orifice plate. If in equation 3(1) we substitute for V_0 the flow through an orifice plate, Q can be expressed as:

$$Q = K\sqrt{\Delta P/SG} \times CV \qquad 3(2)$$

where ΔP is the pressure differential across an orifice plate and SG is the specific gravity of the flowing gas.

The volumetric flow through an orifice plate (V_0) can be expressed as the product of a constant (K) and the square root of the ratio $\Delta P/SG$. Rearranging that relationship for $\sqrt{\Delta P}$ gives the following equation for the orifice pressure drop:

$$\sqrt{\Delta P} = K \times V_0 \times \sqrt{SG} \qquad 3(3)$$

The Wobble Index (WI) measures the ratio between the net calorific value (CV) and the square root of specific gravity (SG):

$$WI = CV/\sqrt{SG} \qquad 3(4)$$

The advantage of detecting the Wobble Index is that it eliminates the need for separately measuring the specific gravity, because the product of Wobble Index [equation 3(4)] and orifice pressure drop [equation 3(3)] results in a value (KQ) which is directly related to the heat flow rate (Q), without necessitating a separate measurement of SG:

$$WI \times \sqrt{\Delta P} = (CV/\sqrt{SG}) \times (K \times V_0 \times \sqrt{SG}) =$$
$$= CV \times K \times V_0 = K \times Q \qquad 3(5)$$

Because Wobble Index can be measured continuously in hazardous areas, this approach provides an on-line method of detecting heat flow rate.

THE BTU FLOWMETER LOOP

The BTU flowmeter for gaseous fuels consists of a heat value measuring instrument and a flow measuring loop with

TABLE 3a
Combustion Constants and Composition of Representative Manufactured and Natural Gases

	Blast Furnace Gas	Coal Gas	Coke Oven Gas	Natural Gas Residual, Follansbee, W. Va.	Natural Gas Sandusky, Ohio	SNG Green Springs, Ohio	LNG Columbia Gulf Coast	NG Columbia Gulf Coast	Refinery Gas	Producer Gas
% Methane, CH_4		34.0	28.5		83.5	98.914	85.136	97.528	27.0	2.6
% Ethane, C_2H_6				79.4	12.5	.01	10.199	1.238		
% Propane, C_3H_8				20.0			3.06	.241		
% Ethylene, C_2H_4		6.6	2.9			.0016			2.7	0.4
% Carbon monoxide CO	26.2	9.0	5.1			.025			10.6	22.0
% Carbon dioxide, CO_2	13.0	1.1	1.4		0.2	.439	.018	.487	2.8	5.7
% Hydrogen, H_2	3.2	47.0	57.4			.61			53.5	10.5
% Nitrogen, N_2	57.6	2.3	4.2	0.6	3.8	.002	.201	.224	3.4	58.8
% Oxygen, O_2			0.5			.007				
% Other*							1.37	.192		
BTU per cu. ft., high (gross) 60°F, 30 in. Hg, satd. H_2O	93	634	536	1868	1047				516	136
BTU per cu. ft., low (net) 60°F, 30 in. Hg, satd. H_2O	91.6	560	476	1711	946				461	128
Flame Temp. °F	2660	3910	3430	3830	3740				3970	3050

*Heavier hydrocarbons and traces of compounds including sulfurs.

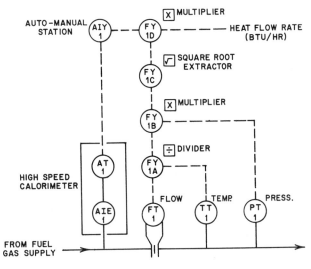

FIG. 3b
BTU flowmeter loop.

an orifice element. When the signals are properly conditioned and multiplied to satisfy equation 3(5), the resulting output provides a BTU flow rate signal as shown in Figure 3b.

APPLICATIONS

As shown in Figure 3b, the BTU measuring loop can be used to generate a true indication of the BTU flow rate for use in combustion control systems. This control system will meter the air in proportion to the BTU flow rate in a feedforward mode as well as respond to the firing rate demand signal of the feedback loop, as depicted in Figure 3c. This feedback signal to the setpoint of FIC-1 modulates the fuel gas control valve to deliver the desired BTU flow rate rather than just maintaining a volumetric gas flow rate.

The illustrated configuration is quite common in combustion optimization or plant updating. In boilers and furnaces that operate on multiple fuels, this control system makes the system easier to control, because by fixing the BTU flow rate

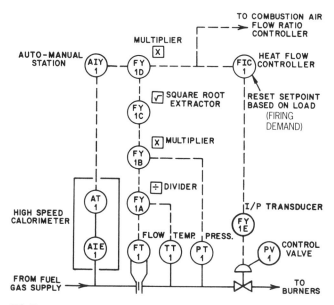

FIG. 3c
Combustion control loop using BTU rate signal.

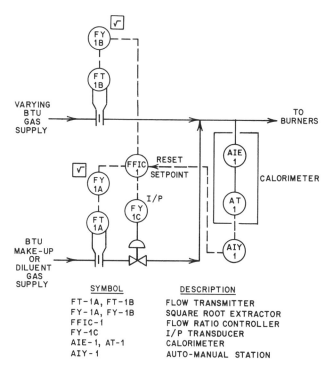

FIG. 3d
Mixing of two gas streams to supply controlled BTU gas.

of the gas fuel, it is feasible to modulate the liquid fuels as a function of the excess air in the flue gas.

Another common application of BTU flowmetering is the blending of an enriching or diluent gas into various fuels to control the BTU of the blend at a constant level. The fuel gas thus produced can be distributed throughout the plant to various boilers, heaters, and gas-fired processes. In this case, the BTU flow rate controller becomes the cascade master of the makeup or diluent flow controller. Figure 3d shows such a flow schematic of a typical blending control loop.

CONCLUSIONS

Using proven, available technology, a BTU flow rate loop can provide a direct signal for use in control applications. The basic elements of such a loop are an orifice flowmeter and a heat value measuring instrument having relatively good speed of response for most burning processes. Measuring BTU flow rate provides a direct and accurate method of controlling and optimizing combustion processes.

Bibliography

Foundos, A.P., "Measuring Heat Release Rate from Fuel Gases," *Instrumentation Technology,* Instrument Society of America, 1977.

Foundos, A.P., "On-Line Optimizing Fuel Gas Composition Variations," *Instrumentation Technology,* Instrument Society of America, 1980.

Hardy, W.M., "Process Analyzers—Predictions for the 1990s and Beyond," Paper #90-0456 at 1990 ISA Conference in New Orleans.

Perry, J., *Chemical Engineers Handbook,* 3rd Edition. New York: McGraw-Hill, 1950, p. 1577.

Reineke, H.F., Patent No. 1,055,259, Federal Republic of West Germany, 1957.

4 Cross-Correlation Flowmetering

B. G. LIPTÁK (1982, 1993)

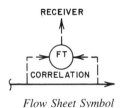

Flow Sheet Symbol

Current Applications:	Pumped paper pulp, pneumatically conveyed coal dust, cement, grain, plastic granules, chalk, and animal food stuffs
Sizes:	½ to 8 in. (13 to 200 mm) diameter pipe with 150# or 300# flanges
Cost:	A 4 in. 150# mass flowmeter with epoxy-resin-lined enameled steel pipe costs $5000
Partial List of Suppliers:	Endress + Hauser Inc. (Germany); Kajaani Electronics Ltd. (Finland)

The oldest and simplest methods of flow measurement are the various tagging techniques. Here a portion of the flowstream is tagged at some upstream point and the flow rate is determined as a measurement of transit time. Variations of this technique include particle tracking, pulse tracking, dye or chemical tracing, including the radioactive types. The advantages of tagging techniques include the ability to measure the velocity of only one component in a multicomponent flowstream without requiring calibration or pipeline penetration. For example, electromagnetic tagging of gas-entrained particles allows for the determination of their speed through the detection of their time of passage between two points that are a fixed distance from each other.

Flowmetering based on correlation techniques[1,2] is similar in concept to the tagging or tracing techniques because it also detects transit time. As illustrated in Figure 4a, any measurable process variable which is noisy (displays localized variations in its value) can be used to build a correlation flowmeter. The only requirement is that the noise pattern must persist long enough to be seen by both detectors "A" and "B" as the flowing stream travels down the pipe. Flow velocity is obtained by dividing the distance (between the identical pair of detectors) by the transit time. In recent years, the required electronic computing hardware, with fast pattern recognition capability, has become available and consequently it is feasible to build on-line flowmeters using this technique.[3]

The following process variables display persistent-enough noise patterns (or local fluctuations) so that correlation flowmeters can be built by using an identical pair of these sensors:

Density
Pressure
Temperature
Ultrasonics
Gamma radiation
Capacitive density
Conductivity

Several of the above process variables, such as temperature,[4,5] gamma radiation, and capacitive density[6] have been investigated as potential sensors for correlation flowmeters. One instrument has been developed utilizing the principle of ultrasonic cross-correlation to measure heavy water flow.[3] Others are available for paper pulp applications using photo-

FIG. 4a
Cross-correlation flowmetering.

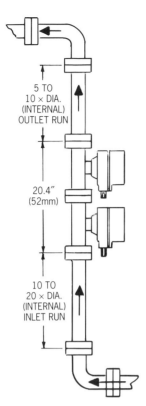

FIG. 4b
Installation requirements of cross-correlation-type solids flowmeter. (Courtesy of Endress + Hauser Inc.)

metric sensors and for solids flow measurement utilizing capacitance detectors (Figure 4b).

When fully developed, correlation flowmetering can extend the ability to measure flow not only into the most hostile process environments, but also into areas of multiphase flow and into three-dimensional flow vectoring.

References

1. Porges, K.G., "On-line Correlation Flowmetering in Coal Utilization Plants," 1980 Symposium on Instrumentation and Control of Fossil Energy Processes, June 9–11, 1980, Virginia Beach, Virginia.
2. Porges, K.G., "Correlation Flowmetering Review and Application," 1979 Symposium on Instrumentation and Control for Fossil Energy Processes, August 20–22, 1979, Denver, Colorado.
3. Flemans, R.S., "A New Non-Intrusive Flowmeter," Transactions, Flow Measurement Symposium, NBS, February 23–25, 1977.
4. Ashton, M.W., and Bentley, P.G., "Design Study For On-Line Flow Measurement by Transit Time Analysis of Temperature Fluctuations," Conference on Industrial Measurement Techniques for On-Line Computers, June 11–13, 1968, London.
5. Boonstoppel, F., Veltman, B., and Vergouwen, F., "The Measurement of Flow by Cross-Correlation Techniques," Conference on Industrial Measurement Techniques for On-Line Computers, June 11–13, 1968, London.
6. O'Fallon, N.M., "Review of the State-of-the-Art of Flow and Analysis Instruments," 1977 Symposium on Instrumentation and Control for Fossil Demonstration Plants, July 13–15, 1977, Chicago, Illinois.

Bibliography

Beck, M.S., Calvert, G., Hobson, J.H., Lee, K.T., and Mendies, P.J., "Flow Measurement in Industrial Slurries and Suspensions Using Correlation Techniques," Transactions of the Institute of Measurements and Control, Volume 4, Number 8, August 1971, England.

Mersh, F., "Speed and Flow Measurement by an Intelligent Correlation System," Paper #90-0632, 1990 ISA Conference in New Orleans.

Robinson, C., "Obstructionless Flowmeters," *InTech,* December 1986.

Spitzer, D.W., "Industrial Flow Measurement," ISA, 1984.

5 Elbow Taps

W. H. HOWE (1969)   **B. G. LIPTÁK** (1982, 1993)

Flow Sheet Symbol

Design Pressure:	Up to 928 psig (64 bar)
Operating Temperature Range:	−330 to +1100°F (−200 to +600°C)
Fluids:	Liquids, vapors, or gases
Differential Pressure:	0 to 10 in. wc (0 to 250 mm)
Sizes:	½ to 20 in. (12 to 500 mm)
Inaccuracy:	±2 to ±10%FS
Cost:	Approximately 1.5 times the cost of the elbow
Partial List of Suppliers:	Turbo Instruments Inc.

A flow measurement using elbow taps depends on the detection of the differential pressure developed by centrifugal force as the direction of fluid flow is changed in a pipe elbow.[1] Taps are located on the inner and outer radii in the plane of the elbow; the diameter which passes through the taps is at either 45 degrees or 22½ degrees from the inlet face of the elbow (Figure 5a).

Elbow taps are easy to implement because most piping configurations already contain elbows in which taps can be located. This guarantees an economical installation and results in no added pressure loss. The measurement introduces no obstructions in the line. Accumulation of extraneous material in the differential pressure connections can plug the elbow taps. Therefore they should be purged if the process fluid is not clean.

As with other head-type primary flow measurement devices, the differential pressure developed by a given flow is precisely repeatable. However, the flow coefficient of an elbow tap calculated from the physical dimensions of the pipe is generally considered reliable to only ±5 to ±10%. This is quite satisfactory for many flow control applications, where repeatability is the primary consideration. If absolute accuracy is desired, a more precise flowmeter should be used or the elbow tap readings should be calibrated, preferably in place and using the working fluid. There is not enough data to establish precise correction factors for effects of upstream disturbances, viscosity, and roughness in pipe and elbow surfaces.

Elbow taps develop relatively low differential pressures. For this reason, they cannot be used for measurement of streams with low velocity. Typically, water flowing at an average velocity of 5 ft/s (roughly 200 gpm in a 4 in. pipe or 45 m^3/hr in a 100 mm pipe) through a conventional elbow with a centerline radius equal to the pipe diameter develops about 10 in. (254 mm) of water differential pressure. This is about the minimum full scale value recommended for reliable measurement. Taps in long radius pipe or tube bends do not develop sufficient differential pressure for good flow measurement at low flow velocities.

The upstream piping is a factor in the installation of elbow taps. It is recommended to provide at least 25 pipe diameters of straight pipe upstream and 10 diameters downstream. The tap holes should be perpendicular to the surface

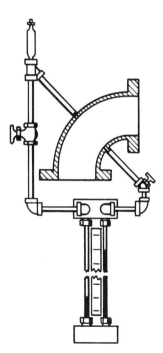

FIG. 5a
Flow detection with elbow taps.

of the elbow and slightly rounded at the pipe surface with no burrs or protrusions. Tap hole diameter should not exceed 1/8 of the pipe diameter. Elbows should be of the flange type with the elbow diameter equal to the pipe diameter. An elbow of smaller diameter than the pipe with a reducer between pipe and elbow has the advantage of higher differential for a given flow. Threaded elbows with the flow section larger than the pipe develop less differential pressure and thereby increase the error. The flow coefficient of a pipe elbow can only be reliably determined if the inside surface of the elbow is smooth. The elbow should be precisely aligned with the pipe making sure that no gaskets are protruding into the flowing stream either at the inlet or outlet of the elbow.

When selecting an elbow for flow measurement purposes, it is preferable to pick one that is located between two horizontal pipe sections. This will guarantee that the pressure taps will be horizontal and material will not accumulate in them. If the elbow were located between a horizontal and a vertical pipe section, the pressure tap on the inner radius would slope upward and the one on the outer radius would slope downward. As the differential pressure instrument is piped to these taps, this piping will not be self-draining. The high and low points in the connecting piping will tend to trap either the vapors on liquid services or the liquid condensate on vapor services. As the total pressure differential to be measured is already low, this interference can make the installation unsatisfactory.

Some tests suggest that the 22.5-degree tap locations provide more stable and reliable readings and are less affected by upstream pipe configuration. On the other hand, the 45-degree tap locations are preferred for bidirectional flow measurement applications.

The mass flow through an elbow can be calculated to an accuracy of about ±10% by the following equation:

$$W = 244 \sqrt{rhD^3\rho} \qquad 5(1)$$

where

- W = mass flow in pounds per hour
- r = elbow center-line radius in inches
- D = elbow (pipe) diameter in inches
- h = differential pressure in inches of water
  Water density is assumed to be 62.32 pounds per cubic foot (one kilogram per liter), corresponding to 68°F (20°C).
- ρ = operating density in pounds per cubic foot

If it is desired to calculate the pressure differential produced by a pipe elbow at a particular flow rate, the following equation can be used:

$$h = W^2/59{,}536 r\rho D^3 \qquad 5(2)$$

In addition to elbow taps, the differential pressure produced by centrifugal forces can also be converted into flow readings by other configurations. These include the Winter-Kennedy taps installed in the scroll case of hydraulic turbines. Another design is the full circle loop with taps located at the midpoint of the loop. It is claimed that this design provides high accuracy and minimum sensitivity to upstream piping configuration.

References

1. Bulletin #289, Wallace M. Lansford, Engineering Experiment Station, University of Illinois, Urbana, Illinois.

Bibliography

Hauptmann, E.G., "Take a Second Look at Elbow Meters for Flow Monitoring," *Instruments and Control Systems,* October 1978, pp. 47–50.

Moore, D.C., "Easy Way to Measure Slurry Flowrates," *Chemical Engineering,* October 2, 1972, p. 96.

6 Flow Switches

B. G. LIPTÁK (1982, 1993)

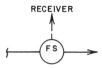

Flow Sheet Symbol

Types of Designs:	A. Bypass or differential pressure B. Capacitance or RF inductance C. Hot wire anemometer D. Paddle, vane, or rotor E. Thermal F. Ultrasonic (including Doppler) G. Valve body H. Variable area or piston I. Microwave J. Surface charge
Types of Services:	0. Liquids (flow or no-flow) 1. Liquids (adjustable high or low flow) 2. Gases, vapors (adjustable high or low flow) 3. Solids (flow/no-flow, some adjustable)
Costs:	Small oil, air, and water switches (types A and D) can be obtained for $50 to $150; paddle type flow switches for air ducts cost $150 and for water service through 1 inch (25 mm) NPT fitting and stainless steel construction, $375; types B, E, and H units range from $150 to $800 in small sizes and standard materials; types C and F types start at around $500 and range to over $1000; solids flow switch types I and J cost around $1000 Type G in 1 inch (25 mm) size and bronze construction is $120, in stainless steel $400; while a 2-inch (50 mm) unit in these materials is $300 and $600 respectively.
Partial List of Suppliers:	ABB Kent Inc. (B-3); AM Sensors Inc. (I-3); W.E. Anderson Div. of Dwyer (A-2, D-1); Aqualarm (D-1); Auburn International Inc. (J-3); Babbit International Inc. (B-3); Brooks Instrument Div. of Rosemount (F-1, H-1,2); Cannonbear Inc. (I-3); Chem-Tec Equipment Co. (D-1,2); Cosense Inc. (F-0); Custom Services (D-0); Delta Controls Corp. (B-0, D-1); Delta M Corp. (E-1,2); H.G. Dietz Co. (D-1,2); Endress + Hauser Instruments (B-0, F-3); ETA Control Instruments (E-1,2); FCI Fluid Components Inc. (E-1,2); Fischer & Porter Co. (H-1,2); Flow & Level Controls (E-1,2,3); GEMS Express Service, Imo Industries (D-1,2); Harwil Corp. (D-1); Hedland Flow Meters (G-1,2); Hydril Co. (E-0); ICC Federated (H-1,2); Intek Inc. (E-1,2); Johnson Controls (D-1,2); Kay Ray (F-1,3); Ketema, Schutte & Koerting Div. (H-1,2); Kobold Instruments Inc. (D-1,2); Krone America Inc. (H-1); Kurz Instruments Inc. (C-2); Magnetrol International (G-1); Meriam Instrument (A-1,2); Midwest Instrument (A-1,2); Monitor Manufacturing (I-3); Oilgear Co. (D-1); Omega Engineering Inc. (D, F, H-1,2); OPW Div. of Dover Corp. (D-1); Orange Research (H-1); Princo Instruments Inc. (B-1,3); Proteus Industries Inc. (D-1); Revere Corp. of America (D-1); Robbins and Myers Inc. Moyno Div. (B-0); SOR Inc. (B, D, H-1); TSI Inc. (C-2); Turbo Instruments Inc. (H-1); Universal Flow Monitors (D, H-1); Zi-Tec Instrument Corp. (F-1,2);

Flow switches are used to determine if the flow rate is above or below a certain value. This value (the setpoint) can be fixed or adjustable. When the setpoint is reached, the response can be the actuation of an electric or pneumatic circuit. When the flow switch is actuated, it will stay in that condition until the flow rate moves back from the setpoint by some amount. This difference between the "setpoint" and the "reactivation point" is called the switch "differential." The differential can be fixed or adjustable. If the differential is small, the switch is likely to cycle its control circuit as the flow fluctuates around its setpoint.

In certain applications a manual reset feature is desirable. This will guarantee that once the switch is actuated, it will not be allowed to return to its preactuation state until manually reset by the operator. This feature is designed to require the operator to review and eliminate the cause of the abnormal flow condition before resetting the switch.

All instruments that can measure flow can also be used as flow switches. On the other hand, if only a flow switch is required for a particular application, the installation of indicating or transmitting devices cannot be economically justified. Therefore, in this section only the direct flow switches will be discussed. Indirect devices, such as differential pressure switches piped around orifice plates, or receiving switches connected to the output signals of transmitters, are not covered.

DESIGN VARIATIONS

The least expensive and therefore the most widely used are the various paddle type devices. At "no flow" the paddle hangs loosely in the pipe in which it is installed. As flow is initiated, the paddle begins to swing upward in the direction of the flow stream. This deflection of the paddle is translated into mechanical motion by a variety of techniques including a pivoting cam, a flexure type, or a bellows assembly. The mechanical motion causes the switch to open or close. If a mercury switch is used the mechanical motion drives a magnetic sleeve into the field of a permanent magnet which trips the switch. A hermetically sealed switch will be directly actuated by the permanent magnet as it moves up or down according to the paddle movement. If a microswitch is used the translated motion will cause direct switch actuation.

The range and actuation point of paddle switches can be changed and adjusted by changing the length of the paddle. For any given pipe size, the actuation flow rate decreases as the paddle length increases.

Paddle-type flow switches are sensitive to pipeline turbulence, pipeline vibration, and installation configuration. For these reasons, it is advisable to provide them with the equivalent of a 10 pipe diameter straight upstream run, to use dampers if pipe vibration or pulsating flow is expected, and to readjust their settings if they are to be mounted in vertical upward flowlines. The conventional paddle-type designs are incapable of distinguishing low flow velocities from no-flow conditions. Therefore, if low flows are to be detected, the folding circular paddle should be used (Figure 6a) which permits the full diameter paddle to fold back upon itself, to minimize pressure drop.

In smaller sized pipelines where it is desired to provide local flow indication, in addition to the flow switch action, the variable-area-type flow switches can be considered. If the vertical upward flow configuration of the rotameter design is not convenient from a piping layout point of view, the circular swinging vane design, illustrated in Figure 6b, can be considered. In existing systems, the clamp-on-type ultrasonic liquid flow switch can be a convenient solution because it does not require process shutdown or pipe penetration. If the purpose of the flow switch is to protect pumps

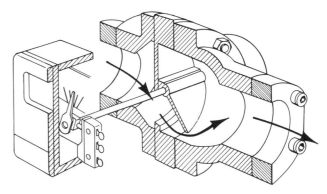

FIG. 6b
Swinging vane flow switch.

from running dry, the wafer-type capacitance insert unit is a good choice.

Flow switch reliability is increased by the elimination of moving parts, so that pipe vibrations or fluid flow pulses will not cause erroneous switch actuation. One of the most popular solid-state designs is the thermal flow switch. All heat-actuated flow switches sense the movement or stoppage of the process stream by detecting the cooling effect (temperature change) on one or more probes. They are available both in the flow-through and in the probe configuration. One design consists of a heater probe and two sensor probes, connected in a Wheatstone bridge. When the flow stops, an inbalance in the bridge circuit occurs, as illustrated in Figure 6c. The main advantage of this design is its ability to detect very low flow velocities (Table 6d). Its main limitation is that it cannot respond instantaneously to flow changes. Depending on switch adjustments and on type of process fluid, the speed of response will vary from 2 seconds to 2 minutes.

The valve body-type flow switches are built into a pipe fitting that resembles the body of a single seated globe valve. A flow disk is allowed to move in a vertical direction within what is normally considered the valve seat. A magnetic sleeve is mounted above the flow disk and as the disk is lifted upward due to initiation of flow, a mercury switch is

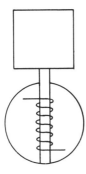

FIG. 6a
Folding paddle switch.

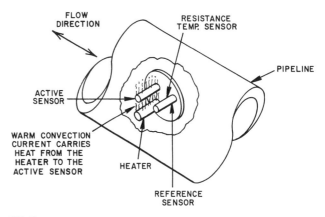

FIG. 6c
Thermal flow switch.

TABLE 6d
Minimum Settings for Flow Switches

Flow Switch Type	Minimum Velocity FPM (m/min)	
	Air	Water
Thermal	10 (3)	0.5 (0.15)
Variable Area	300 (91)	20 (6)
Ultrasonic	—	60 (18)
Paddle	300 (91)	60 (18)

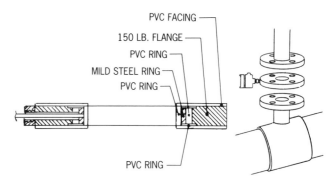

FIG. 6f
Capacitance-type flow/no-flow switch. (Courtesy of Endress + Hauser Inc.)

actuated by the movement of the magnetic sleeve into the field of the externally mounted permanent magnet.

A bypass-type switch (Figure 6e) has an externally adjustable vane that creates a differential pressure in the flow stream. This differential pressure forces a proportional flow through the tubing that bypasses the vane. A piston retained by a spring is in the bypass tubing and will move laterally as flow increases or decreases; the piston's movement actuates a switch. Bypass flow switches can be used for fairly low flow rates and their ability to be externally adjusted is a very desirable feature.

Capacitance-type ring-sensors can detect the absence or presence of liquids or slurries, which usually relate to flow/no-flow conditions. This can prevent pumps from running dry or cavitating. This ring-sensor (Figure 6f) is a capacitance switch that fits between two flat faced flanges in standard pipe sizes between 2 in. and 8 in. (50 to 200 mm). When the ring-sensor contains air (no flow condition), it detects a dielectric constant of 1. When flow starts, it detects the dielectric constant of the flowing fluid, which is greater than 1, which causes a corresponding increase in radio frequency (MHz) current flow. The circuitry detects this

current flow and operates the switch accordingly. An adjustable time-delay of 0 to 20 seconds is provided to protect from premature shutdowns. The materials of construction are usually PVC and steel, with the electronics housed in explosion-proof or water-tight aluminum or thermoplastic housing. The operating temperature range for the standard ring-sensor is 0 to 212°F (-18 to 100°C).

SOLIDS FLOW SWITCHES

In pneumatically conveyed or gravity-flow-type solids handling systems it is important to quickly detect blockages or other abnormal conditions, such as feed loss, bin bridging, cyclone overflow, or the rupture of baghouse filters. The flow of certain powders such as flyash, cement, or alumina are particularly difficult to detect. Abnormal flow conditions have in the past been detected by level sensors; tilt switches; and capacitance, radiation, sonic, and optical devices. In addition to these, one can also measure the flow rate of these solids' flows by the following techniques.

One solids flow switch, "Triboflow," collects on its probe surface the static charges of the solid particles passing over its surface. The resulting current is related to the flow rate of solids. These probes are sensitive enough to detect flow increases as small as what results from baghouse rupture. These probe-type solids flow switches are inexpensive and can be installed in hazardous areas.

Microwave switches detect the flow of solids by detecting motion, or the absence of it. In the microwave-type motion detector the transducer emits a 24 GHz signal into the flowing solid stream and analyzes the reflected frequency (Doppler effect) to determine the speed of the object that reflected it. The sensitivity of the solids flow switch is adjustable, so that it might be used to detect flow/no-flow or trip at a velocity as low as 6 in./min (15 cm/min) when the pipe is full or at a velocity of one particle every 5 seconds in a free-falling gravity flow system. Units are available in aluminum or stainless steel and can be connected to a pipe by a coupling or flange (Figure 6g) or can look through

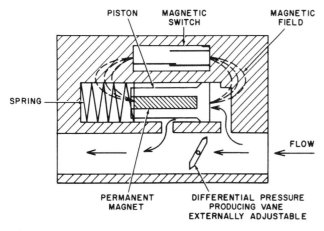

FIG. 6e
Bypass flow switch.

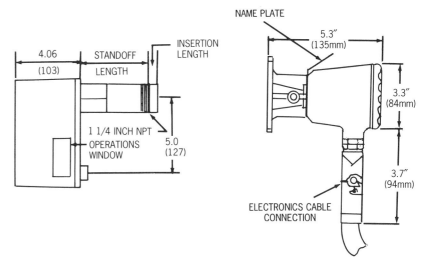

FIG. 6g
Microwave solids flow switches with screwed and flanged connections. (Courtesy of TN CannonBear and Endress + Hauser, respectively)

windows or nonmetallic walls without any openings. The units are intrinsically safe and can be used at working pressures up to 15 PSIG (1 bar). The switch can also observe motion at a distance of several feet from the detector and can tolerate 0.5 in. of nonconductive coating buildup or 0.1 in. of conductive coating buildup.

Bibliography

Clark, W.J., *Flow Measurements,* Pergamon Press, 1967.
"Flow Switches," *Measurements and Control,* June 1991.
Lipták, B.G., "Online Instrumentation," *Chemical Engineering,* March 31, 1986.
Spink, L.K., *Principles and Practices of Flow Meter Engineering,* 9th edition, The Foxboro Co., 1967.

7 Jet Deflection Flow Detectors

D. S. KAYSER (1982)   **B. G. LIPTÁK** (1993)

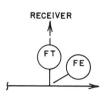

Flow Sheet Symbol

Maximum Process Pressure:	10 PSIG (0.7 bar)
Design Temperature:	450°F (232°C) standard and up to 1200°F (650°C) special
Standard Materials:	316 stainless steel
Connection and Insertion:	Standard connection is 3 in. flanged; insertion depth is adjustable from 0 to 60 in. (0 to 1.5 m)
Air (Nitrogen) Requirement Pressure:	10 to 90 PSIG (0.7 to 6.2 bar) over process pressure
Flow:	2.5 and 5.0 SCFM (71 to 142 l/m)
Velocity Ranges:	0 to 50 fps and 0 to 85 fps (0 to 15 mps and 0 to 26 mps)
Output Differential Range:	0 to 80 in. H_2O and 0 to 130 in. H_2O (0 to 20 kPa and 0 to 32 kPa)
Inaccuracy:	±2% full scale (if sensor is inserted to average velocity point in the duct)
Rangeability:	20:1
Cost:	About $3000; varies with accessories
Partial List of Suppliers:	Fluidynamic Devices Ltd.; Lear Siegler Measurement Controls Corporation

Flow rate may be inferred from a measurement of gas velocity in a pipe or duct. The velocity measurement can be made with the jet deflection flow detector (Figure 7a). Air or another gas that is compatible with the process is expelled from the nozzle, forming a jet with the pressure profile shown. Under conditions of no process flow, the profile is symmetrical in relation to the two receiver ports and the differential pressure between the two ports is zero. As process flow increases, the jet is deflected by an amount that is related to the velocity of the process stream. Deflection of the jet causes increased pressure at the downstream port and decreased pressure at the upstream port. The geometry of the ports is such that the change in differential pressure is linearly proportional to process stream velocity over the useful range of the element. The actual value of the pressure differential is a function of the product of process velocity and the square root of process gas density, but if the density is constant, it varies only with velocity.

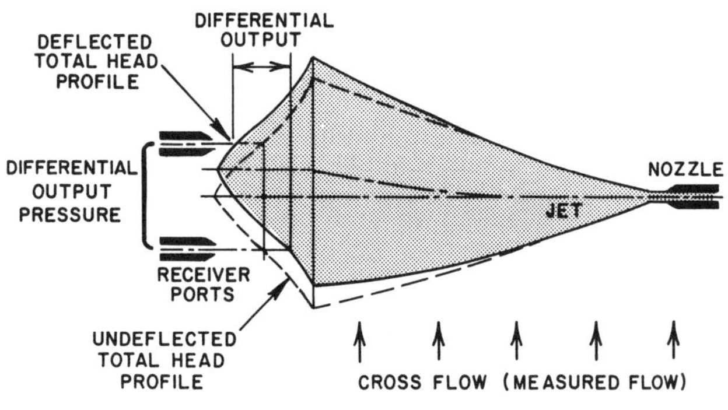

Figure 7a also shows how the pressure profile of the jet shifts as the velocity of the process stream increases.

This flowmeter is similar to the conventional pitot tube in some respects. These include its negligible pressure drop, its high speed of response, and its retractable design, which can be used for wet tapping or to measure flow profiles by traversing the pipe or duct cross section. Features that are superior to those of conventional pitot sensors include the provisions of continuous backpurge and the auxiliary cleaning jets, which keep the receiver ports clean; the fact that the element is heated to a temperature above the dew point, and therefore condensation is avoided; and the linear output signal which is much stronger then that of a conventional pitot. At an air (or nitrogen) supply pressure of 50 PSIG (345 kPa), the output differential pressure signal generated by this design is 100 times that of a pitot tube and amounts to about 1.5 in. of water column per each fps (0.3 mps) of process gas velocity. If the particulate concentration in the process fluid (flue or stack gases) is separately measured, this reading when combined with gas velocity will yield mass emission rate of particulates.

The relationship between stream velocity and flow rate is:

$$Q = 60VA \qquad 7(1)$$

FIG. 7a
Pressure profile of jet deflection flow detector.

where

Q = flow rate, actual cubic feet per minute
V = velocity, feet per second
A = pipe area, square feet

The point velocity measurement obtained in a rectangular or circular duct will not necessarily correspond to the flow rate because the velocity profile is not uniform. In case of laminar flow in a circular duct the profile is parabolic with the maximum velocity at the center and zero velocity at the walls. In this case the maximum velocity is twice the average; therefore, the reading taken at the center will be twice the average for the laminar flow case. In turbulent flow in a circular duct the average velocity point is located at approximately 25% of the radius as measured from the duct wall. The accurate determination of average velocity in rectangular ducts is more complicated and no rules of thumb can be given.

In many cases the point of average velocity will not be where it is expected because of disturbances introduced by the upstream piping configuration. The upstream piping should be straight for at least 20 pipe diameters to allow for the disturbances to smooth out. When this installation requirement cannot be met, an average velocity point can sometimes be established by using the flow element to make a traverse of the pipe, taking readings at a number of points across the cross section.

A traverse should always be made on rectangular ducts if accurate measurement is required. Traversing rectangular ducts is complicated by the need for traversing in two or more planes, and by the fact that average velocity points can shift as flow rates vary in the rectangular (or circular) duct.

HOT-TAPPING

Probe-type instruments, such as the jet deflection element or the pilot tube, can be installed so that they can be removed for inspection without shutting the process down. They can also be hot-tapped into a line, meaning that the installation can be made without shutting the process down. Figure 7b shows the detail of such an installation that enables removal of the probe while the pipe is under pressure. To remove the element, the gland nut is loosened just enough to allow the shaft of the element to be withdrawn until the element is in the chamber outside of the gate valve. The valve is then closed, the chamber is vented, and the gland nut is removed, allowing safe removal of the element. Normally the jet deflection element is installed in low pressure or vacuum piping and thus the risk of having the element blow out during removal is slight. Nonetheless, some users install stop rods or safety chains in order to completely eliminate the possibility of a blowout.

The hot-tapping procedure is shown step-by-step in Figure 7c. First a weld neck flange is welded to the pipe that is to be tapped. Next a flanged gate valve is made up to the flange. The third step is to bolt a hot-tap machine to the downstream side of the gate valve and open the valve. The hot-tap machine is used to drill the opening into the process pipe. The pressure seal on a hot-tap machine is similar to that of a control valve packing box, allowing taps to be made in pipes that contain relatively high pressures. The drilling assembly shown in the figure is hand-operated, but units are also available with pneumatic or electric drives. A scale on the machine enables the operator to know how far he has drilled into the pipe. After the hole is made, the bit is retracted, the valve is closed, and the hot tap machine is removed. The tap is now ready for the installation of the flow element, gland nut, and chamber, as shown in Figure 7b. Rigid safety procedures must be enforced during hot-tapping operations, particularly if the process is flammable or hazardous. Hot-tapping can only be performed if the piping specification does not require that welds be stress-relieved, and if there is flow in the pipe, so that the heat of welding and drilling is removed.

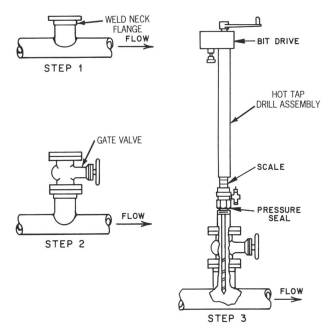

FIG. 7c
Hot-tap procedure.

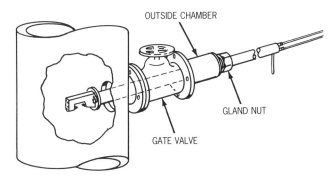

FIG. 7b
Installation that allows removal of the element under pressure.

CONCLUSION

Jet deflection flow detectors are appropriate for flow measurement in low-pressure circular and rectangular ducts. They can be purged or flushed and can be removed for inspection and cleaning; they are suitable for dirty, abrasive, corrosive, or plugging services. The accuracy of ±2% of full scale is generally acceptable for measurements around flare headers, stacks, and air ducts, although it should be remembered that overall installation accuracy is dependent on inserting the element to the point of average velocity.

Bibliography

Brooks, E.F., et al., "Continuous Measurement of Total Gas Flow Rate from Stationary Sources," Washington, D.C.: TRW Systems, U.S. Dept. of Commerce, P.B. 241894.

Federal Register, Vol. 42, No. 20, Monday, January 31, 1977, p. 5936.

Federal Register, Vol. 42, No. 160, Thursday, August 18, 1977, p. 41754.

Spitzer, D.W., *Industrial Flow Measurement,* Instrument Society of America, 1984.

8   Laminar Flowmeters

R. SIEV (1969)   **J. B. ARANT** (1982)   **B. G. LIPTÁK** (1993)

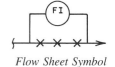

Flow Sheet Symbol

Design Pressure:	Up to 5000 PSIG (34 MPa)
Design Temperature:	Up to 300°F (150°C) normally, but can be higher with special designs
Material of Construction:	Stainless steel, aluminum, or any alloy available in small bore tubing
Fluids:	Liquids and gases
Flow Range:	0.0001 to 2000 scfm for gases (3 cm^3/min to 57 m^3/min) 0.0003 to 10 gpm for liquids (1 cm^3/min to 38 l/min)
Inaccuracy:	0.5% to 1% of actual flow for commercial gas flow elements, if calibrated
Flow Turndown:	10:1 minimum
Flow Characteristic:	Linear to approximately linear
Costs:	A ½ in. (13 mm) stainless steel laminar flow element costs $600; a 2 in. (50 mm) unit costs $1500; and a 16 in. (300 mm) all-stainless unit costs $10,200. The differential pressure readout devices are additional to the above element costs
Partial List of Suppliers:	Meriam Instrument Div. of the Scott and Fetzer Co.; National Instrument Laboratories, Inc. Units are for gas flow only.

Laminar flowmeters fill a special need in flow measurement where the requirements might include low to extremely low flow rates, linear calibration and low noise, the ability to measure high-viscosity liquids, or steady low flow repeatability and control accuracy. Laminar flowmeters are intended for very low flow rates where other types of meters are either marginal in performance or cannot be used at all. Laminar flowmeters can be constructed by various methods, but the most common method is with capillary tubes. Hence, the terms "laminar flowmeter" and "capillary flowmeter" are virtually synonymous. Proprietary commercial units use other matrix shapes and are intended for use with gases (Figure 8a). Where gas is metered, it is preferable to use calibrated commercial units instead of undertaking the design of a laminar flowmeter.

The flowmeter consists of the laminar flow element and a differential pressure measuring instrument. While the flow is theoretically linear with pressure drop, in practice some nonlinearities are often encountered. In most cases, these are of little consequence.

The theory for laminar flowmeters is based upon the Hagen-Poiseuille Law for laminar flow and Reynolds number as a means of defining the type of flow. Both are required to investigate and design a laminar flow element. More detailed explanations and discussions of theory can be found in any standard textbook on fluid mechanics.

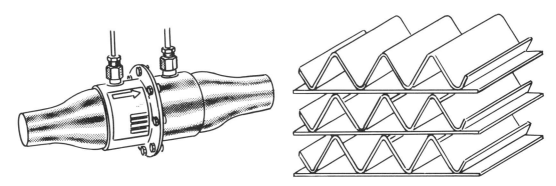

FIG. 8a
The laminar flowmeter and its matrix element with miniature triangular duct passages with under 0.1 mm effective diameters. (Courtesy of Meriam Instrument Div. of the Scott and Fetzer Co.)

THEORY

Fluid flow in pipes and tubes is characterized by a nondimensional number called the Reynolds number. Up to approximately Reynolds number 2000, the flow is called laminar, viscous, or streamline flow. Above 10,000 the flow is called fully developed turbulent. The region between 2000 and 10,000, where the flow is shifting from laminar to turbulent, is not clearly defined but is called transitional. Generally, laminar flow elements are restricted to numbers under 2000 and most commonly well below 1200. There are some methods that will enable a capillary element to be used satisfactorily up to a Reynolds number of 15,000 with a modest sacrifice in error and linearity.

Reynolds number is defined by the following equations:
For liquid flow

$$\text{Re} = \frac{50.7\,\rho Q}{D\mu} \text{ or } \text{Re} = \frac{6.32\,W}{D\mu} \qquad 8(1)$$

where

Re = Reynolds number
ρ = density (lb/ft^3) at flowing temperature
Q = flow rate (gal/min)
D = internal tube diameter (in.)
μ = viscosity of flowing temperature (centipoise)
W = flow rate (lb/hr)

For gas flow

$$\text{Re} = \frac{6.32\,\rho Q}{D\mu} \text{ or } \text{Re} = \frac{6.32\,W}{D\mu} \qquad 8(2)$$

where

ρ = density at standard conditions (lb/ft^3)
Q = flow rate (scfh)
and other units are defined the same as for liquid

TABLE 8b

Gas Properties under the Standard Conditions of 29.92 in. of Mercury and 70°F (760 mm of Mercury and 21°C)

Gas	Density (lb/ft^3)	μ Viscosity Micropoises	Specific Gravity
Air	0.0749	181.87	1.000
Argon	0.1034	225.95	1.380
Helium	0.0103	193.9	0.138
Hydrogen	0.0052	88.41	0.0695
Nitrogen	0.0725	175.85	0.968
Oxygen	0.0828	203.47	1.105
Carbon Dioxide	0.1143	146.87	1.526

For the laminar flowmeter shown in Figure 8a the Reynolds number is limited to a range of 150 to 300 and is calculated as:

$$\text{Re} = 228(SG)(P)(\Delta P)/m\mu \qquad 8(3)$$

where

SG = specific gravity relative to air (Table 8b)
P = flowing gas pressure in inches of mercury absolute
ΔP = differential pressure in inches of water
$m\mu$ = viscosity of the flowing gas in micropoise (Table 8b)

Hagen-Poiseuille Law

Once the tube inside diameter required to give laminar flow according to the Reynolds number calculation has been defined the length of the capillary has to be determined to design the laminar flowmeter system. These equations are as follows:

For liquid flow

$$L = 1.5876 \times 10^3 \frac{\Delta P D^4}{\mu Q} \qquad 8(4)$$

or

$$L = \frac{\Delta P D^4 \rho}{7.86 \times 10^5 \, \mu W} \qquad 8(5)$$

where

L = length of tube (in.)
ΔP = differential pressure drop (in. water)
D = tube internal diameter (in.)
μ = viscosity at flowing temperature (centipoise)
ρ = density at flowing temperature (lbm/ft^3)
Q = flow rate (gal/min)
W = flow rate (lbm/hr)

Equation 8(5) can also be used for calculating a gas flow capillary element provided the value of ΔP is no greater than 10% of the inlet pressure. Otherwise, changes in gas density, specific volume, and flow velocity cause too many complications in the calculations. While the calculation is in weight units, this can be easily converted to read in any desired scale units.

Design Parameters

There are a number of guidelines for successful design of a laminar flowmeter.

1. The differential pressure drop can range from 5 to 800 in. of water (1.24 to 200 kPa).
2. (L/D)/Re should be a minimum of 0.3; for best linearity, a value of 0.6 or greater is preferable. Large L/D ratios and/or lower Reynolds numbers contribute to accuracy. For example, the entrance effect for laminar flow is negligible if (L/D)/Re > 0.3 and Re < 500.
3. The area of the flow conduit preceding the capillary should be a minimum of 20 times the capillary area.
4. The differential pressure instrument pressure connections

should be located 100 to 200 capillary diameters from the capillary ends.

5. A filter capable of removing particles 0.1 in. (2.54 mm) or larger than the capillary internal radius should be installed upstream of the system.
6. The metering system should be sloped up for liquids to permit gas venting and sloped down for gases to permit liquid draining.
7. Examination of the Hagen-Poiseuille equation shows that viscosity is a primary variable; changes in viscosity can result in large flow measurement errors. With a known fluid or composition, the only thing that affects viscosity is temperature. For this reason, the temperature must be known and held essentially constant. This can be done by immersing the metering system and measuring capillary in a constant temperature bath as shown in Figure 8c. If the flow is measured in weight units such as lbs/hr then fluid density must be known. Fluid density also varies with temperature, but controlling the temperature to fix viscosity will also fix density. With some fluids, cooling may be required instead of heating, but the overall principle is the same.

Design Calculations for Liquid Service

Based upon the flow rate and the viscosity of the fluid, select a tube internal diameter that will result in a Reynolds number within the laminar range and preferably less than 1200. Calculate the length of tubing required using the selected tube diameter to ensure that it is a reasonable length and that it meets the (L/D)/Re criteria. By working back and forth between the various equations, the system can be tailored to meet almost any design criteria. For example, let us assume that it is desired to design a capillary flowmeter to measure a small liquid catalyst stream, and the basic data for the catalyst flow is as follows:

Maximum flow capacity: 50 lbm/hr
Viscosity: 20 cps at 100°F
Density: 53.8 lbm/ft^3
Desired instrument ΔP: 100 in. of water

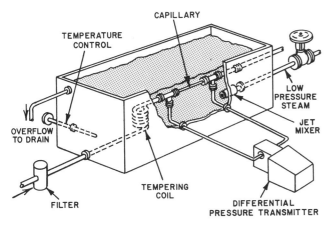

FIG. 8c
Typical capillary with constant temperature bath.

Small diameter standard stainless steel tubing that is readily available should be used. To design as linear and accurate a flowmeter as possible, a tube bore that provides a large (L/D)/Re is desirable. To minimize plugging problems and to enable the use of a filter that won't clog easily, start by looking at a 3/16 × 0.032 in. wall thickness tubing with a nominal internal diameter of 0.1235 in. From Equation 8(1):

$$\mathrm{Re} = \frac{6.32\,W}{D\mu} = \frac{6.32 \times 50}{0.1235 \times 20} = 128 \qquad \mathbf{8(6)}$$

This is well into the laminar range, so the length of the flow element can be calculated to determine if it will make a reasonable design. From Equation 8(5):

$$L = \frac{\Delta P D^4 \rho}{7.86 \times 10^{-5}\,\mu W} = \frac{100 \times 0.1235^4 \times 53.8}{7.86 \times 10^{-5} \times 20 \times 50} = 15.7 \text{ in.} \qquad \mathbf{8(7)}$$

$$(L/D)/\mathrm{Re} = (15.7/0.1235)/128 = 0.993 \qquad \mathbf{8(8)}$$

This is an easy length to work with in fabricating a meter element and a constant temperature bath and looks like a reasonable design based upon the criteria.

ERRORS

Changes in viscosity and density can result in flow measurement errors. Viscosity changes in liquid due to temperature can be substantial, while density changes are more moderate. With gases, the reverse is usually true, with temperature having more influence on density and less on viscosity. The need for careful control of operating temperature to minimize these effects must be emphasized.

From Equations 8(4) and 8(5), it can be seen that internal diameter of the tube is very important, because it is multiplied to the fourth power. While high-quality tubing will be very close to published specifications, manufacturing tolerances will result in variations from these dimensions both laterally and longitudinally. If the actual effective internal diameter of the capillary tube differs by 1% from the value used in the calculation for a given ΔP, an error of about 4% will result. Therefore the laminar flowmeter should be calibrated on a known fluid before use, and appropriate design adjustments should be made as necessary.

To measure the true capillary differential pressure drop per the Poiseuille equation, it would be necessary to put the pressure taps into the capillary at the calculated L dimension. This is impractical because of the small tubing. A pressure tap must be perfectly flush with the inside of the tube and must be clean with no burrs or other projections into the tube. Otherwise, considerable differential pressure measurement error will result. Using practical methods of constructing a capillary flowmeter, there are three additional sources of pressure drop in addition to the capillary loss. These are all additive and will give a greater indicated pressure drop than the capillary flow alone. These three sources of error are inlet loss, exit loss, and capillary entrance loss. These losses also contribute to nonlinearity.

There is very little loss from the entrance fitting into the capillary tube if laminar flow conditions exist. But if the piping cavity ahead of the capillary is extremely large relative to the capillary (approximating a reservoir) and the fluid velocity is thus extremely low (approaching zero), there can be an inlet effect and pressure loss.[1] This is due to the sudden contraction from the large reservoir to the small tube bore, forming a bell-mouth shape approach flow. This loss can be expressed by:

$$\Delta P_i = \frac{2.8 \times 10^{-7} W^2}{D^4 \rho} \qquad 8(9)$$

This equation is derived from Bernoulli's equation for flow out of a reservoir.

When the fluid exits the capillary, the flow path enlarges. If the piping is similar to that described under inlet loss, the loss can be calculated by:

$$\Delta P_e = \frac{5.6 \times 10^{-7} W^2}{D^4 \rho} \qquad 8(10)$$

Entrance loss occurs in addition to the normal capillary pressure drop in the initial fluid path distance, or to state it in another way, for a short distance the pressure drop is higher than that predicted by the Poiseuille equation.[2,3] The additional loss is due to the work expended in the formation of the parabolic velocity distribution profile characteristic of laminar flow. It can be expressed in terms of an equivalent length of capillary, L_{eq}, added to that calculated by the Poiseuille equation. Refer to Figure 8d for determining the L_{eq}.

The following equation can be used for the pressure drop:

$$\Delta P_{en} = \frac{1.96 \times 10^{-7} W^2}{D^4 \rho} \qquad 8(11)$$

Table 8e can be used as a quick guide for judging the design factors that will minimize overall entrance effects. For the conditions given in the table, the error involved will

TABLE 8e
L/D Ratio to Minimize Entrance Effect

Re	10	50	100	500	1000	2000
L/D>	15	75	150	750	1500	3000

be less than 1%. In general, the effect of all of the above errors will be minimized if the Reynolds number is low, the laminar flow element is long, and the pressure drop is high. The overall error can be calculated by this equation as:

$$\text{Percent error} = \frac{\Delta P_i + \Delta P_e + \Delta P_{en} \times 100}{\Delta P} = \frac{0.367 W \times 100}{\mu L} \qquad 8(12)$$

RANGE EXTENSION TECHNIQUES

There are two techniques to expand the range capability of laminar flow elements. One is to use a number of capillary tubes in parallel. The other is to use a tight helical coil capillary. The choice of technique depends upon such factors as desired flow rate, nonlinearity requirements, Reynolds number, capillary length, and system space design limitations.

If the amount of flow desired is greater than can be conveniently handled by a single capillary, the flow can be split into as many smaller units as necessary.[4] Units with matrix elements (Figure 8a) or with over 900 individual capillary tubes have been successfully built and used. The mechanical construction of multi-parallel capillaries can be a problem. Tube packing voids may not affect meter operation, but add considerable difficulty to calculating the meter range. Normally, it is best to eliminate the voids by filling the spaces with solder, braze material, or plastic resin; the filler material chosen will depend upon fluid compatability and operating conditions. Overall, it is a tricky mechanical design.

Coiling a length of straight capillary results in a flow phenomenon called the Dean effect. When a fluid flows through a curved pipe or coil, a secondary circulation of fluid, known as a double eddy, takes place at right angles to the main direction of flow. This circulation accounts for the fact that the pressure drop in curved pipe is greater than in a corresponding length of straight pipe. The Dean effect stabilizes laminar flow and raises the Reynolds number at which turbulent flow starts. It has been established that this will allow properly designed coiled capillaries to be operated up to a Reynolds number of 15,000.[5] The Reynolds number at which laminar flow can be sustained for various coil curvature ratios is called the critical Reynolds number. It is a function of the internal diameter of the tube and the coil tightness or diameter. Table 8f gives the approximate critical Reynolds number at which laminar flow can be sustained for various coil curvature ratios.

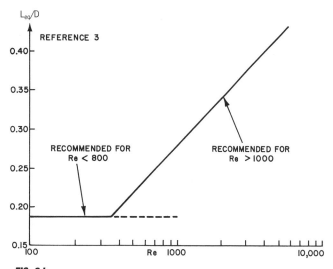

FIG. 8d
Equivalent length of capillary (L_{eq}).

TABLE 8f
Critical Reynolds Number Versus Coil Curvature Ratio

Coil Curvature Ratio (Dc/D)	Critical Reynolds Number (Re)$_c$
Straight Pipe	2100
2000	2700
1000	2900
500	3200
100	4600
50	5700
10	10,000
9	15,000

In this table, D is the tube inside diameter and D_c is the mean coil diameter, centerline to centerline. From a practical viewpoint, the ratio of $D/D_c = 1/9$ is equivalent to the maximum allowable critical Reynolds number of 15,000 and can be used as a safe design in most cases.

The pressure drop of laminar flow through coils can be expressed in terms of an equivalent length, L_e, of straight pipe of the same diameter and shape which will have the same friction loss as the curved pipe. The ratio of the equivalent to actual coil length, L_e/L, is a function of the Dean number, or $Re/(D_c/D)^{1/2}$, as shown in Figure 8g. This curve is accurate to about ±5%.

The equation for calculating the length of a coiled capillary required to meet a specific metering design is expressed by:

$$L = \frac{\Delta P\, D^4\, \rho}{7.86 \times 10^{-5}\, \mu WC} \qquad 8(13)$$

where C = the coil factor correction

The coil factor correction is a function of the term $Re/(D/D_c)^{1/2}$. Refer to Figure 8h for C versus $Re/(D/D_c)^{1/2}$ or to Figure 8i for C versus Re for various D/D_c ratios. In very small capillaries, the coil diameter can be the nominal value, since exact centerline measurement is insignificant.

In laminar flow, the friction factor is a function of Reynolds number only and is independent of surface roughness.

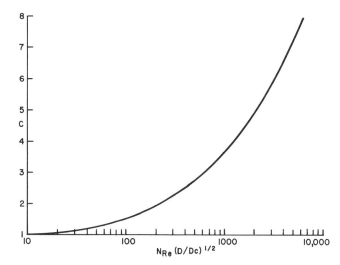

FIG. 8h
Correction factors for coiled capillary flowmeter (data adapted from reference 5).

The friction factor can be expressed as:

$$f = 16/Re \qquad 8(14)$$

Therefore, the Fanning equation 8(15), can be used as an alternate means of calculating the capillary element as shown by:

$$L = \frac{2\,\Delta P\, g_c\, D}{4f\, \rho\, V^2} \qquad 8(15)$$

where

L = capillary length (ft)
ΔP = pressure drop (lbf/ft^2)
g_c = gravity constant 32.17 (ft/sec^2)
D = capillary internal diameter (ft)
ρ = fluid density (lbm/ft^3)
V = fluid velocity (ft/sec)

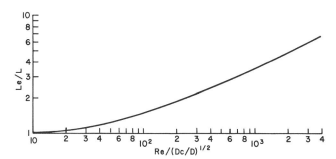

FIG. 8g
Equivalent lengths for curved pipe.

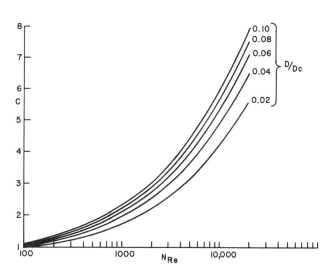

FIG. 8i
Correction factors for coiled capillary flowmeters (data adapted from reference 5).

COMMERCIALLY AVAILABLE UNITS

Laminar flowmeters are used on the measurement of air flow to internal combustion engines, in leak testing, and in fan and blower calibration. Standard units are available in stainless steel construction in pipe sizes ranging from ¼ to 16 in. (6 to 400 mm). The corresponding air flow range of these units ranges from 5 cc/min to 64 m^3/min (2250 SCFM). The maximum pressure drop of these commercially available laminar elements for gas service is 4 to 8 in. of water (102 to 203 mm of water). The recommended installation practice is to provide 10 to 15 diameters of straight pipe upstream to the flow element. A filter at the meter inlet is also recommended. In engine testing a backfire trap is also desirable to prevent carbon deposits on the matrix. The measurement error is usually between 0.5% and 1% of actual flow within a 10:1 range, but it is affected by both the calibration of the system and by the precision of the instrument used to detect the pressure drop.

CONCLUSIONS

Laminar flowmeters are very useful in measuring low flow rates of liquids and gases. Design of these elements is based upon the use of Reynolds number and Poiseuille's Law. Design is relatively simple, but fabrication of a complete measurement system can be complex. It is recommended that the final system be calibrated before use. The calibration can be performed with fluids such as air, nitrogen, or water instead of the actual fluid. The important factor is to set up a calibration procedure that closely approximates the actual in-service Reynolds number.

References

1. Kreith, F., and Eisenstadt, R., Transactions of the ASME, July 1967, pp. 1070–1078.
2. Willoughby, D.A., and Kittle, P.A., *Industrial and Engineering Fundamentals,* Vol. 6, No. 2, May 1967, pp. 304–306.
3. Rivas, M.A., Jr., and Shapiro, A.H., Transactions of the ASME, April 1956, pp. 489–497.
4. Greeff, C.E., and Hackman, J.R., ISA Journal, August 1965, pp. 75–78.
5. Powell, H.N., and Browne, W.G., *The Review of Scientific Instruments,* Vol. 28, No. 2, February 1957, pp. 138–141.

Bibliography

Bowen, LeBaron R., "Designing Laminary-Flow Systems," *Chemical Engineering,* June 12, 1961.
Gann, R.G., *Journal of Chemical Education,* Vol. 51, No. 11, November 1974, pp. 761–762.
Hughes, R.A., "New Laminar Flowmeter," *Instruments and Control Systems,* April 1962.
Instruments and Control Systems, November 1976, pp. 75–76.
Mahood, R.F., and Littlefield, R., Private Communications, March 1952.
Polentz, L.M., "Capillary Flowmetering," *Instruments and Control Systems,* April 1961.
Roth, R.E., "Four Weeks to Fill a Bucket," *Instrumentation,* Vol. 7, No. 6.
Thomas, D.L., "Laminar Flow Elements," *Control,* March 1991.
Welch, J.V., "Trends in Low Flow Metering," *InTech,* February 1991.
Weigand, J., and Lombardo, L., "The Use of Laminar Flow Element in Computerized Flow Measurement," 1989 ISA Conference, Paper #89-0002.

9 Magnetic Flowmeters

J. G. KOPP (1969, 1982) **B. G. LIPTÁK** (1993)

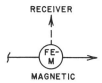

Flow Sheet Symbol

Design Pressure:	Varies with pipe size. For a 4 in. (100 mm) unit the maximum is 285 PSIG (20 bars); special units are available with pressure ratings up to 2500 PSIG (172 bars)
Design Temperature:	Up to 250°F (120°C) with Teflon liners and up to 360°F (180°C) with ceramic liners
Materials of Construction:	Liners: ceramics, fiberglass, neoprene, polyurethene, rubber, Teflon, vitreous enamel, Kynar Electrodes: platinum, Alloy 20, Hastelloy C, stainless steel, tantalum, titanium, tungsten carbide, Monel, nickel, platinum-alumina cermet
Type of Flow Detected:	Volumetric flow of conductive liquids, including slurries and corrosive or abrasive materials
Minimum Conductivity Required:	The majority of designs require 1 to 5 μS/cm. Some probe types require more. Special designs can operate at 0.05 or 0.1 μS/cm
Flow Ranges:	From 0.01 to 100,000 GPM (0.04 to 378,000 l/m)
Size Ranges:	From 0.1 to 96 in. (2.5 mm to 2.4 m) in diameter
Velocity Ranges:	0–0.3 to 0–30 ft/s (0–0.1 to 0–10 m/s)
Error (Inaccuracy):	±1% of actual flow with pulsed DC units within a range of up to 10:1 if flow velocity exceeds 0.5 ft/s (0.15 m/s). ±1% to ±2% full scale with AC excitation
Cost:	The least expensive designs are the probe versions that cost about $1500. A 1-in. (25-mm) ceramic tube unit can be obtained for under $2000. A 1-in. (25-mm) metallic wafer unit can be obtained for under $3000. An 8-in. (200-mm) flanged meter that has a Teflon liner and stainless electrodes and is provided with 4 to 20 mA DC output, grounding ring, and calibrator will cost about $8000. The scanning magmeter probe used in open-channel flow scanning costs about $10,000
Partial List of Suppliers:	ABB Kent-Taylor Inc.; AccuDyne Systems Inc.; Accurate Metering Systems Inc.; ADE-Applied Digital Electronics; Badger Meter Inc.; Baily Controls Co.; Brooks Instrument Div. of Rosemount; Colorado Engineering Experimental Station; Dantec Electronics; H.R. Dulin Co.; Dynasonics Inc. (probe-type); Edinboro Computer Instruments Corp.; Electromagnetic Controls Corp.; Endress + Hauser Instruments; Engineering Measurements Co.; Fischer & Porter Co.; Foxboro Co.; Harwil Corp.; Honeywell, Industrial Controls Div.; Instrumark International Inc.; Johnson Yokogawa Corp.; K & L Research Co. (probe-type); Krone-America Inc.; Marsh-McBirney Inc. (probe-type); Meter Equipment Mfg.; Mine Safety Appliances Co.; Monitek Tech. Inc.; Montedoro Whitney; MSR Magmeter Manufacturing Ltd. (probe-type); Omega Engineering; Rosemount Inc.; Sarasota Measurements & Controls; Schlumberger Industries Inc.; Signet Industrial (probe-type); Sparling Instruments Co.; Toshiba International; Turbo Instruments Inc.; Vortab Corp.; Wallace & Tiernan Inc.; Wilkerson Instrument Co.; XO Technologies Inc.; Yokogawa Electric Corp.

Magnetic-type flowmeters use Faraday's Law of electromagnetic induction for making a flow measurement. Faraday's Law states that when a conductor moves through a magnetic field of given field strength, a voltage level is produced in the conductor that is dependent on the relative velocity between the conductor and the field. This is the concept that is used in electric generators. Faraday foresaw the practical application of the principle to flow measurement, because many liquids are adequate electrical conductors. In fact, he attempted to measure the flow velocity of the Thames River using this principle. He failed because his instrumentation was not adequate, but 150 years later, we can successfully apply the principle in magnetic flowmeters.

THEORY

Figure 9a shows how Faraday's Law is applied in the electromagnetic flowmeter. The liquid is the conductor which has a "length equivalent" to the inside diameter of the flowmeter, D. The liquid conductor moves with an average

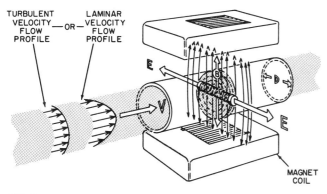

FIG. 9a
Schematic representation of the magnetic flowmeter.

velocity V through the magnetic field of strength B. The induced voltage is E. The mathematical relationship is:

$$E = BDV/C \qquad 9(1)$$

C is a constant to take care of the proper units.

When the pair of magnetic coils is energized, a magnetic field is generated which is in a plane that is mutually perpendicular to the axis of the liquid conductor and the plane of the electrodes. The velocity of the liquid is along the longitudinal axis of the flowmeter body; therefore, the voltage induced within the liquid is mutually perpendicular to both the velocity of the liquid and the magnetic field. The liquid should be considered as an infinite number of conductors moving through the magnetic field with each element contributing to the voltage that is generated. An increase in flow rate of the liquid conductors moving through the field will result in an increase in the instantaneous value of the voltage generated. Also, each of the individual "generators" is contributing to the instantaneously generated voltage. Whether the profile is essentially square (characteristic of a turbulent velocity profile), parabolic (characteristic of a laminar velocity profile), or distorted (characteristic of poor upstream piping), the magnetic flowmeter does an excellent job of averaging the voltage contribution across the metering cross section. The sum of the instantaneous voltages generated is therefore representative of the average liquid velocity because each increment of liquid velocity within the plane of the electrode develops a voltage proportional to its local velocity. The signal voltage generated is equal to the average velocity almost regardless of the flow profile. The magnetic flowmeter detects the volumetric flow rate by sensing the linear velocity of the liquid.

The Equation of Continuity (Q = VA) is the relationship which converts the velocity measurement to volumetric flow rate providing the area is constant. Therefore the area must be known and constant and the pipe must be full in order to obtain a correct measurement.

AC AND DC EXCITATION

AC-type magnetic flowmeters apply line voltage to the magnet coils. The signal generated is a low level AC signal in the high microvolt to low millivolt range. A more recent development is the pulsed DC-type magnetic flowmeter. In this design, the magnet coils are periodically energized. There are many forms of excitation in use, but generally they can be categorized into two families: those which are on-off excitation and those which use plus-minus excitation. In either case, the principle is to take a measurement of the induced voltage when the coils are not energized and to take a second measurement when the coils are energized and the magnetic field has stabilized. Figure 9b shows some of the types of excitation offered by various manufacturers.

In all of the pulsed DC approaches, the concept is to take a measurement when the coils are excited and store (hold) that information, then take a second measurement of the induced voltage when the coils are not excited. The voltage induced when the coils are energized is a combination of both noise and signal. The induced voltage when the coils are not energized is noise only. Subtracting the noise measurement from the signal plus noise yields signal only (see Figure 9c).

The pulsed DC-type systems establish zero during each on-off cycle. This occurs several times every second. Because zero is known, the end result is that pulsed DC systems are potential percent-of-rate systems (see Figure 9d). The AC-type systems must be periodically rezeroed by stopping flow and maintaining a full pipe in order to zero out any voltage present at that time. The noise voltage can change with time, resulting in a potential offset; therefore, AC-type systems normally are percent-of-full scale systems.

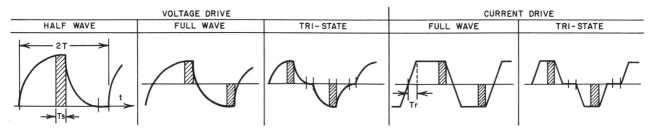

FIG. 9b
Types of pulsed DC coil excitation.

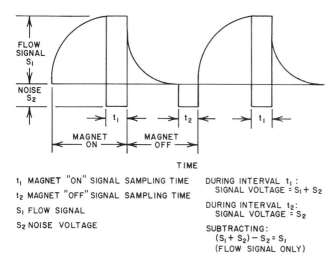

t_1 MAGNET "ON" SIGNAL SAMPLING TIME
t_2 MAGNET "OFF" SIGNAL SAMPLING TIME
S_1 FLOW SIGNAL
S_2 NOISE VOLTAGE

DURING INTERVAL t_1:
SIGNAL VOLTAGE = $S_1 + S_2$

DURING INTERVAL t_2:
SIGNAL VOLTAGE = S_2

SUBTRACTING:
$(S_1 + S_2) - S_2 = S_1$
(FLOW SIGNAL ONLY)

FIG. 9c
Signal development of pulsed DC-type magnetic flowmeter with half wave excitation. As shown, the magnetic field is generated by a square wave which, in function, turns the magnet "on" and "off" in equal increments. When "on," the associated signal converter measures and stores the signal which is a composite of flow plus a variable (non-flow-related) residual voltage. During the "off" period, the converter measures the variable (non-flow-related) residual signal only. Since no field excitation is present, no flow signal will be generated. The converter then subtracts the stored residual signal from the flow developed-plus residual signal, resulting in the display of a pure flow signal.

Dual-Frequency Excitation

Changing the method of excitation from line frequency (AC) to low frequency (DC) provided dramatic improvements in both the accuracy and the zero stability of magnetic flowmeters. Yet it did not represent the end in technological advances. A limitation of low-frequency (DC) designs is their relatively low speed of response (0.2 to 2 seconds) and their sensitivity to measurement noise caused by slurries or low-conductivity fluids.

The idea behind dual-frequency excitation is to apply both and thereby benefit from both of their advantages: the

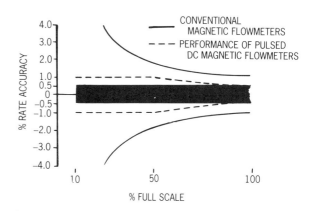

FIG. 9d
Pulsed DC design reduces measurement error to ±0.5% of actual flow over a range of 10:1.

zero stability of low-frequency excitation and the good noise rejection and high speed of response of high-frequency excitation. This is achieved by exciting the magnetic field coils by a current with such a compound wave form as illustrated in Figure 9e. One component is a low-frequency waveform, much below 60 Hz, which guarantees good zero stability. The output generated by the low-frequency signal is integrated via a long time constant to provide a smooth and stable flow signal.

The high-frequency component is superimposed on the low-frequency signal to provide immunity to noise caused by low conductivity, viscosity, slurries, or electrochemical reactions. The output generated by the high-frequency component is sampled at a high frequency and is processed in a differentiating circuit having the same time constant as the integrating circuit. By adding the two signals, the result is an output that is free of "slurry" noise and has good zero stability plus good speed of response.

CONSTRUCTION

Most meters are built with flanged end fittings, although the insert types are also common. Designs are available with

FIG. 9e
Dual-frequency excitation design combines the advantages of both systems.

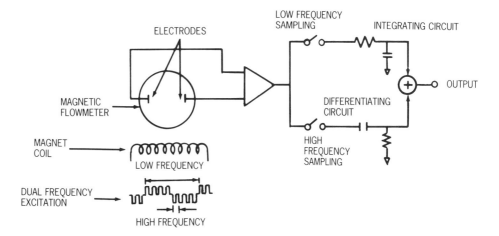

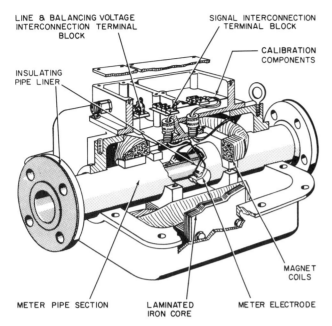

FIG. 9f
Cutaway view of the magnetic flowmeter.

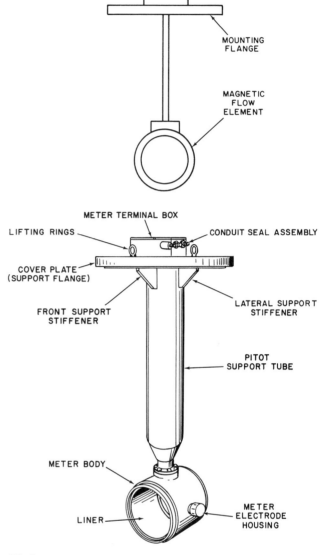

FIG. 9g
Pitot-type magnetic flowmeter.

sanitary-type fittings. In larger pipe sizes, Dresser-type and Victaulic-type end connections are also widely used.

Figure 9f is a cutaway view showing how the principle of electromagnetic induction is employed in a practical flowmeter. The basic elements of the flowmeter are a section of nonconducting pipe such as glass-reinforced polyester or a nonmagnetic pipe section lined with an appropriate electrical conductor such as Teflon, Kynar, fiberglass, vitreous enamel, rubber, neoprene, or polyurethane, among others. On alternate sides of the pipe section are magnet coils which produce the magnetic field perpendicular to the flow of liquid through the pipe. Mounted in the pipe, but insulated from it and in contact with the liquid, is a pair of electrodes which are located at right angles both to the magnetic field and the axis of the pipe. As the liquid passes through the pipe section, it also passes through the magnetic field setup by the magnet coils inducing a voltage in the liquid; the amplitude of the voltage is directly proportional to the liquid velocity. This voltage is conducted by the electrodes to a separate converter which in effect is a precision voltmeter (electrometer) capable of accurately measuring the voltage generated and converting that voltage to the desired control signals. These may be equivalent electronic analog signals, typically 4 to 20 mA DC, or a frequency or scaled pulse output.

Versions of magnetic flowmeters are available for periodic accidental submergence and for continuous submergence in water at depths of up to 30 ft (9 m). An outgrowth of the continuous submergence design is a sampling-type (pitot). The pitot-type magnetic flowmeter samples the flow velocity in large rectangular, circular, or irregularly shaped pipes or conduits. A typical design is shown in Figure 9g. A small size magnetic flowmeter is suspended in the flow stream. The magnet coils are completely encapsulated in the liner material, allowing submersion in the liquid to be measured. The short length of the meter body and the streamlined configuration are designed to minimize the difference between the flow velocity through the meter and the velocity of the liquid passing around the meter. The velocity measurement of the liquid through the meter is assumed to be representative of the pipe velocity. Repeatability of the system is typically 0.25 to 0.5% of full scale. As with any sampling-type flowmeter, the information from the flowmeter is representative only of the flow through the flowmeter. It is the user's responsibility to relate that "sampled" velocity to the average velocity in the pipe, which reflects the total volumetric flow rate. When applying any sampling-type flowmeter, including the pitot-type magnetic flowmeter,

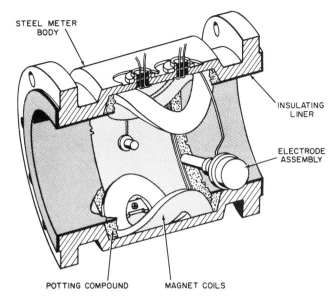

FIG. 9h
The short-form magnetic flowmeter.

substantial errors can occur in applications where the velocity profile can change due to changes in Reynolds number or due to the effects of upstream piping configuration.

Modern magnetic flowmeters are constructed so that the flux density of the magnetic field generated by the magnet coils is constant. The distance between the electrodes is also constant. Induced voltage is dependent only on liquid velocity and is not affected by temperature, viscosity, turbulence, or electrical conductivity, provided that electrical conductivity is above a minimum threshold.

Most manufacturers construct their flowmeters with coils external to the meter pipe section. Some designs place the coils within the flowmeter body, which is made from carbon steel to provide the return path for the magnetic field (see Figure 9h). In this design the meters can be shorter in length, have reduced weight, and require lower power consumption. Lowest power consumption is a result of the pulsed DC design because its coils are energized only part of the time. An additional saving with pulsed DC-types is that the power factor approaches one.

The power consumption of a conventional (high-frequency AC excited) 2 in. (50 mm) flowmeter is about 30 watts and for a 30 in. (76 cm) flowmeter about 300 watts. Low-frequency DC excitation has reduced the power consumption of some magnetic flowmeters to 20 watts, regardless of meter size.

Ceramic Liners

The use of ceramic liners represented a major improvement in the design of magnetic flowmeters, because they not only cost less to manufacture, but they also provide a better meter. Ceramic materials such as Al_2O_3 are ideal liner materials because their casting is inexpensive, they are electrically nonconductive, and they are abrasion- and wear-resistant. In contrast with plastic liners, they can be used on abrasive slurry services (pipelining of minerals or coal) and their inner surfaces can be scraped with wire brushes to remove hardened coatings. Ceramic units are also preferred for sanitary applications because they do not provide any cavities in which bacteria can accumulate and grow. Ceramic meters can also handle higher temperatures (360°F, or 180°C) than Teflon-lined ones (250°F, or 120°C). Because magnetic flowmeters are velocity sensors and in order to convert velocity into volumetric flow rate the pipe cross section has to be constant, the ceramic liners have the added advantage of expanding and contracting less with changes in temperature than do metals or plastics. Ceramic liners are also preferred by the nuclear industry because they are not affected by radiation, while plastics are destroyed by it.

The design of the ceramic insert-type magnetic flowmeter (Figure 9i) also eliminates the possibility of leakage around the electrodes. This perfect seal is produced by allowing a droplet of liquid platinum to sinter through the ceramic wall of the liner. Through this process the ceramic particles and the platinum fuse into a unified whole, providing not only a perfect seal, but also a permanent, rugged, and corrosion-resistant electrode. This electrode can not move, separate, or leak.

While for the above-listed reasons the ceramic insert-type magnetic flowmeter is an improvement, it too has some limitations. One of its limitations has to do with its brittle nature. Ceramic materials are strong in compression but should not be exposed to pipe forces that cause tension or bending. Another possible way to crack the ceramic lining is by sudden cooling. Therefore these elements should not be exposed to downward step-changes in temperature that exceed 90°F (32°C). Another limitation of the Al_2O_3 ceramic liner is that it cannot be used on oxidizing acid or on hot and concentrated caustic applications (over 120°F, or 50°C).

Probe-Type Units

The probe-type magnetic flowmeter is an "inside out" design in the sense that the excitation coil is on the inside of the probe (Figure 9j). As the process fluid passes through the magnetic field generated by the excitation coil inside the probe, a voltage is detected by the electrodes that are imbedded in the probe. The main advantage of this design is its low cost, which is not affected by pipe size, and its retractable nature, which makes it suitable for wet-tap installations. The probe-type magmeter is also suited for the measurement of flow velocities in partially full pipes or in detecting the currents in open waters. When water flow is not constrained by a pipe, flow velocity has to be expressed as a three-dimensional vector. By inserting three magmeter probes parallel with the three axes, one can detect that vector.

The main disadvantage of the magmeter probe is that it detects the flow velocity in only a small segment of the cross-sectional area of the larger pipe. Therefore, if the

44 Flow Measurement

1 CABLE CONNECTION
2 HOUSING
3 METERING ELECTRODES (SINTERED)
4 FLANGE GASKET
5 CONNECTION BOX
6 GASKET
7 MAGNETIC COILS
8 MAGNETIC STRIP
9 METERING SECTION MADE FROM Al_2O_3 CERAMIC LINER

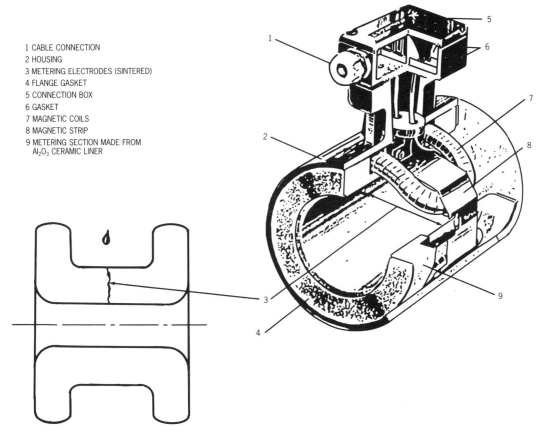

FIG. 9i
The ceramic insert-type magnetic flowmeter.

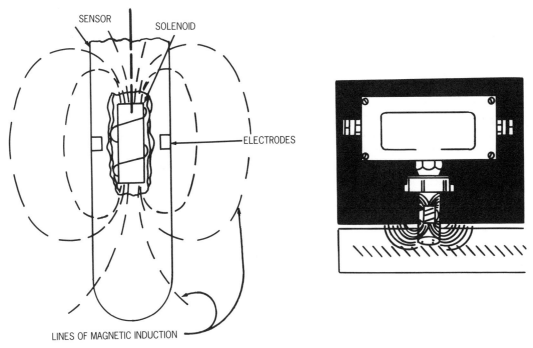

FIG. 9j
The probe-type magnetic flowmeter.

Magnetic Flowmeters 45

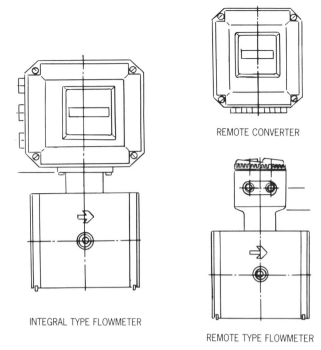

FIG. 9k
The converter/transmitter can be integral or remote. (Courtesy of Yokogawa Corp.)

flowing velocity in that location is not representative of the rest of the cross section, a substantial error can result.

Magmeter Electronics and Intelligence

Each magnetic flowmeter requires electronics to convert the electrode output into a standardized analog or digital signal. The electronics can be mounted locally, directly on the flowmeter, or remotely (Figure 9k). Integral mounting simplifies the installation, reduces its cost, and eliminates the noise and other problems associated with the transmission of the low-level magmeter signal over some distance. The advantages of remote mounting include the reduced headroom requirement for the meter, accessibility, operator convenience, and the distancing of the sensitive electronics from the high-temperature or otherwise undesirable environment of the flowmeter. If shielded and twisted wires are used the electronics can be 200 ft (67 m) from the meter.

The housings of the electronics can be designed for indoor or outdoor use and for general purpose or hazardous environments. The converters can serve several flowmeters simultaneously and provide for interfacing with computers. The displays can provide flow rate or total flow indication. "Smart" magmeters provide the added features of self-diagnostic and detection of coil/converter/metering tube failure or of empty-pipe, as well as switching, alarming, flow integration, and preset batching functions. They can also detect pipe blockage; signal erroneous settings; or change the range, engineering units, dampening times (63% response time settable from 0.1 to 100 seconds), or even the flow direction of metering.

CAPACITY AND RANGE

Figure 9l is a nomograph for magnetic flowmeter capacities. Magnetic flowmeters have an excellent operating range, at least 100:1. For AC types, typical inaccuracy is ±1% of full scale; to improve performance, range is usually divided into two portions and automatically switched between the two. Pulsed DC-types have typical inaccuracy of ±1% of rate applicable to a 10:1 range or ±0.5% of rate over a 2:1 or 5:1 range, and at flow rates below 10% of maximum it is on the order of ±0.1% of full scale. The converter can be set for 20 mA output at any flow between 10% and 100% of meter capacity and still have at least a 10:1 operating range. This ability to field set or reset the meter for the actual operating conditions provides optimum performance.

APPLICATIONS

Most liquids or slurries are adequate electrical conductors to be measured by electromagnetic flowmeters. If the liquid conductivity is equal to 20 microSiemens per centimeter or greater, most of the conventional magnetic flowmeters can be used. Special designs are available to measure the flow of liquids with threshold conductivities as low as 0.1 mi-

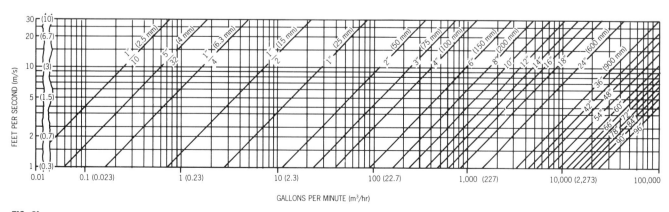

FIG. 9l
Magnetic flowmeter capacity nomograph.

croSiemens per centimeter. Some typical electrical conductivities are:

Liquid (at 25°C except where noted)	Conductivity MicroSiemens/centimeter
Acetic acid (up to 70% by weight)	250 or greater
Ammonium nitrate (up to 50% by weight)	360,000 or greater
Molasses (at 50°C)	5,000
Ethyl alcohol	0.0013
Formic acid (all concentrations)	280 or greater
Glycol	0.3
Hydrochloric acid (up to 40% by weight)	400,000 or greater
Kerosene	0.017
Magnesium sulphate (up to 25% by weight)	26,000 or greater
Corn syrup	16
Phenol	0.017
Phosphoric acid (up to 87% by weight)	50,000 or greater
Sodium hydroxide (up to 50% by weight)	40,000 or greater
Sulphuric acid (up to 99.4% by weight)	8,500 or greater
Vodka (100 proof)	4
Water (potable)	70

The effect of conductivity changes above the threshold conductivity is minimal, but the effect of liquid operating temperature upon the threshold conductivity should be considered. Most liquids have a positive temperature coefficient of conductivity. Liquids which are marginal at one temperature and can become sufficiently nonconductive at a lower temperature so as to impair metering accuracy. At a higher temperature, the same liquid may be metered with good results. There are a few liquids that have a negative temperature-coefficient; these should be carefully checked for their minimum conductivity before applying magnetic flowmeters.

Magnetic flowmeters are not affected by viscosity or consistency (referring to Newtonian and non-Newtonian fluids respectively). The changes in flow profile due to changes in Reynolds numbers or due to upstream piping do not greatly affect the performance of magnetic flowmeters. The voltage generated is a summation of the incremental voltages across the entire area between the electrodes, resulting in a measure of the average fluid velocity. Nevertheless, it is recommended to install the meter with 5 diameters of straight pipe before and 3 diameters of straight pipe following the meter.

Magnetic flowmeters are bidirectional. Manufacturers offer converters with output signals for both direct and reverse flows.

The magnetic flowmeter must always be full to assure accurate measurement. If the pipe is only partially full, the electrode voltage, which is proportional to the fluid velocity, will still be multiplied with the full cross section and the reading will be high. Similarly, if the liquid contains entrained gases, the meter will measure them as liquid and the reading will be high.

The meter's electrodes must remain in electrical contact with the fluid being measured, and should always be installed in the horizontal plane. In applications where a buildup or coating of the inside wall of the flowmeter occurs, periodic "flushing" or cleaning is recommended. Coatings can have conductivities that are the same, lower, or higher than the liquid. These effects are significantly different. Where the conductivity of the coating is essentially the same as that of the liquid, there is no effect on the accuracy of the measurement except for the effect of a reduced cross-sectional area. This can be viewed as a specific profile condition and the meter will average the velocity to give the correct value for the particular flow rate. Fortunately, this is the most common coating condition. If the conductivity of the coating is significantly lower than the liquid being measured, the electrically insulating coating can disable the meter. If periodic cleaning is not possible, mechanical, ultrasonic, thermal, and other electrode cleaning techniques can be applied. Manufacturers also offer specifically shaped protruding electrodes to take advantage of the self-cleaning effect of the flow at the electrode. If the conductivity is higher than that of the process fluid, no corrective measure is needed.

Grease and other nonconductive electrode coatings introduce an error in the measurement, because the voltage generated by the conductive fluid is measured by the magmeter electronics as a voltage drop across its input impedance (R_{fm} in Figure 9m). When there is an electrically resistant coating on the electrodes ($R_c/2$ in Figure 9m), some of the voltage generated by the conductive liquid drops across the coating and less of it remains to be detected by the input impedance. The resulting error percentage can be calculated as:

$$E = 100R_c/(R_{fm} + R_c) \qquad 9(2)$$

Coating resistances (R_c) can reach 10^7, and if the input impedance (R_{fm}) is similar, substantial errors will result. In some of the newer designs the input impedance of the flowmeter has been increased to $R_{fm} = 10^{11}$, which even at a coating impedance of $R_c = 10^7$ limits the coating error to 0.01%. With such high-impedance electronics the need for electrode cleaning is minimized or eliminated.

Although magnetic flowmeters are not affected by temperature or pressure changes, there are specific limits of both temperature and pressure for particular designs (see feature summary at the beginning of this section). Temperature limitations for the particular lining materials require special consideration.

Special meters for measuring sewage sludge flow are designed to prevent the buildup and carbonizing of sludge

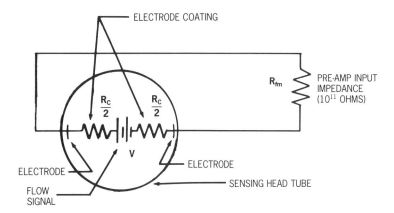

FIG. 9m
Increased flowmeter impedance (R_{fm}) reduces coating error.

on the meter electrodes. They use self-heating to elevate the metering body temperature to prevent sludge and grease accumulation.

Special systems are available for measuring slurries that contain permeable solids such as magnetic oxides and magnetic sulfides. Additional circuitry compensates for signal changes induced by the presence of permeable material that alters the slurry's magnetic field distribution.

The magnetic flowmeters described were of the alternating current-type. A DC magnetic flowmeter is available from MSA which has been designed specifically for handling liquid metals such as liquid potassium and liquid sodium.

The very low head loss of bypass-type magmeters makes them suitable for use in measuring bypass flow around a mainline orifice, as shown in Figure 9n.

Mass flowmetering systems are available using magnetic flowmeters in association with gamma radiation density gauges. The signal developed by the magnetic flowmeter relates to volumetric flow; the output signal from the gamma radiation density gauge is a signal proportional to density. The two signals are multiplied to obtain an output signal proportional to mass flow. Some designs incorporate the flowmeter and the density gauge into the same housing for installation ease and simplicity.

INSTALLATION

The signal detected by magnetic flowmeter electrodes is in the high microvolt to low millivolt range. Proper electrical installation and grounding is mandatory. Individual manufacturer recommendations for installation are the result of extensive experience and should be scrupulously followed (see Figure 9o).

AC-type magnetic flowmeters occasionally shift their no-flow indication after some operating time, requiring a zero reset. One of the most important installation considerations with electromagnetic flowmeters is a proper "bonding" of the flowmeter to the adjacent piping to minimize zero shifts. The intent of this bonding, or "jumpering," is to prevent stray currents from passing through the flowmeter near the electrodes. Magnetic flowmeters are lined with an electrically insulating material; generally, this lining covers the flange face of the meter, making the meter an electrical discontinuity in the system. The flange bolts should not be used for bonding since rust, corrosion, paint, or other insulating materials can create an insulating barrier between the bolts and the flanges. Manufacturers supply and insist upon the installation of copper braid jumpers from the meter flange to the pipe flange at either end of the flowmeter. The jumpers provide a continuous path for the stray currents, which guarantees a more stable zero. It is also essential to install a ground strap to a grounded piece of structural steel, a grounding rod, or a cold water pipe.

The above-described installation process, which involves labor-intensive drilling, tapping, and strapping of adjacent pipe flanges in metallic pipes or the installation of expensive grounding rings in lined or nonconductive pipes, can be eliminated if the magmeter is provided with built-in grounding electrodes. When installing the flowmeter, the grounding electrode must always be at the bottom and must be connected to the third wire ground of the power input.

The conservative installation of magnetic flowmeters requires 3 to 5 diameters of straight pipe, the same size as the

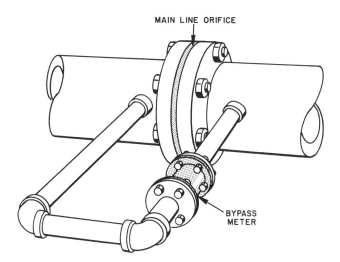

FIG. 9n
Magnetic flowmeter used to measure bypass flow around an orifice plate to infer main line flow rate.

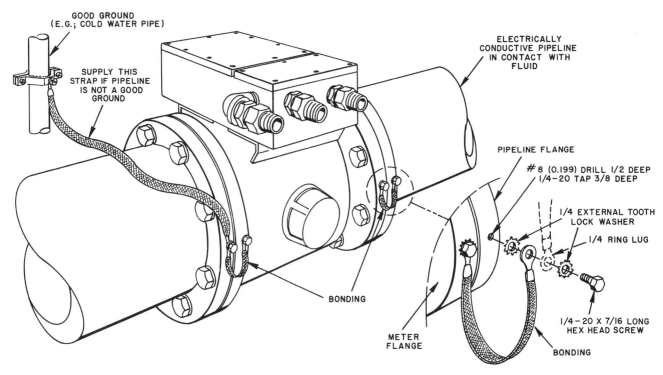

FIG. 9o
Typical bonding and grounding procedure.

flowmeter, to be installed upstream to the meter, and 2 or 3 diameters downstream. Meters can be installed in horizontal pipelines, vertical pipelines, or sloping lines. It is essential to keep the electrodes in the horizontal plane to assure uninterrupted contact with the liquid or slurry being metered. In gravity feed systems, the meter must be kept continually full; therefore, the meter should be installed in a "low point" in horizontal lines (Figure 9p) or, preferably, in a vertical upflow line.

ADVANTAGES

1. The magnetic flowmeter is totally obstructionless and has no moving parts. Pressure loss of the flowmeter is no greater than that of the same length of pipe. Pumping costs are thereby minimized.

2. Electric power requirements can be low, particularly with the pulsed DC-types. Electric power requirements as low as 15 or 20 watts are not uncommon.

3. The meters are suitable for most acids, bases, waters, and aqueous solutions because the lining materials selected are not only good electrical insulators but also are corrosion-resistant. Only a small amount of electrode metal is required, and stainless steel, Alloy 20, the Hastelloys, nickel, Monel, titanium, tantalum, tungsten carbide, and even platinum are all available.

4. The meters are widely used for slurry services not only because they are obstructionless but also because some of the liners, such as polyurethane, neoprene, and rubber, have good abrasion or erosion resistance.

5. Magmeters are capable of handling extremely low flows. Their minimum size is less than 1/8 in. (3.175 mm) inside

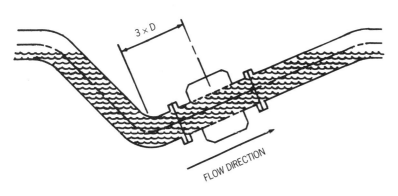

FIG. 9p
Magmeter installation in horizontal pipeline.

diameter. The meters are also suitable for very high volume flow rates with sizes as large as 10 ft (3.04 m) offered.
6. The meters can be used as bidirectional meters.

LIMITATIONS

The meters do have some specific application limitations:

1. The meters work only with conductive fluids. Pure substances, hydrocarbons, and gases, cannot be measured. Most acids, bases, water, and aqueous solutions can.
2. The conventional meters are relatively heavy, especially in larger sizes. Ceramic and probe-type units are lighter.
3. Electrical installation care is essential.
4. The price of magnetic flowmeters ranges from moderate to expensive. Their corrosion resistance, abrasion resistance, and accurate performance over wide turndown ratios can justify the cost. Ceramic and probe-type units are less expensive.
5. To periodically check the zero on AC-type magnetic flowmeters, block valves are required on either side to bring the flow to zero and keep the meter full. Cycled DC-units do not have this requirement.

Magnetic flowmeters are the first to be considered for very corrosive applications and for applications involving measurement of abrasive and/or erosive slurries. They are widely used in pulp and paper stock measurement and other non-Newtonian applications. They can be used for very low flow rates; pipe inside diameters as small as 0.1 in. (2.5 mm) are offered which can handle flow ranges as low as 0.01 to 0.1 gpm (0.038 to 0.38 l/m). Magnetic flowmeters are also available in pipe sizes up to 120 in. (3 m).

Bibliography

Bean, H.S., ed., "Fluid Meters—Their Theory and Application," 6th edition, ASME Research Committee on Fluid Meters, New York: ASME, 1971, pp. 125–128.

Bonfig, K.W., and Feith, B., "Magnetic Inductive Flow Measurement at Open Channels," *Mess Pruef,* Jan–Feb 1982.

Dellerson, A.N., "Magmeters: Leaders in Flow Technology," *InTech,* July 1989.

Eastman, P.C., Brodie, D.E., and Sawyer, D.J., "A Magnetic Flowmeter with Concentric Electrodes," *Instrumentation Technology,* June 1970, pp. 52–55.

"Electromagnetic Flowmeters," *Measurements and Control,* April 1991.

Elrod, H.J., and Fouse, R.R., "An Investigation of Electromagnetic Flowmeters," *Transactions ASME,* Vol. 74, May 1952, pp. 589–594.

Head, V.P., "Electromagnetic Flowmeter Primary Elements," *ASME Journal of Basic Engineering,* Paper 58-A-126, 1958.

Kolin, A., "An Alternating Field Induction Flowmeter of High Sensitivity," *Review of Scientific Instruments,* Vol. 16, May 1945, pp. 109–116.

Lilla, R., "Ceramic-Liner Electromagnetic Flowmeters," *Measurements and Control,* April 1991.

Ostling, H., "Dual-Frequency Excitation Sets New Magmeter Performance Standard," *Control Engineering,* September 1990.

Robinson, C., "Obstructionless Flowmeters," *InTech,* December 1986.

Shercliff, J.A., "Experiments on the Dependence of Sensitivity of Velocity Profile on Electromagnetic Flowmeters," *Journal of Scientific Instruments,* Vol. 32, 1955, pp. 441–442.

Shercliff, J.A., *The Theory of Electromagnetic Flow Measurement,* Cambridge University Press, 1962.

Webb, A.S., "Electromagnetic Flowmetering," *Instrumentation Technology,* March 1974, pp. 29–33.

10 Mass Flowmeters—Coriolis

CATHY APPLE (1993)

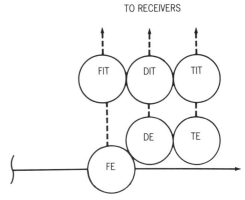

Flow Sheet Symbol

Sizes:	1/16 to 6 in. (1.5 to 150 mm)
Flow Range:	0 to 25,000 lb/m (0 to 11,340 kg/m)
Fluids:	Liquids, slurries, compressed gases, liquified gases; not gas-liquid mixtures or gases at below 150 psig (10.3 bars)
Output Signal:	Linear frequency, analog, digital, scaled pulse, display
Detector Types:	Electromagnetic, optical, capacitive
Operating Pressure:	Depends upon tube size and flange rating: 1800 psig (124 bars) typical standard; 5000 psig (345 bars) typical high-pressure
Pressure Drop Required:	From under 10 psig (0.7 bars) to over 100 psig (6.9 bars) as a function of viscosity and design
Operating Temperature:	Depends upon design: −100 to 400°F (−73 to 204°C) typical standard; 32 to 800°F (0 to 426°C) high-temperature
Materials of Construction:	Stainless steel, Hastelloy, titanium, NiSpan C as standard; tantalum, Tefzel-lined as special
Inaccuracy:	±0.15 to 0.5% of rate (see Fig. 10e) $$\pm 0.15\% \text{ of rate } \pm \frac{\text{zero offset}}{\text{mass flow rate}} \times 100\%$$ Zero offset depends upon size and design of the flowmeter; for a 1 in. (25 mm) meter with a typical maximum flow rate of 400 to 1,000 lb/m (180 to 450 kg/m), the zero offset is typically in the range of 0.03 to 0.1 lb/m (0.014 to 0.045 kg/m) which is under 0.01%
Repeatability:	±0.05 to ±0.2% of rate
Rangeability:	20:1 calibration range (typical)
Cost:	Depends on size and design: 1/16 in. (1.5 mm)—$3950; 6 in. (150 mm)—$21,000; typical 1 in. (25 mm) meter, with full scale flow rate of 400 to 1000 lb/m (180 to 450 kg/m)—$5300
	A typical flowmeter comes standard with one pulse or frequency output that represents flow rate; one analog output configurable for flow rate, density, or temperature; and a display or digital output which provides flow rate, density, temperature, and flow total. In addition, most devices provide standard alarm outputs. The number and type of outputs vary from one manufacturer to another. Additional analog, frequency, pulse, and digital outputs are often provided as options
Partial List of Suppliers:	Bailey Controls; Danfoss A/S (Denmark); Endress & Hauser Instruments; Exac Corp.; Fischer & Porter Co.; The Foxboro Co.; Heinrichs, K-Flow; Krohne, Bopp & Reuther; Micro Motion Inc.; Neptune Measurement Co.; Schlumberger Industries; Smith Meter Inc.;

Since the appearance of the first commercial meters in the late 1970s, Coriolis flowmeters have become widely used. The ability to measure mass flow directly with high accuracy and rangeability and to measure a wide variety of fluids is making Coriolis flowmeters the preferred flow measurement instrument for many applications. Coriolis flowmeters are also capable of measuring the process fluid density and temperature. Since Coriolis flow measurement is a relatively new technology, many of the subtleties of its operation are still being investigated.

THEORY

The operation of a Coriolis flowmeter can be understood by examining an element of fluid moving through a tube that is rotating about a fixed axis perpendicular to the centerline of the tube. Refer to Figure 10a Part A. A fluid element, of mass m, is moving outward, at a constant speed v, away from the rotational axis of the tube, which is rotating at a constant angular velocity of ω. The rotation of the tube imparts a Coriolis acceleration to the moving fluid element. The Coriolis acceleration consists of two acceleration components. The first component is the result of the rotation of the tube, which causes a directional change in the fluid element velocity (v). The magnitude of this acceleration component is the product of the fluid element velocity (v) and the angular velocity (ω) of the tube: $a_d = v\omega$. The angular velocity (ω) of the tube turns the velocity vector (v) of the fluid element toward the left as shown in Figure 10a Part A, so the acceleration vector (a_d) is directed to the left.

The second acceleration component is the result of the change in the linear velocity (v_t) of the tube with changing distance (r) from the rotational axis. The linear velocity (v_t) at any point on the tube is the product of the angular velocity (ω) and the distance (r) from the rotational axis: $v_t = r\omega$. At the axis the linear velocity (v_t) of the tube and the fluid element is zero. As the fluid element moves outward, r increases and the linear velocity (v_t) increases, causing the fluid element to be accelerated. The magnitude of this acceleration is the rate of increase in the magnitude of v_t, which is the result of the rate of change in the magnitude of r. The rate of change in the distance r is the fluid element velocity, v. Therefore the magnitude of the acceleration is defined by: $a_t = v\omega$. Since v_t is increasing, the acceleration is in the same direction as the velocity vector v_t and is directed to the left.

The Coriolis acceleration is the sum of these two acceleration components and is defined by the equation $a_c = 2v\omega$. If the fluid element is at rest, $v = 0$ and the Coriolis acceleration is also zero. The Coriolis force is the product of the fluid mass and the Coriolis acceleration: $F_c = 2mv\omega$. The rotation of the tube exerts a force, the Coriolis force (F_c), on the fluid element. This force pushes the fluid element to the left. The Coriolis force (F_c) is opposed by the inertial force (F_i) of the fluid, which pushes to the right against the tube. This inertial force (F_i) resists the rotation of the tube. Due to the elasticity of the tube, the inertial force (F_i) of the fluid element causes the tube to *deflect slightly to the right*.

Now examine a fluid element, of mass m, moving inward at a constant speed v toward the axis of the tube, which is still rotating at a constant angular velocity of ω (as shown in Part B of Figure 10a). Since the fluid element is moving in the opposite direction, the angular velocity (ω) turns the fluid element velocity vector (v) toward the right. This causes the directional acceleration component (a_d) to be directed to the right. Since the fluid element is moving inward, r is decreasing and the linear velocity, v_t, is decreasing, causing the fluid element to be decelerated. The acceleration (a_t) is in the opposite direction of the velocity vector v_t. Both acceleration components are now directed to the right. Changing the direction of the fluid element velocity (v) results in a Coriolis acceleration (a_c) of the same magnitude but opposite in direction. This also results in a Coriolis force (F_c) which is in the opposite direction. The Coriolis force (F_c) pushes the fluid element to the right. This Coriolis force (F_c) is opposed by the inertial force (F_i) of the fluid which pushes to the left against the tube. The inertial force (F_i) assists the rotation of the tube and causes the tube to *deflect slightly to the left*.

In the practical application of a Coriolis flowmeter, the tube carrying the fluid is not rotated but rather oscillated or vibrated. The direction of the angular velocity (ω) alternates, but the Coriolis forces are still generated in the same manner as described above. A driving mechanism is typically placed at the center of the tube, which oscillates the tube back and forth. The ends of the tube are fixed, as shown in Figure 10a Part C. This structure is analogous to the rotating tube systems described above. The first half of the flowmeter contains fluid elements which are moving from a fixed axis toward the point of maximum rotation. The second half of the flowmeter contains fluid elements which are moving away from the maximum rotation toward the fixed axis. The fluid elements moving toward the tube driver *oppose the motion of the rotating tube*, while the fluid element moving away from the tube driver *assist the rotation of the tube*. The effect of the opposing Coriolis forces on the tube carrying the fluid is shown in Part C. These oppositely directed reaction forces create small scale distortions in the flow tube, which are shown (greatly exaggerated) in Figure 10a Part C.

The Coriolis force is determined from the equation $F_c = 2mv\omega$. From this equation it is apparent that the Coriolis force is proportional to the mass of the fluid element and to the fluid element velocity. Some substitutions enable the Coriolis force to be related directly to the mass flow rate, $\dot{m}$, of the fluid element: $F_c = \dot{m}2l\omega$, where l is the fluid element length. Conversely, if the Coriolis force or opposing inertial force can be measured, then the mass flow rate of the fluid element can be determined from the following equation:

$$\dot{m} = F_c/2\omega l \qquad \mathbf{10(1)}$$

Direct measurement of the Coriolis force or inertial force is not practical, but measurement of the effect of the Coriolis force on the flow tube can be readily accomplished. The driver oscillates the flow tube in a sinusoidal manner. This sinusoidal motion is depicted in Figure 10a Part D. When fluid flows through the tube, there is a slight time or phase difference between the two sides of the tube, resulting from the inertial forces. The mass flow rate can be determined by placing sensors or detectors on the flow tube, one on either side of the driver, and evaluating the difference between the two signals. Since the motion of the inlet portion of the flow tube lags behind the motion of the outlet portion of the flow tube, the right detector sine wave will always lead the left detector sine wave. Mathematically, the mass flow rate mea-

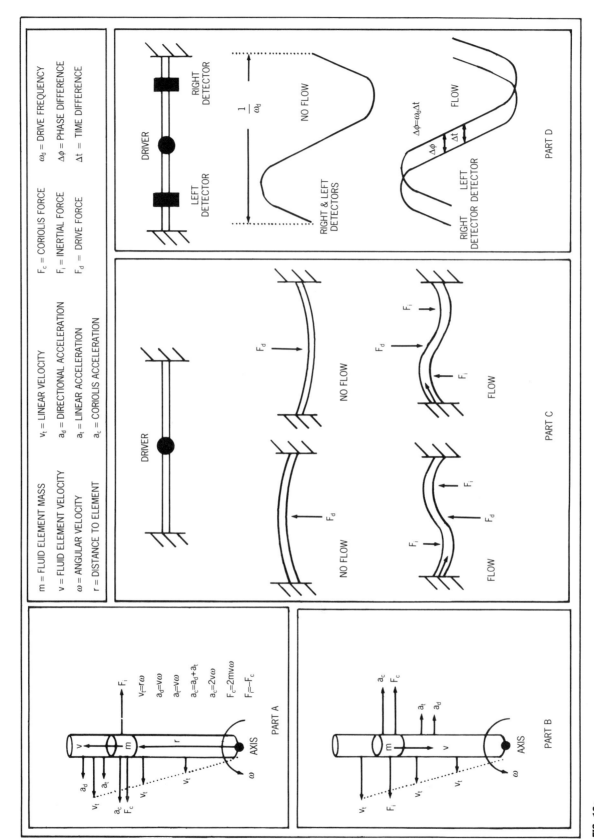

FIG. 10a *Principle of operation.*

surement can be expressed by the following idealized equations, which do not take into consideration any secondary effects associated with the construction of the flow tube:

$$\dot{m} = K\Delta\phi/\omega_d \quad \text{or} \quad \dot{m} = K\Delta t \qquad 10(2)$$

The variables in equation 10(2) are defined in Figure 10a. The value of the constant K depends on the geometry and material of construction of the flow tube. The mass flow rate can be determined either by measuring the time difference between the two detector signals or by measuring the phase difference and the frequency of oscillation. If the fluid is not flowing, there is no time or phase difference between the two detector signals. As the mass flow rate of the fluid increases, the Coriolis forces increase, causing an increase in the time or phase difference between the two detector signals.

CONSTRUCTION

A Coriolis flowmeter consists of two primary components: the flow tube assembly, which is typically called the sensor, and an electronics assembly. The sensor is commonly interfaced to the electronics with a cable that permits the electronics to be located remotely from the sensor. The function of the sensor is to react to the Coriolis forces produced by the fluid flowing through the oscillating tubes. The electronics provides the energy to oscillate the flow tubes, processes the signals from the flow detectors mounted on the flow tubes, and produces outputs which represent the mass flow rate of the fluid.

Sensor

A typical mass flow sensor is shown in Figure 10b Part A. Its main components are the flow tubes, a driving mechanism for oscillating the flow tubes, and flow detectors mounted on either side of the driving mechanism for measuring the time or phase difference between the two sides of the flow tube. The sensor depicted in Figure 10b has two flow tubes. Although the previous discussion on the theory of operation only described a single tube flowmeter, all of the principles which were explained are directly applicable to a dual tube flowmeter. Since both tubes have fluid flowing through them, they are both being deformed by the reaction to the Coriolis forces. As shown in Figure 10b Part B, the tube sections *between the inlet and the driving mechanism will always resist* the driver, and the tube sections *between the driving mechanism and the outlet will always assist* the driver. The flow detectors measure the total influence of the fluid flow on both tubes. The majority of Coriolis flowmeters use a dual tube design, or a single tube bent to have mechanical properties similar to a dual tube design.

All of the commercially available Coriolis flowmeters use a coil and a magnet as the driving mechanism to vibrate the flow tubes. Typically, the coil is mounted on one tube and the magnet is mounted on the other tube. The flow tubes are generally vibrated in opposition at their natural frequency, because this mode of vibration requires the least amount of energy to excite. There are a few designs which vibrate the sensor at frequencies other than the natural frequency.

Mechanical Properties
The flow tube is a vibrating system and as such its behavior is affected by the mass of the system, the elastic properties of the materials of construction, and the damping properties of the system. The mass of the system is the combination of the mass of the fluid flowing through the tube, the tube mass, and the mass of any items attached to the tubes such as the driver and flow detectors. In order for the flowmeter to have stable vibrational properties, the masses of the two tubes should be balanced. The addition of the drive mechanism and the flow detectors to the tubes must be done in such a way that maintains the overall mass balance of the tubes. Changes in the mass of the system will cause the natural frequency of the system to change, which is commonly observed when the density of the fluid changes.

The elastic properties of the flow tube have a direct influence on the mass flow rate measurement. The value of K, from equation 10(2), is directly proportional to the elastic modulus of the flow tube material. The elastic modulus will vary with changing temperature. As temperature increases, the tube material becomes more elastic. As temperature decreases, the tube material becomes stiffer. If the temperature of the tube increases while the mass flow rate through the tube remains constant, the tube will deflect a greater amount. This would be interpreted by the flow detectors to be an increase in the mass flow rate. Coriolis flowmeters have a temperature measuring device, such as an RTD, mounted to the flow tube to measure the temperature of the flow tube material. The temperature indication is input into the electronics where it is used in the calculation of the actual mass flow rate.

Damping variations have a very low-level influence on the flowmeter performance, and are often insignificant. The flow tubes are typically fabricated from materials that have very little damping. Additional damping is introduced by the adhesives and tapes which are used to mount the RTD and to route the signal wires from the driver and flow detectors down the flow tubes.

Tube Geometries
A variety of Coriolis flowmeter designs are currently available, as shown in Figure 10c. Each design has unique performance characteristics which are a function of its geometry. The key design aspects of the flow tube are the diameter, the wall thickness, the material of construction, the overall length, and the shape. The majority of designs which have been developed try to employ a geometry which will magnify the effect of the Coriolis forces on the flow tube to generate a larger time or phase difference between the flow detectors.

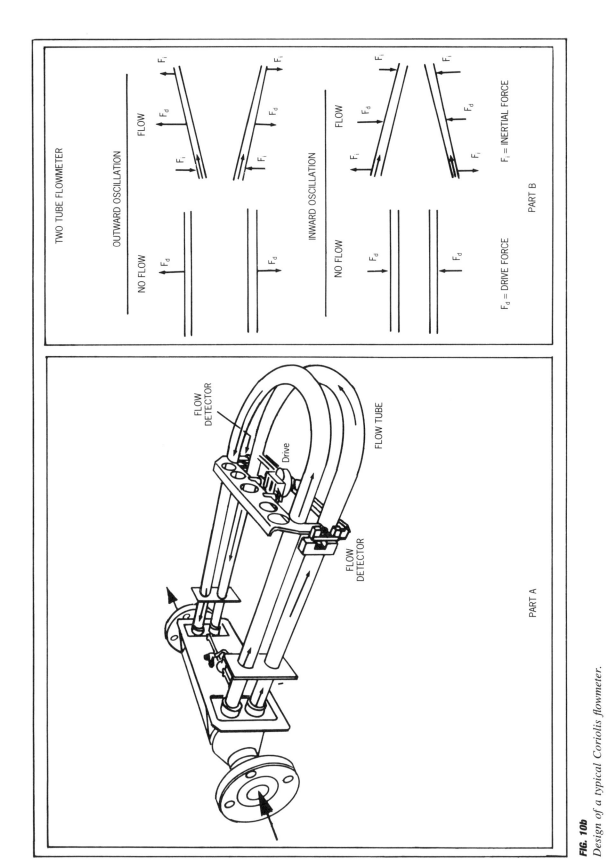

FIG. 10b
Design of a typical Coriolis flowmeter.

FIG. 10c
Geometries of various Coriolis flowmeters.

Flow Detectors The flow detectors can be any type of measurement sensor that will represent the motion of the flow tubes. They can measure position, velocity, or acceleration. The devices which are currently employed as flow detectors are electromagnetic sensors, which measure velocity; and optical sensors and capacitive sensors, which measure position. Electromagnetic sensors are the most common.

Electromagnetic sensors use a magnet mounted on one of the tubes and a coil mounted on the other tube. As the tube vibrates the magnet and coil change their position with respect to one another, which causes a change in the magnetic field in the coil. In a properly designed electromagnetic sensor, the voltage output from the coil is directly proportional to the rate of change of the magnetic field in the coil and reflects the sinusoidal motion of the tube. The optical and capacitive sensors operate in a fashion similar to the electromagnetic sensors, also generating a sinusoidal voltage that reflects the motion of the flow tubes relative to one another.

The signal from the two flow detectors mounted on either side of the driver are sent to the electronics. Both detectors output an electrical signal which varies sinusoidally, but there is a slight time or phase difference between the two signals which results from the deflection of the tube in response to the Coriolis forces.

Electronics

A typical block diagram of the electronics of a Coriolis flowmeter is shown in Figure 10d. The electronics is comprised of three main sections: the signal interfaces between the sensor and electronics, the signal processing section, and the outputs to external devices and displays. The interfaces between the electronics and the sensor are the sensor driving mechanism, the flow detectors, and the flow tube temperature measuring device. The electronics provides the energy for the driver. In most designs, the tubes are vibrated at their natural frequency by minimizing the amount of energy input into the driver while maintaining enough tube displacement to produce sufficient Coriolis force. The electronics applies an alternating current to the drive coil, which is mounted on one of the flow tubes. This generates an alternating magnetic field in the coil. This alternating magnetic field causes the fixed magnet mounted on the other tube to be repelled and attracted, forcing the tubes alternately away from and toward one another in a sinusoidal manner. The oscillation of the flow tubes is reflected by the sinusoidal voltage signal generated by the flow detectors. The sinusoidal voltages from the flow detectors are input into the electronics, along with the signal from the temperature measuring device.

Signal Processing The sine waves from the two flow detectors are compared to determine the time difference (Δt) or phase difference ($\Delta \phi$) between the two signals. Also, the tube frequency, ω_d, is typically measured. The time difference, or the phase difference and tube frequency, along with the tube temperature and the sensor calibration factor are input into an electronic device that determines the mass flow rate. This device is commonly a microprocessor; however, analog circuitry can also be used.

A Coriolis flowmeter can also be used to determine the volumetric flow rate. The electronics typically measures the tube frequency. Since the frequency of tube vibration is proportional to the density of the fluid, the electronics can compute the volumetric flow rate from the mass flow rate and the fluid density using the equation: $Q = \dot{m}/\rho$.

Outputs The primary output from a Coriolis flowmeter is the mass flow rate; however, a majority of the electronics designs are also capable of providing temperature, density, and volumetric flow rate outputs. Many of the Coriolis flowmeter electronics have totalizers to provide mass or volume totals.

The electronics typically provides both frequency and analog (4 to 20 mA) outputs to represent the flow rate of the fluid. The frequency output can be configured to produce a selected number of pulses for every unit of mass, or volume, that the flowmeter measures. The analog outputs can typically be configured to represent one or more of the following variables: mass flow rate, volumetric flow rate, density, or temperature. Most of the electronics are equipped with configurable alarm outputs. Many of the Coriolis flowmeter electronics have a built-in display to directly display the process information. Also, many of the electronics are equipped with digital communications capability that allow for RS485/RS422 or RS232 interfaces. Some of the electronics provide digital communication with hand-held terminals which are typically used to configure the flowmeter's calibration information.

CALIBRATION

Coriolis flowmeters are unique in that each flowmeter is individually calibrated by the manufacturer. Individual calibrations are performed because the flow detector time or phase difference, generated by the Coriolis forces, is different for each flowmeter design. Calibration factors for meters of the same physical size are slightly different because of small variations in manufacturing processes. Each flowmeter is installed in a calibration facility in which fluid flows through the meter into a tank placed on a precision weigh scale. The mass readings from the flowmeter and the weigh scale are compared, and the flowmeter calibration factor is adjusted until the flowmeter reading coincides with the weigh scale reading.

Before a flowmeter can be calibrated it must first be zeroed. Zeroing a Coriolis flowmeter is a process that nulls out the inherent signal differences between the flow detectors. The zero is typically initiated with a push button. This zeroing operation should not be confused with the resetting or zeroing of a flow totalizer. In order to zero the flowmeter, the sensor must be full of the process fluid, but the flow of

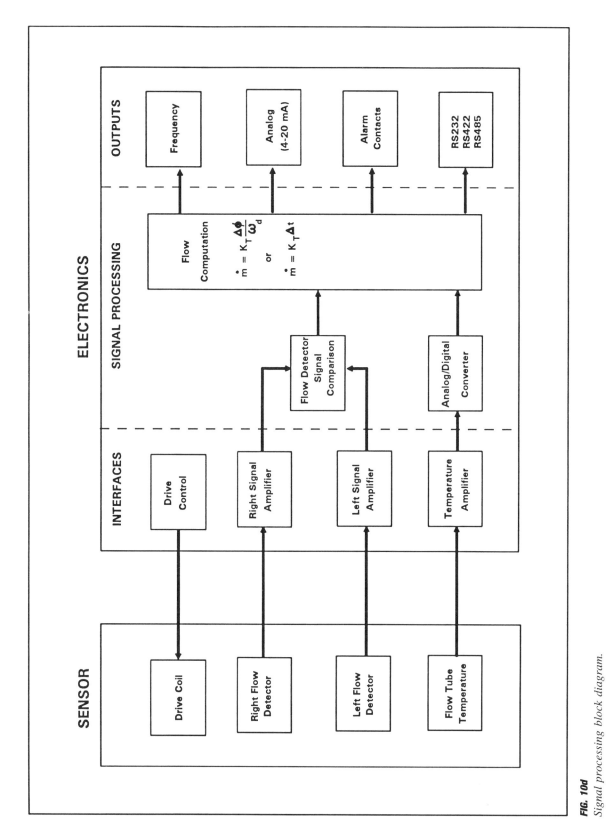

FIG. 10d
Signal processing block diagram.

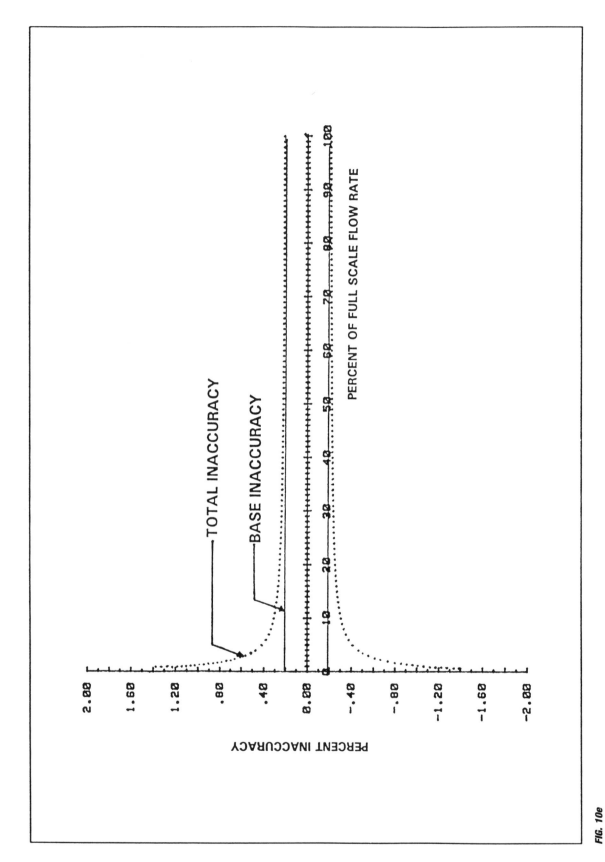

FIG. 10e
Typical Coriolis flowmeter inaccuracy.

fluid through the sensor must be stopped. The signals from the flow detectors are measured and compared to one another. Even though there is no fluid flow, there will typically be a time or phase difference between the two signals. This difference is the result of mechanical variations, or asymmetry, from one side of the sensor to the other. Component variations in the electronics will also contribute to this difference. The electronics measures and stores the difference or "zero," and subtracts this value from subsequent time or phase measurements to eliminate this offset.

Zero Offset or Zero Stability

Even though the flowmeter has been properly zeroed, a slight offset in the zero measurement will usually exist. This offset is due to limitations in the electronics' ability to sample and to measure these small signal levels adequately, and asymmetry in the flow detector signals that is independent of changes in the mass flow rate. This slight offset is commonly defined as the "zero stability" or "zero offset" of the flowmeter. The value of the zero offset is dependent upon the design of the sensor and electronics. Each manufacturer has characterized the maximum amount of offset associated with its flowmeter. Zero offset will only impact the accuracy of the flow measurement at low flow rates, relative to the rated full scale flow rate of the meter.

Inaccuracy

The effect of the zero offset on the inaccuracy of the flowmeter can be more easily understood if the basic inaccuracy equation for Coriolis flowmeters is examined:

$$\text{total inaccuracy \%} = \pm \left[\text{base inaccuracy \%} + \frac{\text{zero offset}}{\text{mass flow rate}} \times 100\% \right] \quad 10(3)$$

For a constant zero offset, decreasing the mass flow rate will increase the magnitude of the zero offset component in the total inaccuracy equation. This relationship is depicted graphically in Figure 10e, which represents the maximum inaccuracy boundaries for a Coriolis flowmeter. When evaluating the inaccuracy of a Coriolis flowmeter for a specific application, the effect of the mass flow rate on the zero offset component must be considered.

FLUID PROPERTIES

When Coriolis flowmeters were introduced, it was thought that the accuracy of the meter was independent of variations in fluid properties: temperature, density, pressure, and viscosity. These beliefs were based on the fact that the Coriolis force depends only on the mass flow rate and is independent of changes in fluid properties. However, experience has shown that the deformation of the flow tube in response to the Coriolis force is influenced by the process conditions to which the tube is subjected. Each geometric shape will behave somewhat differently when fluid properties are changed. Fluid property changes will usually affect one or more of the three calibration parameters: 1) the calibration factor for geometric and material properties, 2) the temperature correction for material elasticity properties, and 3) the zero offset which represents flow detector asymmetry. With the maturing of Coriolis flowmeter technology, manufacturers have become aware of these influences and are producing designs which minimize or eliminate their effects. The performance of the Coriolis flowmeter will continue to improve as investigations of the small scale influences continue.

Temperature

The effect of temperature on the elastic properties of the flow tube depends upon the material of construction. As long as the flow tube temperature is measured and the effects of temperature on the tube stiffness have been characterized, the variations in the mass flow rate measurement can be adequately compensated for. Changes in temperature can also affect the damping properties of the tapes and adhesives used to attach components to the flow tubes. The influence of damping is a low-level effect which can cause changes in the zero offset of the sensor.

Density

Variations in the density of the process fluid cause the mass of the flow measurement system to change, which can alter the mass balance of the sensor. If the flow tube design does not provide for proper mass balance with changing fluid density, measurement errors will arise which will require different calibration factors at different fluid densities. Proper geometric design and mass loading of the flow tube can minimize the influence of density changes on the calibration factor to the point where it is negligible. Changes in the mass balance can also result in asymmetry between the flow detectors, which will cause a change in the zero offset of the flowmeter.

Pressure

The direct effect of increasing pressure causes the flow tubes to be deformed outward. This results in a stiffening of the flow tube which will tend to reduce the generated time or phase difference, even though the mass flow rate has not changed. However, the effect of pressure on the sensor is complicated by the tube geometry. In the bent tube designs the fluid pressure generates significant forces normal to the tube walls which act to offset the outward deformation of the flow tube. The net effect of pressure on the sensor is a combination of several different mechanisms. The flow tube wall thickness, diameter, geometry, and material of construction determine the magnitude of the pressure influence on the sensor. Many designs exhibit insignificant variation with changes in pressure. If a flowmeter design is sensitive to pressure changes a correction method similar to the temperature correction for material elasticity can be employed.

Otherwise, the meter calibration factor must be determined at the operating pressure of the process fluid.

Viscosity

Very little documented information is available on the effect of fluid viscosity on the accuracy of Coriolis flowmeters. Although viscosity influences have been reported, no documented test data have been produced which confirm these claims. Coriolis flowmeters are currently used on a wide variety of viscous products and exhibit excellent accuracy. In one documented test report, on a single sensor geometry, viscosity influences on the flowmeter calibration factor were not observed.[1] However, the zero offset did appear to change. This behavior is likely because changes in fluid viscosity will change the damping properties of the system, which could affect zero offset.[2] It has not been established if fluid viscosity has any influence on the calibration factor or the zero offset.

Fluid property changes which affect the zero offset of the flowmeter impact the accuracy at lower flow rates. Zero offset errors due to fluid property changes can be eliminated by zeroing the flowmeter at operating conditions.

Pressure Drop, Plugging, Erosion, and Corrosion

Pressure Drop Pressure drop varies significantly from one Coriolis flowmeter design to another. The highest drop designs approach a 100 psig drop, at the maximum rated flow of the meter, with water as the flowing media. The pressure drop becomes significantly greater with viscous products. Some designs have high pressure drop because they require high mass flow rates or fluid velocities (sometimes as great as 100 ft/s at maximum rated flow) in order to achieve measurable time or phase difference between the detectors mounted on the flow tubes. It is a common misconception that the high pressure drop in Coriolis flowmeters results predominantly from the bends in the flow tubes. The pressure drop depends upon both the effective length of the tube, which accounts for the tube bends, and the tube inner diameter. Straight-tube flowmeter designs typically have high pressure drop even though there are no tube bends, because they tend to use smaller diameter tubes than the equivalent bent tube designs. Better understanding of the relationship between tube geometry and the time or phase difference produced by the flow detectors has resulted in a variety of low pressure drop designs. Designs are currently being produced that have pressure drops of 10 psig or less, with water, at the rated full scale flow. The manufacturer's literature gives the pressure drop for specific applications.

Plugging Due to their ability to handle many difficult fluid applications, Coriolis flowmeters are often applied to products which have a tendency to harden, freeze, or crystallize if the flow of fluid through the meter is stopped. In some of these applications the product in the piping leading to and from the flowmeter will still be in a liquid state while the product in the flowmeter has hardened. This condition usually results when the flow tube diameter is much smaller than the diameter of the process piping. The smaller diameter flow tubes cause the process fluid to be exposed to an increased surface area, which assists heat loss and product crystallization. Dual tube flowmeters can be particularly troublesome in applications prone to plugging. The dual tube meters use two flow tubes which individually are smaller than an equivalent single tube flowmeter, and therefore have a greater surface area. Also, when trying to unplug a dual tube meter one of the flow tubes will often become clear while the other tube remains plugged. When this condition occurs it is very difficult to unplug the other tube, and the flow rate through the system will be significantly diminished due to the reduced flow through the meter. Even though one of the tubes is plugged, the flowmeter will usually measure flow properly, as long as the plugged tube remains full of process fluid. For a process fluid that has tendencies to solidify, a meter should be selected that provides a flow tube inner diameter as close to the inner diameter of the process pipe as possible. Designs that employ a single tube are preferred over dual tube designs, due to the lower surface area and because a single tube is easier to unplug than are two tubes. Straight tube meters have an advantage in that they are easier to clean once plugging has occurred. Extreme care should be used when cleaning a plugged meter so that the flow tube walls are not scratched or damaged in any way, which can lead to flow measurement errors or tube failure. Some designs can be heat traced to maintain a constant fluid temperature to prevent solidification.

Erosion Coriolis flowmeters can be used to measure solid/liquid mixtures that contain extremely abrasive solids. However, caution must be exercised to avoid erosion of the flow tubes. Erosion will reduce the flow tube wall thickness, which affects the sensors response to the Coriolis forces, and will lead to calibration shifts. Severe erosion can result in failure of the flow tube. In order to minimize the effects of erosion the fluid velocity inside the flow tubes should be kept below 7 ft/s (2.1 m/s) when measuring abrasive materials. Select flow tubes with the largest inner diameter that is practical for the measurement application. Designs with sharp bends are not as desirable as designs with gradual bends or straight tubes.

Corrosion The vibration of the flow tubes results in alternating stresses continuously being applied to the tubes. The presence of these alternating stresses causes the flow tubes of a Coriolis flowmeter to be subject to corrosion fatigue failure. A Coriolis flowmeter may fail in an environment that would not be predicted from general corrosion data. An example of this is a stainless steel flow tube exposed to a process fluid that contains free halogen ions. Halogen ions cause a breakdown of the protective oxide layer of stainless steels and causes pits to form. In an environment where stresses are low the halogen ions would not cause a tube failure. However, if the vibration of the tube results in stress levels that exceed the fatigue limit of the material, a crack might form at a pit. Once a crack has begun it quickly widens and causes failure of the flow tube. In this type of corrosion, additional wall thickness is of little benefit and will not greatly extend the life of the flow tube. In order to minimize or eliminate these types of failures, low stress levels must be maintained in the flow tubes. The fatigue limit for flow tube materials varies with process fluid and temperature. Coriolis flowmeters are commonly available in 316L stainless steel. Some manufacturers provide titanium, Hastelloy, tantalum, and Tefzel-lined meters for use in more aggressive environments.

ENVIRONMENTAL CONDITIONS

Temperature

Ambient temperature changes will affect the elastic properties and zero offset of the sensor in the same manner as changes in the process fluid temperature. The temperature measurement device mounted on the flow tube accounts for both process fluid and ambient temperature changes.

The ambient temperature can also affect the performance of the flowmeter electronics due to the inherent temperature drift characteristics of electronic components. The drift of these components will usually lead to changes in the zero offset of the flowmeter. The electronics are designed to minimize drift effects on critical measurement components. Installation of the electronics in a temperature-controlled environment will eliminate component drift.

Humidity

Humidity does not usually have a significant detrimental effect on the performance of the flowmeter. The sensor is typically enclosed in a sealed case that isolates it from atmospheric contaminants. In designs that do not have adequate case sealing, extremely humid environments can create condensation on the flow detector coils, which may lead to corrosion and failure of these components.

The flowmeter electronics is commonly enclosed in a housing that provides protection against humid conditions. Some manufacturers require that their electronics be installed in a control room to isolate them from the effects of the atmosphere.

Vibration

Since a Coriolis flowmeter is a vibrating device it is commonly assumed that the flowmeter will not function properly in an environment which is subject to vibration. This assumption is incorrect and the flowmeters are commonly installed in pipelines where they are exposed to external vibrations and provide excellent performance. However, it is not advisable to install a Coriolis flowmeter without adequate vibration isolation in an environment where piping and other process equipment have experienced vibration-related failure. The flowmeters are designed to withstand vibrational amplitudes associated with good pipeline practices. In severely vibrating pipelines the sensor should be isolated from the vibration with flexible piping and vibration isolating pipe supports. Another precaution is to ensure that the pipeline vibrational frequency is not the same as the flowmeter operating frequency or one of its harmonics.

If the sensor is externally excited at its operating frequency, measurement errors will occur. It is not often that a pipeline transmits enough vibrational energy at the operating frequency of the sensor to affect the performance of the flowmeter. In the event that the pipeline frequency affects the operation of the meter, rigid clamps may be applied to the piping to change its frequency. In some cases when multiple flowmeters are installed in the same pipeline a problem can arise because they can operate at very similar frequencies. The sensors can transmit enough vibrational energy through the pipeline to excite one another, which can lead to measurement errors. In order to prevent this from occurring, the sensors should be isolated from one another by vibration isolators.

Pipe Stress

Pipe stresses can be applied to a Coriolis flowmeter if piping is misaligned or when the temperature of the piping changes. These conditions cause compression, tension, or shear forces to be applied to the flowmeter. These forces can affect the alignment of the flow tubes with respect to one another, which will lead to asymmetry between the flow detectors. Whenever a flowmeter is installed or relocated, it should be rezeroed in order to nullify these effects. If the pipeline is severely misaligned, it may not be possible to zero out the effects of the pipe stress. Changes in the piping temperature will produce thermal expansion forces and stresses on the sensor. Many flowmeter designs employ a heavy flow manifold in the axis of the pipeline in order to minimize the effects of changing pipe stresses on the flow tubes. Meters which do not provide adequate isolation from changes in pipe stresses are susceptible to changes in the zero offset. Straight tube meters are particularly susceptible to this effect and may require thermal expansion isolators in the pipeline.

INSTALLATION

Each of the different Coriolis flowmeter designs has its own unique installation requirements to minimize the influences of pipe stress and pipeline vibration. The manufacturers' specific recommendations should therefore be adhered to when installing the flowmeter. There are some general recommendations that are applicable to all Coriolis flowmeters. The sensor should remain full of the process fluid. Mixtures of gas and liquid should be avoided. When measuring liquids, do not install the sensor at the highest point in the system, because gas may accumulate in the sensor. When measuring gases, do not install the sensor at the lowest point in the system, because condensation may accumulate in the sensor. If the sensor is mounted in a vertical orientation, fluid should always be pumped upward through the sensor; downward flow can result in incomplete filling of the sensor. A shutoff valve downstream of the sensor should always be installed to allow flow to be stopped to enable zeroing of the flowmeter. Pipeline supports *should not be* attached directly to the sensor, and the sensor should never be used to support process piping. The sensor should not be installed near equipment which generates strong magnetic fields, which can interfere with the normal operation of the drive mechanism and the flow detectors.

APPLICATIONS

Coriolis flowmeters are currently being used in many industrial areas that require flow measurement: chemical, petroleum, petrochemical, food and beverage, pulp and paper, and pharmaceutical. Due to the versatility of these flowmeters they are being used for process control, batching, inventory, precision filling of containers, and custody transfer. An overview of some of the applications for Coriolis flowmeters is presented in Table 10f.

ADVANTAGES

1. A Coriolis flowmeter is capable of measuring a wide range of fluids that are often incompatible with other flow measurement devices. The operation of the flowmeter is independent of Reynolds number; therefore, extremely viscous fluids can also be measured. A Coriolis flowmeter can measure the flow rate of Newtonian fluids, all types of non-Newtonian fluids, and slurries. Compressed gases and cryogenic liquids can also be measured by some designs.
2. Coriolis flowmeters provide a direct mass flow measurement without the addition of external measurement instruments. While the volumetric flow rate of the fluid will vary with changes in density, the mass flow rate of the fluid is independent of density changes.
3. Coriolis flowmeters have outstanding accuracy. The base accuracy is commonly on the order of 0.2%. In addition, the flowmeters are extremely linear over their entire flow range.
4. The rangeability of the flowmeters is usually on the order of 20:1 or greater. Coriolis flowmeters have been successfully applied at flow rates 100 times lower than their rated full scale flow rate.
5. A Coriolis flowmeter is capable of measuring mass flow rate, volumetric flow rate, fluid density and temperature—all from one instrument.
6. The operation of the flowmeter is independent of flow characteristics such as turbulence and flow profile. Therefore, upstream and downstream straight run re-

TABLE 10f
Common Coriolis Flowmeter Applications

Food & Beverage	Chemical & Petrochemical		Petroleum Products
Beer	Adhesives	Hydrogen peroxide	Asphalt
Chocolate	Alcohol	Latex	Bunker C
Fruit juice	Ammonia	Nitric acid	Crude oil
Honey	Catalysts	Phosgene	Diesel fuel
Ice cream	Caustic	Phosphoric acid	Fuel oil
Margarine	Cyclohexane	Polyol	Gasoline
Milk	Ethylene	Propylene	Hydraulic oil
Molasses	Formaldehyde	Resins	Jet fuel
Peanut butter	Freon	Solvents	Kerosene
Pet food	Glycerine	Styrene	Lube oil blending
Tomato paste	Glycol	Sulfuric acid	Oil/water emulsion
Vegetable oil	Hydrochloric acid	Toluene	Tar

Pharmaceutical	Pulp & Paper	Other
Alcohols	Anti-foaming agents	Compressed gases: nitrogen, helium, carbon dioxide, CNG
IV bag filling	Black liquor	
Palm oil	Cellulose slurry	
Perfume	Paper pulp	Dyes
Pill coatings	Red liquor	Ink
Soap	Titanium dioxide	Liquidifed gases: carbon dioxide, LPG, LNG
Sodium methylate		
Talc powder		
Vitamins		Magnetic tape coating
		Paint
		Photographic emulsion
		Wax

quirements and flow conditioning are not necessary. They can also be used in installations that have pulsating flow.
7. Coriolis flowmeters do not have internal obstructions which can be damaged or plugged by slurries or other types of particulate matter in the flow stream. Entrained gas or slugs of gas in the liquid will not damage the flowmeter. There are no moving parts which will wear out and require replacement. These design features reduce the need for routine maintenance.
8. The flowmeter can be configured to measure flow in either the forward or the reverse direction. In reverse flow there will still be a time or phase difference between the flow detector signals, but the relative difference between the two detector signals will be reversed.
9. Coriolis flowmeter designs are available which allow for use in sanitary applications, and for the measurement of shear sensitive fluids. Materials are available to permit the measurement of corrosive fluids.

LIMITATIONS

1. Coriolis flowmeters are not available for large pipelines. The largest Coriolis flowmeter that is currently available has a maximum flow rating of 25,000 lb/min (11,340 kg/min), and is equipped with 6 in. (15 cm) flanges. When larger flow rates must be measured, two or more flowmeters mounted in parallel are required.
2. Some flowmeter designs require extremely high fluid velocities in order to achieve a significant amount of time or phase difference between the flow detector signals. This can result in extremely high pressure drops across the flowmeter.
3. Coriolis flowmeters are expensive. However, the cost of a Coriolis meter is often comparable to (or below) the cost of a volumetric meter plus a densitometer used together to determine the mass flow rate.
4. Coriolis flowmeters have difficulty measuring the flow rate of low-pressure gas. Applications with pressures less than 150 psig are marginal with the flowmeter designs that are currently available. Low-pressure gases have low density, and their mass flow rate is usually very low. In order to generate enough mass flow rate to provide sufficient Coriolis force to be measured, the gas velocity must be extremely high. This in turn may lead to prohibitively high pressure drops across the meter.

References

1. Bug, D.J., and Cork, S.P., "Evaluation of Mass Flowmeter, Model 2100," SIREP Evaluation Report E 1640 S 89, Sira Ltd., October 1989.
2. Keita, N.M., "Contribution to the Understanding of the Zero Shift Effects in Coriolis Mass Flowmeters," *Flow Measurement Instrumentation,* Vol. 1, October 1989.

Bibliography

Babb, M., "New Coriolis Meter Cuts Pressure Drop in Half," *Control Engineering,* October 1991.
Babb, M., "New Mass Flowmeter Design Claims Improved Reliability," *Control Engineering,* May 1990.
Blickley, G., "Flowmeter Calibration Maintained by Design," *Control Engineering,* May 1990.
Blickley, G., "Mass Flow Measurement Aided by Coriolis Methods," *Control Engineering,* April 1991.
Bugher, G., "Coriolis Flowmeters," *Measurement and Control,* September 1990.
Eibl, C., "Is Mass Flow Measurement Increasing Too Rapidly," *Measurement and Control,* September 1990.
"Flowmeter Measures Mass of Fluid Motion," *Chemical Processing,* Mid-November 1982.
Ginesi, D., "Flow a Performance Review," *Chemical Engineering,* June 22, 1987.
Hall, J., "Coriolis Mass Flowmeter Reduces Pressure Loss," *Instruments and Control Systems,* October 1989.
"Mass Flow Control Ups Polyurethane Production," *Instruments and Control Systems,* December 1981.
Plache, K., "Coriolis/Gyroscopic Flow Meter," *Mechanical Engineering,* March 1979.
Robinson, C., "Obstructionless Flowmeters," *InTech,* December 1986.
Schietinger, M., "Mass Flow vs. Volumetric Flow," *Measurement and Control,* September 1990.
Young, A., "Coriolis Mass Flow Measurement," *Measurements and Control,* September 1988.

11 Mass Flowmeters—Miscellaneous

R. SIEV (1969) **K. O. PLACHE** (1982) **B. G. LIPTÁK** (1993)

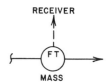

Flow Sheet Symbol

Types:	A. Flow computers using inputs from volumetric flow sensors and densitometers, or pressure/temperature transmitters B. Magnetic flowmeter and radiation-type densitometer combinations C. Angular momentum and other similar principles D. Linear hydraulic Wheatstone-type
Applications:	Gas, liquid, solids
Design Pressures and Temperatures:	Magmeter/radiation units available up to 255 PSIG (17.5 bars) and 300°F (150°C)
Pipe Sizes Available:	A. Unlimited B. 2 to 36 in. (50 mm to 0.9 m) D. Under 1 in. (25 mm), for very low flows only
Errors (Inaccuracy):	Varies with the designs, generally around ±1% full scale
Costs:	A. Computers without the sensors cost about $2000 B. A 6 in. (150 mm) magmeter/radiation combination costs about $13,000 C. Costs vary substantially with size and materials of construction D. Starts at about $2000
Partial List of Suppliers:	Alltru Mass Flowmeter & Equipment Co. (C); Contrec Inc. (A); EACS Div., Advanced Monitors Inc. (A); Eldec Corp. (C); Helicoid Instruments, Div. Bristol Babcock (A); ITT Barton Instruments Co. (A); Johnson Yokogawa Corp. (A); Kay-Ray Inc. (B); Kessler-Ellis Products Co. (A); Ohmart Corp. (B); Omega Engineering Inc. (A); Pierburg Instruments Inc. (Flo-Tron) (D); Pierburg Luftfahrtgerate Union, GMBH (D); Preso Industries (A); Promac USA Inc. (A); Schlumberger Industries, Transducer Div. (A); Texas Nuclear Corp. (B); West Coast Research Corp. (A)

The knowledge of mass flow rates is necessary in combustion fuel control, in reactor recipe formulations, and in many other applications, from mining and dredging to food, pulp and paper, pharmaceuticals, and the chemical industry. The various weighing systems, solids flowmeters, and the more frequently used liquid/gas mass flowmeters (such as the Coriolis and the thermal types) are discussed in other sections of this volume. In this section some of the other techniques of mass flow detection are covered.

RADIATION-TYPE MASS FLOWMETERS

One of the earliest methods of mass flow determination was to install two separate sensors—one to measure the volumetric flow, the other to detect the density of the flowing stream—and then use the two transmitter signals as inputs into a mass flow computing module. While feasible, this approach required coordination between the products of different suppliers and corrections for such process variables as temperature, pressure, viscosity, particle sizes, and velocity profile changes. The introduction of the density/mass flow systems has made it easier to use this technique. The key working component in these combination designs is the multiple input transmitter (Figure 11a), which in addition to a radiation-type density input also accepts a flow measurement signal from any volumetric flowmeter. Based on these two inputs, the microprocessor-based transmitter generates an output signal, which relates to mass flow.

A further improvement occurred in the design of these density/mass flow systems when the density and volumetric flow sensors were combined into a single package (Figure 11b). These units are usually comprised of a magnetic flowmeter and a gamma-radiation-based densitometer, all in a single unit with a microcomputer. This mass flow unit does not require compensation for changes in process variables and is installed as a single, obstructionless mass-flow sensor. Its features and materials of construction are similar to those of a magnetic flowmeter, except that it is bulkier and more expensive, due to the incorporation of the radiation-type densitometer. Because the flow sensor is a magnetic flowmeter, the unit is also limited to being used on process fluids having at least 3.0 μS/cm conductivity.

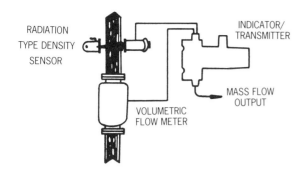

FIG. 11a
Combination mass flow system.

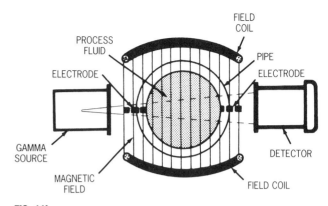

FIG. 11b
Mass flowmeter combining a magnetic flowmeter and a radiation-type densitometer in a single unit.

ANGULAR-MOMENTUM-TYPE MASS FLOWMETERS

The principle of angular momentum can best be described by referring to Newton's second law of angular motion and the definition of angular momentum, using the following notation:

H = angular momentum (lbf-ft-sec)
I = moment of inertia (lbf-ft^2)
ω = angular velocity (rad/sec)
α = angular acceleration (rad/sec^2)
Y = torque (ft-lbf)
r = radius of gyration (ft)
m = mass (slugs)
t = time (sec)

Newton's second law of angular motion states that

$$Y = I\alpha \qquad 11(1)$$

and defines that

$$H = I\omega \qquad 11(2)$$

But, since by definition

$$I = mr^2 \qquad 11(3)$$

Equation 11(1) becomes

$$Y = mr^2\alpha \qquad 11(4)$$

and Equation 11(2) becomes

$$H = mr^2\omega \qquad 11(5)$$

Since

$$\alpha = \frac{\omega}{t} \qquad 11(6)$$

Equation 11(4) becomes

$$Y = \frac{m}{t}r^2\omega \qquad 11(7)$$

Solving for mass flow rate, $\frac{m}{t}$, (lbm/sec), we get

$$\frac{m}{t} = \frac{Y}{r^2\omega} \qquad 11(8)$$

Also, dividing both sides of Equation 11(5) by t:

$$\frac{H}{t} = \frac{m}{t}r^2\omega \qquad 11(9)$$

Since torque is in terms of force, the right-hand side of equation 11(8) must be multiplied by g (32.2 ft/sec^2, or 9.8 m/sec^2) to obtain a dimensionally correct equation. Therefore, since r^2 is a constant for any given system, the mass flow of fluid can be determined if an angular momentum is introduced into the fluid stream and measurements are made of the torque produced by this angular momentum and of the fluid's angular velocity.

Impeller-Turbine

The impeller-turbine-type mass flowmeter uses two rotating elements in the fluid stream, an impeller and a turbine (see Figure 11c). Both elements contain channels through which the fluid flows. The impeller is driven at a constant speed by a synchronous motor through a magnetic coupling and im-

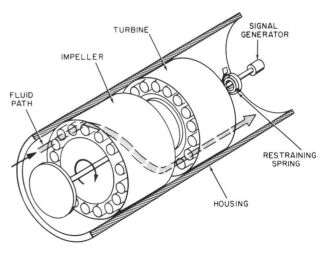

FIG. 11c
Impeller-turbine mass flowmeter. (Illustration reproduced by permission of the General Electric Co.)

parts an angular velocity to the fluid as it flows through the meter. The turbine located downstream of the impeller removes all angular momentum from the fluid and thus receives a torque proportional to the angular momentum. This turbine is restrained by a spring which deflects through an angle which is proportional to the torque exerted upon it by the fluid, thus giving a measure of mass flow.

Constant-Torque-Hysteresis Clutch

Another angular-momentum type mass flowmeter eliminates the necessity of making a torque measurement after imparting a constant torque to the fluid stream. The relationship between mass flow and torque is

$$\frac{m}{t} = \frac{Y}{r^2 \omega} \qquad 11(8)$$

Therefore, if Y is held at a constant value, and since r^2 is a physical constant of any given system,

$$\frac{m}{t} = \frac{k}{\omega} \qquad 11(10)$$

This relationship is used in designing a mass flowmeter as follows: A synchronous motor is placed in the center of the flowmeter assembly. This motor is magnetically coupled to an impeller which is located within the flowing process stream. The magnetic coupling between the motor and the impeller is provided by means of a hysteresis clutch which transmits a constant torque from the motor to the impeller. Thus, a measurement of the rotational speed of the impeller is inversely proportional to the mass flow rate.

Twin-Turbine

Another angular-momentum-type device is the twin-turbine mass flowmeter. In this instrument, two turbines are mounted on a common shaft (see Figure 11d). They are connected with a calibrated torsion member. A reluctance-type pickup coil is mounted over each turbine and a strong magnet is located in each turbine within the twin-turbine assembly.

Each turbine is designed with a different blade angle; therefore there is a tendency for the turbines to turn at different angular velocities. However, since the motion of the turbines is restricted by the coupling torsion member, the entire assembly rotates in unison at some average velocity, and an angular phase shift is developed between the two turbines. This angle is a direct function of the angular momentum of the fluid. As was previously shown, angular momentum can be measured by torque, and angular momentum is a function of mass flow. In the twin-turbine assembly, the turbines are not restrained by a spring, but the torsion member which holds them together is twisted. This torsion member has a well-established torsion-spring rate (ft-lbf/rad). Therefore, the angle developed between the two turbines is a direct function of the twist or torque exerted by the system.

This angle is measured by a unique method. As each turbine magnet passes its own pickup coil, the coil generates a pulse. The pulse from the upstream turbine is used to open a so-called electronic gate, while the pulse from the downstream turbine closes this gate. An oscillator is placed in the electronic circuit, and the oscillations are counted while the gate is opened. The number of oscillations is thus a function of the angle between the two turbines. Knowledge of the angle gives the value of torque which, in turn, is proportional to mass flow rate.

Coriolis

The classic Coriolis-type mass flowmeter (see Figure 11e) consists of a centrifugal-pump impeller wheel and a vaned sensing wheel which acts as a turbine wheel to extract the angular momentum imparted to the fluid by the impeller. The sensing (or turbine) wheel is contained in the same housing as the impeller and is attached to the latter by a strain gauge; the combination is driven at a known constant speed. The power applied to the impeller is merely that required to overcome the frictional drag of the system.

The torque measured is that required to impart to the fluid stream a Coriolis acceleration, and is given by the expression

$$Y = \omega (R_2^2 - R_1^2) \frac{m}{t} \qquad 11(11)$$

where R_2, R_1 = outer and inner radii (ft).

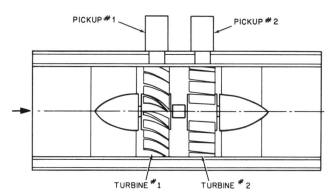

FIG. 11d
Twin-turbine mass flowmeter.

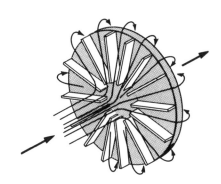

FIG. 11e
Classical Coriolis mass flowmeter.

Comprehensive analysis and laboratory testing of this type of flowmeter are reported in Reference 1.

Gyroscopic

Another angular momentum mass flowmeter (Figure 11f) operates on the principle of a gyroscope. It consists of a pipe shaped in the form of a circle or a square. A motor introduces an oscillating vibration at a constant angular velocity ω about the A axis. When the fluid passes through the loop, a precession-type moment is produced about the B axis and is measured by the deflection of a sensing element. This deflection can be shown to be directly proportional to mass flow.

The gyroscopic mass flowmeter can handle slurries in the medium pressure and temperature ranges but its industrial use is very limited due to its high cost and inability to handle high flow rates.

The gyroscopic and Coriolis flowmeters described above were not successful on the market. A newer combination of the two principles has resulted in the highly successful mass flowmeter, that was described in the previous section.

LINEAR

The linear mass flowmeter is, in principle, a hydraulic equivalent of the electrical Wheatstone Bridge. Four matched orifices make up the bridge and an integral constant flow recirculating pump establishes the internal reference flow. Sensing unbalance generated by external flow through the meter, the hydraulic bridge produces an output of differential pressure which is both linear and proportional to the true mass liquid flow.

This flowmeter has a wide rangeability and is unaffected by changes in process temperature, density, or viscosity. It is a fast responding flowmeter which can detect very low flows at very low pressure drops. This meter is widely used in the automotive industry and whenever engines and fuel systems are checked, such as in the manufacturing of fuel injectors.

CONCLUSIONS

The most successful mass flowmeters have been described in the previous section (Section 10), but the devices covered in this section also have their applications. For example, the integrated magnetic flowmeter/radiation densitometer package is the best solution to the mass flow measurement of large slurry streams, in size ranges of 12 in. (300 mm) and above.

Accurate mass flow detection can frequently increase the efficiency of processes or allow for optimization which results in energy or fuel conservation in combustion systems.

Reference

1. Plache, K.O., "Coriolis/Gyroscopic Flow Meter," *Mechanical Engineering*, March 1979, p. 36.

Bibliography

Albertz, T., "Mass Flow Measurements in Dust," Krupp-Koppers, Essen, Germany, April 1980.

ASME Research Committee on Fluid Meters, "Fluid Meters, Their Theory and Application," 5th edition, 1959.

Babb, M., "New Mass Flowmeter Design Claims Improved Reliability," *Control Engineering*, May 1990.

Bugher, G., "Coriolis Flowmeters," *Measurements and Control*, September 1990.

Close, D.L., "Cryogenic Mass Flowmeter," ISA Symposium in Pittsburgh, May 1971, Paper No. 1-6-34.

Eibl, C., "Is Mass Flow Measurement Increasing Too Rapidly," *Measurements and Control*, September 1990.

"Flowmeter Measures Mass of Fluid Motion," *Chemical Processing*, Mid-November 1982.

Hall, J., "Coriolis Mass Flowmeter Reduces Pressure Loss," *Instruments and Control Systems*, October 1989.

Harrie, P.M., "Mass Flow and Density," *Measurements and Control*, April 1991.

Hayward, A.J., "Choose the Flowmeter Right for the Job," *Processing Journal*, 1980.

Laskaris, E.K., "The Measurement of Flow," *Automation*, 1980.

Li, Y.T., and Lee, S.Y., "Fast-Response True-Mass-Rate Flowmeter," *Trans ASME*, Vol. 75, July 1953, pp. 835–841.

Lomas, D.J., "Selecting the Right Flowmeter," *Instrumentation Technology*, 1977.

"Mass Flowmeters," *Measurements and Control*, September 1991.

Schietinger, M., "Mass Flow vs. Volumetric Flow," *Measurements and Control*, September 1990.

Sheriff, D., "Mass Flow Control," *Measurements and Control*, September 1990.

Snell, C.C., "Relative Volume Fraction Measurement of Gas/Liquid or Solid/Liquid Flow with a Rotating Field Conductance Gauge," 1979 ISA Symposium, ISBN 87664-432-9.

Spink, L.K., "Principles and Practice of Flow Meter Engineering," The Foxboro Co., 1958.

Watson, G.A., "Flowmeter Types and Their Usage," *Chartered Mechanical Engineer Journal*, 1978.

Zapolin, R.E., "New Ways to Meet User Needs for SCADA Mass Flow Monitoring," 1990 ISA Conference, Paper #90-0633, New Orleans.

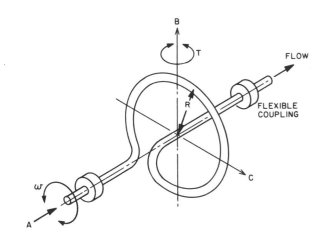

FIG. 11f
Gyroscopic mass flowmeter.

12 Mass Flowmeters—Thermal

R. SIEV (1969, 1982) **B. G. LIPTÁK** (1993)

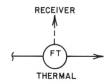

Flow Sheet Symbol

Types:	A. Heat-transfer-type mass flowmeters B. Combination units with orifice or linear element in main line and heat-transfer-type flowmeter in bypass C. Heated-element-type mass flowmeters (Note: Thermal flow switches are discussed in Section 6)
Design Temperature:	A. Up to 350°F (176°C), higher with special designs B. Some limited to 105°F (40°C); others can operate at up to 300°F (150°C) C. Standard units operate at up to 140°F (60°C), special ones up to 930°F (500°C)
Design Pressure:	A. Up to 1200 PSIG (83 bars), higher with special designs B. Up to 150 PSIG (10.3 bars) C. Low-pressure designs up to 15 PSIG (1 bar); others up to 1000 PSIG (69 bars) in smaller sizes
Pressure Drop:	A & C. Usually only a few inches of water B. Up to 45 PSIG (3 bars)
Process Fluids:	A. Air, gas, liquids, and slurries B. Gases only C. Gases and liquids
Flow Range:	A. From 0.5 sccm to 40,000 lbm/h (10,000 kg/h) B. Flow control units from 0–10 sccm to 0–7 scfm (0–30 slm); special units up to 0–500 scfm (0–14 scmm) C. From 0–50 sccm to 0–2500 scfm (0–70 scmm)
Error (Inaccuracy):	±1% to ±2% of full scale, some better
Rangeability:	10:1 to 100:1
Materials of Construction:	Stainless steel, glass, Teflon, Monel, etc.
Cost:	A type "B" thermal mass-flow control loop for low gas flows costs about $1500; purged probes for stack velocity measurement cost about $4,500 each
Partial List of Suppliers:	Aalborg Instruments & Controls Inc. (gas); Air Monitor Corp. (air); Brooks Instrument Div. of Rosemount (gas, liquid); Dresser Industries Inc., Datametrics (gas); Edwards High Vacuum International (gas); Eldridge Products (gas); ETA Control Instruments (flow switch); FCI Fluid Components (gas, liquid); Flow and Level Controls (flow switch); Hydril Co. (flow switch); Integrated Control Concepts Inc. (gas); Intek Inc., Rheotherm Div. (gas, liquid); Kurz Instruments Inc. (gas); Matheson Gas Products Inc. (gas); MKS Instruments Inc. (gas); M-Tek (liquid); Phase One Instruments Inc. (gas); Porter Instrument Co. (gas); Precision Flow Devices Inc. (gas); Saratoga Automation (gas); Scott Specialty Gases Inc. (gas); Sierra Instruments Inc. (gas); Teledyne Hastings-Raydist (gas); Thermal Instrument Co. (gas, liquid, slurry); TSI Inc. (gas, liquid); Tylan General (gas); Unit Instruments (gas); Wahl Instruments Inc. (gas)

Thermal flowmeters can be divided into two categories:

1. Flowmeters that measure the rise in temperature of the fluid after a known amount of heat has been added to it. They can be called heat transfer flowmeters.
2. Flowmeters that measure the effect of the flowing fluid on a hot body. These instruments are sometimes called hot wire probes or heated-thermopile flowmeters.

Both types of flowmeters can be used to measure flow rates in terms of mass, a very desirable measurement, especially on gas service.

HEAT TRANSFER FLOWMETERS

The operation of the heat transfer flowmeter is based on:

$$Q = Wc_p(T_2 - T_1) \quad 12(1)$$

where

Q = heat transferred (BTU/hr or Cal/hr)
W = mass flow rate of fluid (lbm/hr or kgm/hr)
c_p = specific heat of fluid (BTU/lbm°F or cal/kgm °C)
T_1 = temperature of the fluid before heat is transferred to it (°F or °C)

T_2 = temperature of the fluid after heat has been transferred to it (°F)

Solving for W we get

$$W = \frac{Q}{C_p(T_2 - T_1)} \qquad 12(2)$$

A simple flowmeter based upon this equation is shown schematically in Figure 12a. Heat is added to the fluid stream with an electric immersion heater. The power to the heater equals the heat transferred to the fluid (Q) and is measured by a wattmeter. T_1 and T_2 are thermocouples or resistance thermometers. Since we know the fluid, we also know the value of its specific heat. Thus, by measuring Q, T_1, and T_2, the flow rate (W) can be calculated. T_1 and T_2 do not have to be separately detected; they can be connected together so that the temperature difference ($T_2 - T_1$) is measured directly.

A flowmeter of this type of construction has many limitations. The temperature sensors and the heater must protrude into the fluid stream. Thus, these components (particularly the heater) are easily damaged by corrosion and erosion. Furthermore, the integrity of the piping is sacrificed by the protrusions into the fluid stream, increasing the danger of leakage.

To overcome these problems, the heater and the upstream and downstream temperature sensors can be mounted outside of the piping (see Figure 12b). In this type of construction the heat transfer mechanism becomes more complicated and the relationship between mass flow and temperature difference becomes nonlinear. Figure 12c illustrates this nonlinear shift in ΔT in a heated-tube-type flowmeter, where the asymmetricity of the temperature distribution increases with flow.

To understand the principle of operation of this flowmeter we must review the effects of fluid mechanics and heat transfer. When a fluid flows in a pipe (turbulent or laminar) a thin layer (film) exists between the main body of the fluid and the pipe wall. When heat is passing through the pipe wall to the fluid, this layer resists the flow of heat. If the heater is sufficiently insulated and if the piping material is a good heat conductor, the heat transfer from the heater to the fluid can be expressed as:

$$Q = hA(T_{wall} - T_{fluid}) \qquad 12(3)$$

where

h = film heat transfer coefficient [BTU/(hr × ft² × °F)]
A = area of pipe through which heat is passing (ft²)
T_{wall} = temperature of wall (°F)
T_{fluid} = temperature of fluid (°F)

The film heat transfer coefficient h can be defined in terms of fluid properties and tube dimensions—for both laminar and turbulent flow.[1]

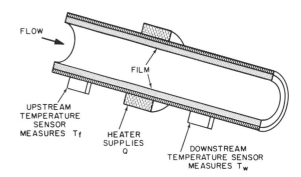

FIG. 12b
Thermal flowmeter with external elements and heater.

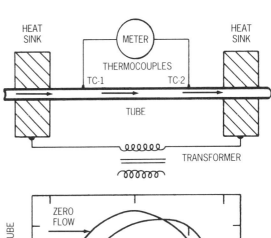

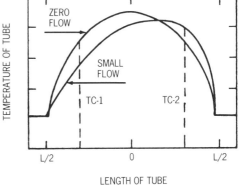

FIG. 12c
Heated-tube-type mass flowmeter.

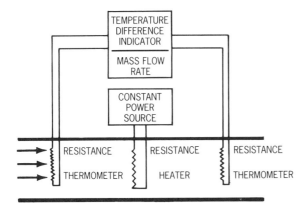

FIG. 12a
Heat transfer flowmeter.

$$h_{turbulent} = \frac{0.023 K^{0.6} c_p^{0.4} W^{0.8}}{D^{1.8} \mu_1^{0.4}} \quad 12(4)$$

$$h_{laminar} = \frac{1.75 K^{0.67} c_p^{0.33} W^{0.33}}{D L^{0.33}} \quad 12(5)$$

where

- K = thermal conductivity of the fluid [BTU/(hr × ft × °F)]
- c_p = specific heat of the fluid [BTU/(lbm × °F)]
- D = pipe diameter (ft)
- L = heated length (ft)
- μ_1 = absolute viscosity of the fluid (lbf/hr-ft²)
- W = flowrate (lbm/hr)

Using the turbulent flow condition as an example, and solving Equations 12(3) for h and 12(4) for W:

$$h = \frac{Q}{A(T_{wall} - T_{fluid})} \quad 12(6)$$

$$W^{0.8} = \frac{h D^{1.8} \mu_1^{0.4}}{0.023 K^{0.6} c_p^{0.4}} \quad 12(7)$$

Substituting Equation 12(6) into 12(7):

$$W^{0.8} = \frac{Q D^{1.8} \mu_1^{0.4}}{0.024 K^{0.6} c_p^{0.4} A(T_{wall} - T_{fluid})} \quad 12(8)$$

Therefore the mass flow (W) will vary with $T_{wall} - T_{fluid}$ if the fluid properties (μ, K, and c_p) and the meter design parameters (Q, D, and A) are all held constant. Letting all these constants = X:

$$W^{0.8} = \frac{X}{T_{wall} - T_{fluid}} \quad 12(9)$$

The downstream temperature sensor is located near the heater so that it measures T_{wall}. The upstream temperature sensor is located where the wall and fluid temperatures are in equilibrium with each other. Thus, flow rate is obtained by measuring ΔT, if the geometry of the flowmeter, the thermal conductivity, the thermal capacity and viscosity of the fluid, and the heater power are constant. This type of flowmeter can also be operated by keeping the ΔT constant and measuring the required power to the heater.

When building and/or using a flowmeter based upon heat transfer principles, the instrument engineer must tread with caution. One must be sure that the values which have been assumed to be constants are truly that. One must also understand that relationships such as those given in Equations 12(4) and 12(5) are limited to a range of Reynolds numbers, L/D ratios, etc. Finally, it is highly recommended that this type of instrument should be calibrated, either by the manufacturer or by the user, under conditions that duplicate as nearly as possible its actual application.

These types of flowmeters are best suited for the measurement of homogeneous gases and are not recommended for applications where the process fluid composition or moisture content is variable. In order for these flowmeters to be useful in a system, both the thermal conductivity and the specific heat of the process fluid must be constant.

Bypass-Type Designs

In order to make the heat-transfer-type flowmeter suitable for the measurement of larger flow rates, the bypass designs have been introduced (Figure 12d). The thermal flowmeter tubes in these bypass units are small capillary tubes, usually under 0.125 in. (3 mm) diameter. They ensure laminar flow over the full operating range of the meter. Their small size is advantageous in minimizing the electric power requirement and also in increasing their speed of response, but it necessitates the use of upstream filters to protect against plugging. Some of the units also require up to 45 PSIG (3 bars) pressure drop in order to develop the laminar flow conditions at which they operate.

These units are also available as complete control loops, with sensor controller, and automatic control valve all assembled into a single unit (Figure 12e). The cost of these units is very competitive, and therefore, if it is sufficient to control the flow of small gas streams within ±2% of full scale, these units represent a good selection. The control valve used in these small units is usually the thermal expansion type, operated without seals and therefore without friction or wear. The stem of a thermal expansion valve is heated by the controller and the expanding stem moves the

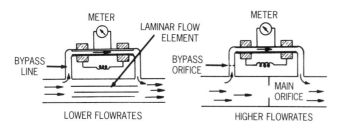

FIG. 12d
Bypass-type thermal mass flowmeters.

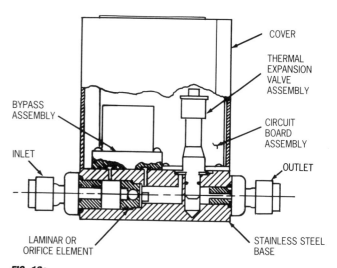

FIG. 12e
Complete flow control loop for small gas flows.

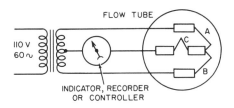

FIG. 12f
Hot wire flow-sensing probe.

valve plug into the seat, thereby closing the valve. Due to the type of valve actuator used it takes 5 to 30 seconds to bring the flow to setpoint.

HOT WIRE PROBES

In this design two thermocouples (A and B) are connected in series forming a thermopile. A schematic of this type of flowmeter is shown in Figure 12f. This thermopile is heated by passing an alternating current through it. A third thermocouple (C) is placed in the direct current output circuit of the thermopile. Alternating current does not pass through this thermocouple, and it is therefore not electrically heated. This assembly is inserted into the process fluid (usually gas) stream. The gas cools the heated thermopile by convection. Since the AC input power to the thermopile is held constant, the thermopile will attain an equilibrium temperature and produce an emf that is a function of the gas temperature, velocity, density, specific heat, and thermal conductivity. The third, unheated thermocouple (C) generates an emf that is proportional to the gas temperature. This cancels the effect of the ambient gas temperature on the output signal of the heated thermopile. A and B are the heated thermocouples; C is the unheated one.

The output signal (voltage) of this instrument is given by the equation derived by King[3]:

$$e = \frac{C}{2(\pi K c_p \rho d v)^{1/2} + K} \qquad 12(10)$$

where

e = voltage generated
C = instrument constant
K = thermal conductivity of fluid (BTU/hr-ft-°F)
c_p = specific heat of fluid (BTU/lbm-°F)
d = diameter of heated thermocouple wire (ft)
v = velocity of fluid (ft/hr)
ρ = density of the fluid (lbm/ft^3)

Once the instrument has been calibrated for a certain gas, change in the gas temperature will have little effect on the gas properties, and thus on the output signal. For example, the properties of air, over a wide range of temperatures, are as follows:

Temp (°F)	K (BTU/hr-ft-°F)	c_p (BTU/lbm°F)	ρ (lbm/ft^3)	$(Kc_p\rho)^{1/2}$
70	0.0150	0.243	0.0753	0.0165
500	0.0246	0.245	0.0416	0.0159
1000	0.0359	0.263	0.0274	0.0161

Since K (in the denominator) is very small, and since the term $(Kc_p\rho)^{1/2}$ remains constant over a wide range of temperatures, this type of instrument can be used to measure the mass flow rate of gases.

The hot-wire-type sensors are also used as air velocity sensors. These devices are called "Anemometers." A major limitation of the hot-wire-type mass flowmeters is similar to the limitations of all pitot-type flowmeters, namely that they do not detect the mass flow across the full cross section of the pipe, but only at the sensor. Therefore, if the sensor is installed in a nonrepresentative location across the velocity profile, the resulting reading will be in error. The more recent developments in the design of this type of flowmeter include a more rugged mass flow sensor element, an integral (usually 10-diameter-long) pipe section to ensure smooth velocity profiles, and a conditioning nozzle which eliminates

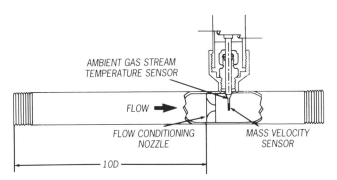

FIG. 12g
Complete mass flow sensor assembly. (Courtesy Kurz Instruments Inc.)

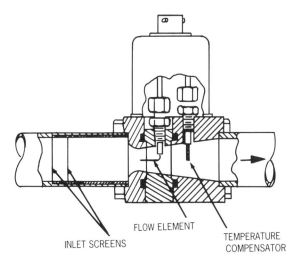

FIG. 12h
Venturi-type thermal mass flowmeter. (Courtesy TSI Inc.)

boundary layer effects and concentrates the flow onto the sensor (Figure 12g). These sensors can also be provided with controllers and control valves to result in a complete flow control loop.

Other designs place the heated mass flow sensor at the throat of a venturi and add a screen upstream to make the flow profile more uniform (Figure 12h). These units are offered for both liquid and gas service. Yet other designs are of the insertion probe type. Their flow ranges are only a function of the size of the pipe into which they are inserted, while their performance is similarly influenced by the correctness of the insertion depth (as are all pitot tubes).

References

1. McAdams, W.H., "Heat Transmission," 3rd edition, equations 9–10a and 9–23, New York: McGraw-Hill, 1954.
2. Louis V. King, "On the Convection of Heat from Small Cylinders in a Stream of Fluid," Philosophical Transactions of the Royal Society of London, Series A, Vol. 214 (1914), pp. 373–432.

Bibliography

Baker, W.C., "Flow Without Fouling," *Measurements and Data,* Nov.–Dec. 1974.

Baker, W.T., "Flowmeters Using Temperature," *Measurements and Control,* September 1991.

Benson, J.M., Baker, W.C., and Easter, E., "Thermal Mass Flowmeter," *Instruments and Control Systems,* February 1970.

Hayward, A.J., "Choose the Flowmeter Right for the Job," *Processing Journal,* 1980.

Laskaris, E.K., "The Measurement of Flow," *Automation,* 1980.

LeMay, D.B., "A Practical Guide to Gas Flow Control," *Instruments and Control Systems,* September 1977.

Lomas, D.J., "Selecting the Right Flowmeter," *Instrumentation Technology,* 1977.

Reason, J., "Thermal Dispersion Elements Measure Flue-Gas Recirculation," *Power Magazine,* June 1987.

Watson, G.A., "Flowmeter Types and Their Usage," *Chartered Mechanical Engineer Journal,* 1978.

Welch, J.V., "Trends in Low Gas Flow Metering," *InTech,* February 1991.

13 Metering Pumps

R. SIEV (1969, 1982)   **B. G. LIPTÁK** (1994)

Types:	A. Peristaltic
	B. Piston or plunger types (provided with packing glands)
	C. Diaphragm or glandless types (mechanical, hydraulic, double-diaphragm, and pulsator designs)
Capacity:	A. 0.0005 ccm to 20 gpm (90 lpm)
	B. 0.001 gph to 280 gpm (0.005 lph to 1250 lpm)
	C. Mechanical diaphragms: from 0.01 to 50 gph (0.05 to 3.7 lpm); mechanical bellows: from 0.01 to 250 gph (0.05 to 18 lpm); others: from 0.01 to 800 gph (0.05 lph to 60 lpm); pulsator pumps: from 30 to 1800 gph (2 to 130 lpm)
Error (Inaccuracy):	A. ±0.1 to ±0.5% of full scale over a 10:1 range
	B & C. ±0.25 to ±1% of full scale over a 10:1 range; can be as good as ±0.1% full scale at 100% stroke and tends to drop as stroke is reduced
Maximum Discharge Pressure:	A. 50 PSIG (3.5 bar)
	B. 50,000 PSIG (3450 bars)
	C. Mechanical bellows: up to 75 PSIG (5 bars); mechanical diaphragm: up to 125 PSIG (8.5 bars); hydraulic Teflon diaphragm: 1500 PSIG (104 bars); pulsator pumps: up to 5000 PSIG (345 bars); hydraulic metallic diaphragms: up to 40,000 PSIG (2750 bars)
Maximum Operating Temperature:	A. 70 to 600°F (−57 to 315°C)
	B. Jacketed designs: up to about 500°F (260°C)
	C. Units containing hydraulic fluids can handle from −95 to 360°F (−71 to 182°C), Teflon and Viton diaphrams are limited to 300°F (150°C), and neoprene and Buna N are limited to 200°F (92°C); the metal bellows and the remote head designs can operate from cryogenic to 1600°F (870°C)
Materials of Construction:	A. Neoprene, Tygon, Viton, silicone
	B. Cast iron, steel, stainless steel, Hastelloy C, Alloy 20, Carpenter 20, Monel, nickel, titanium, glass, ceramics, Teflon, PVC, Kel-F, Penton, polyethylene, and other plastics
	C. Polyethylene, Teflon, PVC, Kel-F, Penton, steel, stainless steel, Carpenter 20, Monel, Hastelloy B & C
Cost:	A. $200 to $800
	B. $1000 to $6000
	C. $1000 to $12,000
Partial List of Suppliers:	American LEWA Inc. (A,B,C); Barnant Co. (A); Blue White Industries; Bran & Luebbe Inc.; Clark-Cooper Corp. (B,C); Cole-Parmer Instrument Co.; Flo-Tron Inc. (B); Fluorocarbon Co.; Gerber Industries; Hydroflow Corporation; LDC Analytical; Leeds & Northrup, Unit of General Signal; Liquid Metronics Inc. Milton Roy Div. (B); Plast-O-Matic Valves Inc.; Ruska Instrument Corp.; S J Controls Inc.; Valcor Scientific; Wallace & Tiernan Inc. (B,C);

A metering pump is a positive displacement pump providing a predictable and accurate rate of process fluid flow. Normally the application-design, specification, and use of pumps are the concern of mechanical engineers and machinery designers. Metering pumps, however, are sometimes used to actually measure or control flow rate, and in many cases they are the final control elements in an instrumentation loop. Therefore, the instrument engineer should be familiar with their operation and application.

Some controlled-volume pumps (as metering pumps are sometimes called) are designed to meet the needs of just one particular application, such as adding sodium hypochlorite to a swimming pool or providing chemical reagents to a chromatograph. Thus, each industry has its own particular types of metering pumps which could be classified by their application. A better way of classifying metering pumps, however, is to distinguish them by their design. Any positive displacement pump, due to its volumetric mode of fluid

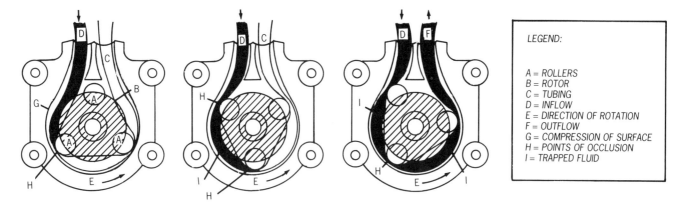

FIG. 13a
The three-roller design prevents backflow and eliminates the need for check valves. (Masterflex® is a registered trademark of Cole-Parmer Instrument Co.)

transfer, can be used as a metering pump. In practice, however, only those positive displacement pumps that have none or only very little internal and/or external leakage can provide the precision and accuracy that are normally required of a metering pump.

The three basic metering pump designs are the peristaltic, piston, and diaphragm types. They are all positive displacement devices used for the precise charging of corrosive, radioactive, toxic, flammable, or otherwise difficult fluids such as slurries, melts, liquified gases, or liquid metals. They are used either in a controlled-volume metering mode or in a mixing-proportioning mode. The piston or plunger designs are provided with a packing gland, while the diaphragm or bellows designs are also referred to as glandless. The pulsator pump is a special variety of the glandless design. The actuators can adjust either the speed or the stroke, and the controls used for stroke adjustment include micrometers, positioners, and reversing motors with slide-wire feedback.

PERISTALTIC PUMPS

Peristaltic action is similar to the operation of the intestines and other hollow muscular structures, where the successive contraction of the muscular fibers of their walls forces their content forward. In the peristaltic pump the fluid is moved forward by progressively squeezing a flexible container from the entrance to discharge. This container is usually a tube that can be made out of any material that possesses sufficient resiliency to allow it to recover to its original shape immediately after compression. There are a variety of methods employed for squeezing the tube (or container) to produce flow, including:

1. Rollers that are connected to a rotating body squeeze the tubing against a circular housing (see Figure 13a).
2. Rollers that are driven by a chain drive squeeze tubes against a flat plate.
3. Cam-operated fingers successively squeeze the tubing against a flat surface (see Figure 13b).
4. A rotating wobbling cam squeezing a tube against a flat plate.

The plastic hose or tubing provides an external, tight, sanitary, and easily cleanable and replaceable container. It must be remembered that the tube is the only component of the pump that comes into contact with the fluid. A plastic material can usually be found that will be suitable for even the most corrosive and abrasive application. However, the use of plastic tubing also places severe limits on the capability of the peristaltic pump. These pumps can furnish only low flow rates and low pressure heads.

The peristaltic pump has found particularly large acceptance in the medical and biochemical fields where high accuracy, low flow rates, inherent enclosure of the fluid, and sterilization are prime requirements. The flow rate of the peristaltic pump can be adjusted by changing the speed of the squeezing mechanism. Power to these pumps is usually provided by 60-Hz, 110-volt electrical motors. Air-operated and electrical explosion-proof motors are also available.

Another peristaltic pump employs a plastic liner separating a rotating cam from the pumped fluid. This allows the use of a cam-type positive displacement pump for metering service, but prevents any external leakage. Higher pressures and flow rates can be obtained from this pump, but the advantages of the flexible tubing are lost.

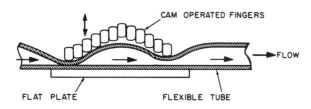

FIG. 13b
Peristaltic pump with cam-operated fingers.

PISTON PUMPS

The piston pump employs a piston or plunger which moves with a reciprocating motion within a chamber. A fixed volume of liquid is delivered with each stroke. The flow rate

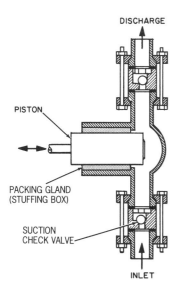

FIG. 13c
Piston pump schematic.

is a function of piston diameter, chamber length, and piston speed. Check valves located at the pump inlet and outlet are required to prevent backflow. A schematic of a typical piston pump is shown in Figure 13c. The piston produces pressure in only one direction; therefore, the flow produced by plunger pumps (as piston pumps are sometimes called) is pulsating. If the pulsating flow characteristics are undesirable, a dampening reservoir (accumulator) should be installed in the discharge line of the pump. Another method available to reduce pulsation is to use a pump that employs more than one chamber/piston combination in parallel. Pumps having as many as four chambers (cylinders) are commercially available. These multiple-piston pumps are called duplex pumps if they have two pistons; triplex, if they have three; etc. (Figure 13d).

The construction materials of the piston pump components must be selected with care since its housing, piston, piston packing, valve body, and valve seat all come into intimate contact with the process fluid.

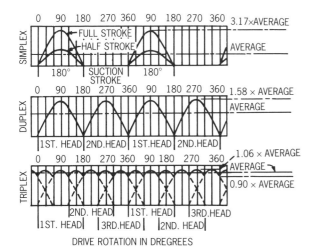

FIG. 13d
Multiple pistons tend to dampen pressure fluctuations.[1]

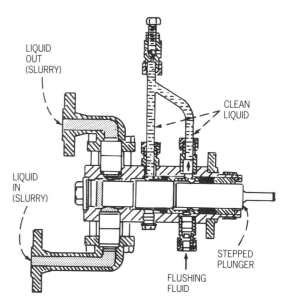

FIG. 13e
Flushing fluid keeps the packing free of solids on slurry services.[1]

The displacement of the pump is the area of the plunger multiplied by the stroke length. This is not the volume that is actually delivered, because some of the fluid slips back into the cylinder while the check valves are closing. Worn valves and fluid compressibility at high discharge pressures further reduce the volumetric efficiency (usually around 95%) of plunger pumps. Check valve leakage can be reduced by installing two check valves in series at both the pump suction and at its discharge. Volumetric efficiency should not be taken as metering accuracy, because leakage is repeatable and can be zeroed out by calibration. Therefore, in well-maintained and calibrated metering pumps the measurement error will be much less than the leakage.

The plunger packing must be carefully selected, not only to minimize leakage and wear, but also for lubrication, cooling, sterilization, or flushing. This pump design can deliver both large flow rates and high discharge pressures. The smaller the plunger diameter, the greater the possible discharge pressure.

When metering abrasive slurries such as kaolin, diatomaceous earth, or metal-based catalysts, it is necessary to introduce a clean flushing fluid. This flushing fluid flows from the piston-cylinder cavity into the process fluid (Figure 13e), thereby keeping the piston and its packing clean during both the suction and discharge strokes. This flushing action prevents the slurry from reaching the plunger packing. The metering action can still take place, because the mix proportion between the slurry and the flushing liquid is kept constant.

DIAPHRAGM PUMPS

The diaphragm or membrane pump uses a flexible member to transmit a pulsating force to the pumped fluid without

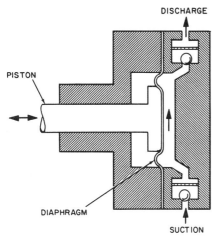

FIG. 13f
Direct-driven diaphragm pump.

allowing external leakage such as what might occur past the piston pump's packing. Like the piston pump, inlet and exit check valves are used to direct the flow. The diaphragm may be soft, made of Teflon, neoprene, etc., or hard, made of metal. The diaphragm may be moved directly by a piston as in a reciprocating piston pump (see Figure 13f). This type, which may also employ a bellows instead of a diaphragm (Figure 13g), generally has a pressure limitation of about 125 PSIG (8.5 bars).

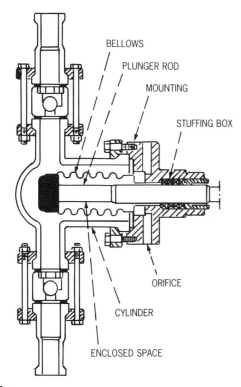

FIG. 13g
For extreme temperature applications the bellows-type, direct, mechanically actuated metering pump is a good choice.[1]

The glandless metering pumps can handle toxic, corrosive, radioactive, high-purity, odorous, volatile, or abrasive materials. The direct-actuated diaphragm designs are the least expensive and are suited for low-pressure, low flow rate services. The mechanically actuated bellows design can deliver more flow and is well suited for vacuum or extreme temperature services. For higher temperatures and corrosion resistance, metal bellows are used.

Hydraulic-Actuated Metering Pumps

In these designs the direct mechanical forces on the diaphragm are replaced by the forces delivered by hydraulic oil which is pumped by a reciprocating piston pump. The other side of the diaphragm is exposed to the process fluid (Figure 13h); therefore, the diaphragm is inherently balanced, as it is exposed to equal pressures on both of its sides. The support plates serve to keep the diaphragm deflections well within the endurance limits of the diaphragm material. This guarantees long useful life.

Metallic diaphragms can be damaged by dirt particles, while plastic diaphragms cannot handle high pressures and temperatures. Where a reliable seal is essential (for example, for pharmaceutical or liquid chlorine services), the double diaphragm design is recommended (Figure 13i). In these designs the diaphragm rupture can be visually observed if a sight glass is installed in the intermediate chamber, which is filled with a liquid of a preselected pH reference color. If the process side diaphragm is damaged, the leakage of process fluid into the intermediate chamber changes the pH and thereby changes the color of the intermediate fluid. The materials exposed to the pumped fluid—diaphragm, housing and valves—must be carefully selected for the application.

As with the piston pump, the flow is pulsating but may be smoothed by multiple diaphragms and/or by the use of a dampening reservoir. Check valve leakage provides the

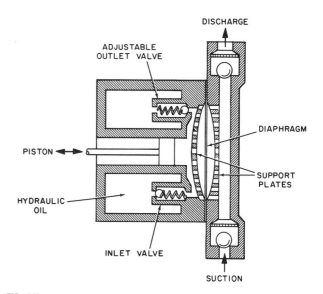

FIG. 13h
Oil-driven diaphragm pump.

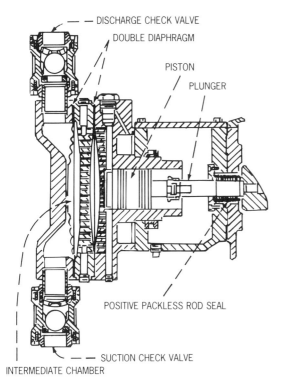

FIG. 13i
Double diaphragm design provides improved seal.[1]

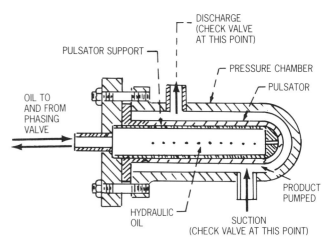

FIG. 13j
The pulsator-type metering pump.[1]

same problem for the diaphragm pump as it did for the piston pump.

The slave fluid's pulsation rate is varied by adjusting the piston pump's stroke length, altering the eccentricity of the crank, or changing the duration of the stroke by diverting a portion of each stroke to idle motion by mechanical or hydraulic means. Since slave hydraulic fluid leakage does not affect the metered rate of flow, many variations of the standard reciprocating piston pump can be used.

When it is desired to keep the metering pump away from hot processes, a remote head design is frequently used in which the check valve assembly remains in the process pipe while the diaphragm and pump are located at some distance away. The connecting pipe is filled with a column of the process fluid and a temperature gradient through this column of liquid isolates the diaphragm and its hydraulic fluid from the hot process.

Pulsator-Head Pumps

These pumps can be valuable when very difficult fluids are being metered, such as boiling sulfuric acid, pyrophoric fluids, fluorinated hydrocarbons, or slurries suspended in Freon. In this design, hydraulic oil is pumped through the hollow cavity of the pulsator support into two hermetically sealed diaphragms, called pulsators. As the pulsator elements expand, they displace the process fluid from the pressure chamber. When the hydraulic oil is exhausted from the pulsator, fresh process fluid is drawn into the pressure chamber through the suction port (Figure 13j). While the illustration shows only one pulsator, normally two are used to smooth out the flow. The alternate expansion and contraction of the pulsators causes the pumping action. The inner pulsator is usually Buna N, the outer is selected to match the fluid being pumped.

The pulsator pump produces a substantially pulsation-free, infinitely adjustable flow rate, which is set by the variable-volume oil pump. In hazardous areas the hydraulic fluid can be pumped from a remote location, thereby removing all electrical components from the process area. The pump can be easily sterilized, because only the outside of the pulsator is exposed to contamination. Consequently, the pump can be cleaned by removing just the pressure chamber.

PROPORTIONING PUMPS

Proportioning pumps are used when several process streams need to be mixed in some preset proportion. These can be continuous mixing processes or batch processes in which the ingredients are simultaneously fed. In digital batch blending, it is the cycle time and the pump size that sets the number of pump strokes or pulses per batch. For example, if the cycle time is 30 minutes and the amount to be charged is 60 gallons, a pump capacity of 2 gpm is required. If this pump operates at a speed of 60 strokes per minute, then the 60 gallons of batch charge would be represented by 1800 strokes or pulses. The pump pulse transmitter transmits the pulses, and when the 1800 count is reached, the batch controller stops the pump.

The proportioning of several process streams can be achieved in many different ways, depending on the hardware used. One possible solution is to operate several metering pumps in parallel. When using piston or diaphragm pumps, it is also possible to place multiple heads on the same drives, which will also achieve the same proportioning goal. On idle motion pumps the proportioning adjustments can be achieved by varying the stroke duration.

The peristaltic pumps are natural proportioning devices.

One can lay as many as 23 tubes of various sizes on a flat plate and the finger or chain-driven rollers of the peristaltic pump will squeeze all tubes simultaneously and at the same speed. In this configuration the tube sizes set the ratios between flows while the total pumping rate for the mix is set by the motor speed.

All proportioning and digital batch blending controls use modern, computer-compatible, microprocessor-based controls.

CONCLUSIONS

Each metering pump discussed has its particular application. Piston pumps are used to deliver high pressures. They require check valves and produce pulsating flows that can be dampened. Diaphragm pumps are utilized in the medium pressure range. The membrane serves as a moving partition between the mechanical or hydraulic drive and the process fluid. Rotary and pulsator pumps furnish pulsation-free high flow rates, and are suitable for high-viscosity service. The accuracy of rotary pumps is a function of the clearances between the rubbing surfaces. This generally results in low precision and therefore the rotary pumps are not considered to be metering devices. Peristaltic pumps are very accurate; they can handle extremely small flows, are self-priming, and require no seals or check valves.

If one is responsible for the operation and maintenance of a metering pump, one must be aware that a pump differs in many aspects from other flowmeters. For example, the pump motor must be lubricated periodically, and the pump must not be operated without liquid in it. The inlet piping must be designed to prevent cavitation. Running a pump dry or cavitating it will cause damage.

A metering pump should be calibrated, not only prior to first use, but also periodically during its operation. The calibration should duplicate fluid properties, suction and discharge pressures, and inlet and outlet piping configuration.

Reference

1. Hernandez, L.A., "Controlled Volume Pumps," *Chemical Engineering,* October 21, 1966.

Bibliography

Greene, "Pumping Machinery," Wiley.
Karassik, Krutzsch, Fraser, Messing, *Pump Handbook,* McGraw-Hill.
"Metering Pump Survey," *Instruments and Control Systems,* April 1971, pp. 103–109.
Payne, D.C., "Reliable Viscous Output with Positive-Displacement Pumps," *Chemical Engineering,* February 23, 1970.
Poyton, "Metering Pumps," Marcel Dekkel, Hydraulic Institute Standards.

14 Orifices

W. H. HOWE (1969) **J. B. ARANT** (1982)
B. G. LIPTÁK (1993)

Reviewed by S. Rudbäck (1993)

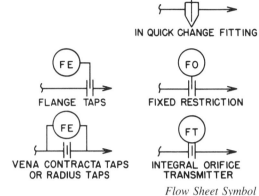

Flow Sheet Symbol

Design Pressure:	For plates it is limited by readout device only. Integral orifice transmitter to 1500 PSIG (10.3 MPa)
Design Temperature:	Function of associated readout system, only when the differential pressure unit must operate at the elevated temperature. For integral orifice transmitter the standard range is −20 to 250°F (−29 to 121°C)
Sizes:	Maximum size is pipe size
Fluids:	Liquids, vapors, and gases
Flow Range:	From a few cc/min using integral orifice transmitters to any maximum flow, limited only by pipe size
Materials of Construction:	There is no limitation on plate materials. Integral orifice transmitter wetted parts can be obtained in steel, stainless steel, Monel, nickel, and Hastelloy
Inaccuracy:	The orifice plate, if the bore diameter is correctly calculated and prepared, can be accurate to ±0.25 to ±0.5% of actual flow. When a conventional d/p cell is used to detect the orifice differential, that will add a ±0.1 to ±0.3% of full scale error. The error contribution of "smart" d/p cells is only 0.1% of actual span
Intelligent d/p cells:	Inaccuracy of ±0.1%, rangeability of 40:1, built-in PID algorithm
Rangeability:	If one defines rangeability as the flow range within which the combined flow measurement error does not exceed ±1% of actual flow, then the rangeability of conventional orifice installations is 3:1. When using intelligent transmitters with automatic switching capability between the "high" and the "low" span, the rangeability can approach 10:1
Cost:	A plate only is $50 to $300, depending on size and materials. For steel orifice flanges from 2 to 12 in. (50 to 300 mm) the cost ranges from $200 to $1000. For flanged meter runs in the same size range the cost ranges from $400 to $3000. The cost of electronic or pneumatic integral orifice transmitters is between $1500 and $2000. The cost of d/p transmitters ranges from $900 to $2000, depending on type and "intelligence."
Partial List of Suppliers:	ABB Kent-Taylor Inc. (incl. integral orifices); Crane Manufacturing Inc.; Daniel Flow Products Inc. (orifice plates and plate changers); Fischer & Porter Co. (incl. integral orifices); Fluidic Techniques, a Div. of FTI Industries; Foxboro Co. (incl. integral orifices); Honeywell Industrial Div.; Lambda Square Inc.; Meriam Instrument, Div. Scott & Fetzer (orifice plates); Rosemount Inc.; Vickery-Simms, a Div. of FTI Industries. In addition, orifice plates, flanges and accessories can be obtained from most other major instrument manufacturers

HEAD METERS

Head-type flow meters comprise a class of devices for fluid flow measurement including orifice plates, venturi tubes, weirs, flumes, and many others. They change the velocity or direction of the flow, creating a measurable differential pressure or "pressure head" in the fluid.

Head metering is one of the most ancient of flow detection techniques. There is evidence that the Egyptians used weirs for measurement of irrigation water in the days of the Pharaohs, and that the Romans used orifices to meter water to households in Caesar's time. In the 18th century, Bernoulli established basic relationships between pressure head and velocity head, and Venturi published on the flowtube bearing his name. However, it was not until 1887 that Clemens Herschel developed the commercial venturi tube. Work on the conventional orifice plate for gas flow measurement was commenced by Weymouth in the United States in 1903. Recent developments include improved primary elements, refinement of data, more accurate and versatile test and calibrating equipment, better differential pressure sensors, and many others.

Theory of Head Meters

Head-type flow measurement derives from Bernoulli's Theorem which states that in a flowing stream, the sum of the pressure head, the velocity head, and the elevation head at one point is equal to their sum at another point in the direction of flow plus the loss due to friction between the two points. Velocity head is defined as the vertical distance through which a liquid would fall to attain a given velocity. Pressure head is the vertical distance which a column of the flowing liquid would rise in an open-ended tube as a result of the static pressure.

This principle is applied to flow measurement by altering the velocity of the flowing stream in a predetermined manner, usually by a change in the cross-sectional area of the stream. Typically, the velocity at the throat of an orifice is increased relative to the velocity in the pipe. There is a corresponding increase in velocity head. Neglecting friction and change of elevation head, there is an equal decrease in pressure head (Figure 14a). This difference between the pressure in the pipe just upstream of the restriction and the pressure at the throat is measured. Velocity is determined from the ratio of the cross-sectional areas of pipe and flow nozzle, and the difference of velocity heads given by differential pressure measurements. Flow rate derives from velocity and area. The basic equations are:

$$V = k \sqrt{\frac{h}{\rho}} \qquad 14(1)$$

$$Q = kA \sqrt{\frac{h}{\rho}} \qquad 14(2)$$

$$W = kA \sqrt{h\rho} \qquad 14(3)$$

where V is velocity, Q is volume flow rate, W is mass flow rate, A is cross-sectional area of the pipe, h is differential pressure between points of measurement, ρ is the density of the flowing fluid, and k is a constant which includes ratio of cross-sectional area of pipe to cross-sectional area of nozzle or other restriction, units of measurement, correction factors, etc., depending on the specific type of head meter. (For a more complete derivation of the basic flow equations, based on considerations of energy balance and hydrodynamic properties, consult References 1, 2, and 3.)

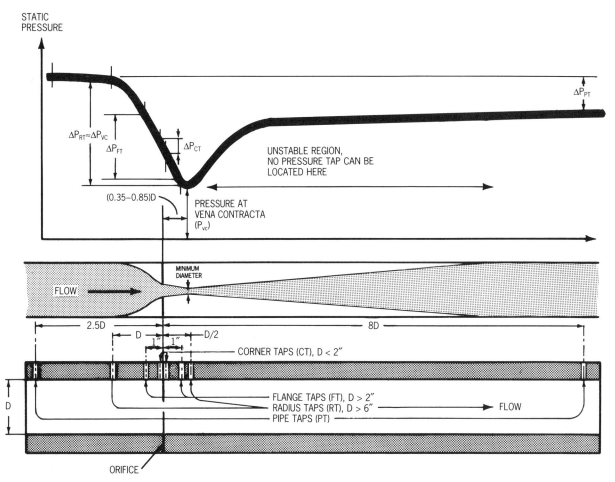

FIG. 14a
Pressure profile through an orifice plate and the different methods of detecting the pressure drop.

Head Meter Characteristics

Two fundamental characteristics of head-type flow measurements are apparent from the basic equations. First is the square root relationship between flow rate and differential pressure. Second, the density of the flowing fluid must be taken into account both for volume and for mass flow measurements.

The Square Root Relationship This relationship has two important consequences. Both are primarily concerned with readout. The primary sensor (orifice, venturi tube, etc.) develops a head or differential pressure. A simple linear readout of this differential pressure expands the high end of the scale and compresses the low end in terms of flow. Fifty percent of full flow rate produces 25% of full differential pressure. At this point, a flow change of 1% of full flow results in a differential pressure change of 1% of full differential. At 10% flow the total differential pressure is only 1% and a change of 1% of full scale flow (10% relative change) results in only 0.2% full scale change in differential pressure. Both accuracy and readability suffer. Readability can be improved by a transducer which extracts the square root of the differential pressure to give a signal linear with flow rate. However, errors in the more complex square root transducer tend to decrease overall accuracy.

For a large proportion of industrial processes, which seldom operate below 30% capacity, a device with pointer or pen motion linear with differential pressure is generally adequate. Readout directly in flow can be provided by a square root scale. Where maximum accuracy is important, it is generally recommended that the maximum-to-minimum flow ratio shall not exceed 3 to 1 or at the most 3½ to 1 for any single head-type flow meter. The high repeatability of modern differential pressure transducers permits a considerably wider range for flow control, where constancy and repeatability of flow rate are the primary concern. However, where flow variations approach 10 to 1, use of two primary flow units of different capacities, or two differential pressure sensors with different ranges, or both, is generally recommended. It should be emphasized that the primary head meter devices produce a differential pressure which corresponds accurately to flow over a wide range. Difficulty arises in the accurate measurement of the corresponding extremely wide range of differential pressure—for example, a 20 to 1 flow variation results in a 400 to 1 variation in differential pressure.

The second problem with the square root relationship is that some computations require linear input signals. This is the case when flow rates are integrated or when two or more flow rates are added or subtracted. This is not necessarily true for multiplication and division; specifically, flow ratio measurement and control do not require linear input signals. A given flow ratio will develop a corresponding differential pressure ratio over the full range of the measured flows.

Density of the Flowing Fluid Fluid density is involved in the determination of either mass flow rate or volume flow rate. In other words, head-type meters *do not* read out directly in either mass or volume flow (weirs and flumes are an exception, as discussed in Section 29). The fact that density appears as a square root gives head type metering an actual advantage, particularly in applications where measurement of mass flow is required. Due to this square root relationship, any error which may exist in the value of the density used to compute mass flow is substantially reduced—a 1% error in the value of the fluid density results in a ½% error in calculated mass flow. This is particularly important in gas flow measurement, where the density may vary over a considerable range and where operating density is not easily determined with high accuracy.

β (Beta) Ratio Most head meters depend on a restriction in the flow path to produce a change in velocity. For the usual circular pipe and circular restriction, the β ratio is the ratio between the diameter of the restriction and the inside diameter of the pipe. The ratio between the velocity in the pipe and the velocity at the restriction is equal to the ratio of areas or β^2. For noncircular configurations, β is defined as the square root of the ratio of area of the restriction to area of the pipe or conduit.

Reynolds Number

The basic equations of flow assume that the velocity of flow is uniform across a given cross section. In practice, flow velocity at any cross section approaches zero in the boundary layer adjacent to the pipe wall, and varies across the diameter. This flow velocity profile has a significant effect on the relationship between flow velocity and pressure difference developed in a head meter. In 1883, Sir Osborne Reynolds, an English scientist, presented a paper before the Royal Society, proposing a single, dimensionless ratio now known as Reynolds number, as a criterion to describe this phenomenon. This number, Re, is expressed as

$$\mathrm{Re} = \frac{\mathrm{VD}\rho}{\mu} \qquad 14(4)$$

where V is velocity, D is diameter, ρ is density, and μ is absolute viscosity. Reynolds number expresses the ratio of inertial forces to viscous forces. At a very low Reynolds number, viscous forces predominate, and inertial forces have little effect. Pressure difference approaches direct proportionality to average flow velocity and to viscosity. At high Reynolds numbers, inertial forces predominate and viscous drag effects become negligible.

At low Reynolds numbers, flow is laminar and may be regarded as a group of concentric shells; each shell reacts in a viscous shear manner on adjacent shells; the velocity profile across a diameter is substantially parabolic. At high Reynolds numbers, flow is turbulent, with eddies forming between the boundary layer and the body of the flowing fluid, and propagating through the stream pattern. A very complex, random pattern of velocities develops in all directions. This tubulent, mixing action tends to produce a uniform average axial velocity across the stream. Change from the laminar flow pattern to the turbulent flow pattern is gradual, with no distinct transition point. For Reynolds numbers above 10,000, flow is definitely turbulent. The coefficients of discharge of the various head-type flowmeters changes with Reynolds number (Figure 14b).

The value for k in the basic flow equations includes a Reynolds number factor. References 1 and 2 provide tables and graphs for Reynolds number factor. For head meters, this single factor is sufficient to establish compensation in coefficient for changes in ratio of inertial to frictional forces, and the corresponding changes in flow velocity profile; a gas flow with the same Reynolds number as a liquid flow has the same Reynolds number factor.

Compressible Fluid Flow

Density in the basic equations is assumed to be constant upstream and downstream from the primary device. For gas or vapor flow, the differential pressure developed results in a corresponding change in density between upstream and downstream pressure measurement points. For accurate calculations of gas flow, this is corrected by an "expansion factor" which has been empirically determined. Values are given in References 1 and 2. When practical, the full scale differential pressure should be less than 0.04 times normal minimum static pressure (differential pressure, stated in inches of water, should be less than static pressure stated in PSIA). Under these conditions, the expansion factor is quite small.

Choice of Differential Pressure Range

The most common differential pressure range for orifices, venturi tubes, and flow nozzles is 0 to 100 in. of water (0 to 25 kPa) for full scale flow. This range is high enough to minimize errors due to liquid density differences in the connecting lines to the differential pressure sensor or in seal chambers, condensing chambers, etc., caused by temperature differences. Most differential pressure responsive devices develop their maximum accuracy in or near this range, and the maximum pressure loss—3.5 PSI (24 kPa)—is not serious in most applications. (As shown in Figure 27f, the pressure loss in an orifice is about 65% when a β ratio of 0.75 is used.) The 100 in. range permits a 2:1 flow rate change in either direction to accommodate changes in operating conditions; most differential pressure sensors can be modified to cover the range from 25 to 400 in. of water (6.2 to 99.4 kPa) or more, either by a simple adjustment, or by a relatively minor structural change. Applications, where the pressure loss up to 3.5 PSI is expensive or is not available, can be handled either by selection of a lower differential pressure range or by the use of a venturi tube or other primary element with high-pressure recovery. Some high-velocity flows will develop more than 100 in. differential pressure with maximum acceptable ratio of primary element effective diameter to pipe diameter. For these applications, a higher differential pressure is indicated. Finally, for low static pressure (less than 100 PSIA) gas or vapor, a lower differential pressure is recommended to minimize the expansion factor.

Pulsating Flow and Flow "Noise"

Short period (1 second and less) variation in differential pressure developed from a head-type flow meter primary element arises from two distinct sources. First, reciprocating pumps, compressors, and the like may cause a periodic fluctuation in the rate of flow. Second, the random velocities inherent in turbulent flow cause variations in differential pressure even with a constant flow rate. Both have similar results and are often confused. However, their characteristics and the procedures used to cope with them are distinct.

Pulsating Flow The so-called pulsating flow from reciprocating pumps, compressors, etc., may significantly affect the differential pressure developed by a head-type meter. For example, if the amplitude of instantaneous differential pressure fluctuation is 24% of the average differential pressure, an error of ±1% can be expected under normal operation conditions. For the pulsation amplitudes of 24, 48, and 98% values, the corresponding errors of ±1, ±4, ±16% can be expected. The Joint ASME-AGA Committee on Pulsation reported that the ratio between errors varies roughly as the square of the ratio between differential pressure fluctuations.

For liquid flow, there is indication that the average of the square root of the instantaneous differential pressure (essentially average of instantaneous flow signal) results in a lower error than the measurement of the average instantaneous differential pressure. However, for gas flow, extensive investigation has failed to develop any usable relationship between pulsation and deviation from coefficient beyond the estimate of maximum error.[4]

Operation at higher differential pressures is generally advantageous for pulsating flow. The only other valid ap-

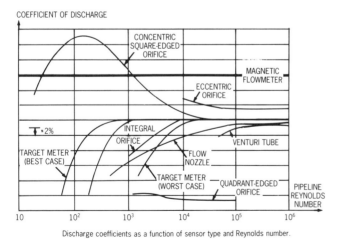

FIG. 14b
Discharge coefficients as a function of sensor type and Reynolds number.

proach to improve accuracy of pulsating gas flow measurement is the location of the meter at a point where pulsation is minimized.

Flow "Noise" Turbulent flow generates a complex pattern of random velocities. This results in a corresponding variation or "noise" in the differential pressure developed at the pressure connections to the primary element. The amplitude of the noise may be as much as 10% of the average differential pressure with a constant flow rate. This noise effect is a complex hydrodynamic phenomenon, and is not fully understood. It is augmented by flow disturbances from valves, fittings, etc., both upstream and downstream from the flow meter primary element and, apparently, by characteristics of the primary element itself.

Tests based on average flow rate accurately determined by static weight/time techniques, compared to accurate measurement of differential pressure including continuous, precise averaging of noise, indicate that the noise, when precisely averaged, introduces negligible (less than 0.1%) measurement error when average flow is substantially constant (change of average flow rate is not more than 1% per second).[5] It should be noted that average differential pressure, not average flow (average of the square root of differential pressure) is measured, since the noise is developed by the random and not the average flow.

Errors in determination of true differential pressure average will result in corresponding errors in flow measurement. For normal use, "damping" in one form or another in devices responsive to differential pressure is adequate. Where accuracy is a major concern, there must be no elements in the system that will develop a bias rather than a true average when subjected to the complex noise pattern of differential pressure.

Differential pressure noise can be reduced by use of two or more pressure sensing taps connected in parallel for both high and low differential pressure connections. This provides major noise reduction. Only minor improvement results from additional taps. Piezometer rings, a form of multiple connections, are frequently used with venturi tubes but seldom with orifices or flow nozzles.

THE ORIFICE METER

The orifice meter is the most common head-type flow measuring device. An orifice plate is inserted in the line, and the differential pressure across it is measured (Figure 14a). This section is concerned with the primary device (the orifice plate, its mounting, and the differential pressure connections).

The orifice in general, and the conventional thin, concentric, sharp-edged orifice plate in particular, has important advantages, including inexpensive manufacture to very close tolerances and ease of installation and replacement. Orifice measurement of liquids, gases, and vapors under a

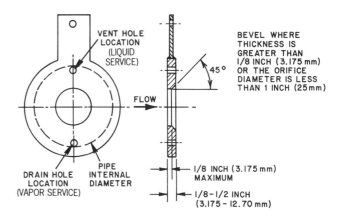

FIG. 14c
Concentric orifice plate.

wide range of conditions enjoys a high degree of confidence based on a great deal of accurate test work.

The standard orifice plate itself is a circular disk; usually stainless steel, from ⅛ to ½ in. thick (3.175 to 12.70 mm), depending on size and flow velocity, with a hole (orifice) in the middle, and a tab projecting out to one side which is used as a data plate (Figure 14c). The thickness requirement of the orifice plate is a function of line size, flowing temperature, and differential pressure across the plate. Some helpful guidelines are:

By Size: 2–12 in. (50–304 mm): ⅛ in. (3.175 mm) thick 14 in. (355 mm) and larger: ¼ in. (6.35 mm) thick
By Temperature ≥600°F (316°C): 2–8 in. (50–203 mm): ⅛ in. (3.175 mm) thick 10 in. (254 mm) and larger: ¼ in. (6.35 mm) thick

All plates thicker than ⅛ in. (3.175 mm) should be beveled 45° on the downstream side to a ⅛ in. edge thickness. In addition, any orifice hole 1 in. (25 mm) or less in size should be beveled 45° to a maximum ⅛ in. edge thickness (¹⁄₁₆ in. or 1.6 mm in a ⅛ in. thick plate).

Flow Through the Orifice Plate

The orifice plate inserted in the line causes an increase in flow velocity and a corresponding decrease in pressure. The flow pattern shows an effective decrease in cross section beyond the orifice plate, with a maximum velocity and minimum pressure at the vena contracta (Figure 14a). This location may be from .35 to .85 pipe diameters downstream from the orifice plate, depending on β ratio and on Reynolds number.

This flow pattern and the sharp leading edge of the orifice plate (Figure 14d) which produces it are of major importance. The sharp edge results in an almost pure line contact between the plate and the effective flow, with negligible fluid-to-metal friction drag at this boundary. Any nicks, burrs, or rounding of the sharp edge can result in surprisingly large errors in measurement.

When the usual practice of measuring the differential pressure at a location close to the orifice plate is followed,

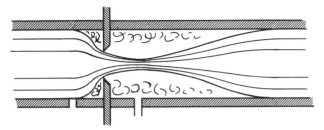

FIG. 14d
Flow pattern with orifice plate.

friction effects between fluid and pipe wall upstream and downstream from the orifice are minimized so that pipe roughness has minimum effect. Fluid viscosity, as reflected in Reynolds number, does have a considerable influence, particularly at low Reynolds numbers. Since the formation of the vena contracta is an inertial effect, a decrease in the ratio of inertial to frictional forces (decrease in Reynolds number), and the corresponding change in the flow profile, results in less constriction of flow at the vena contracta and an increase of the flow coefficient. In general, the sharp edge orifice plate should not be used at pipe Reynolds numbers under 2000 to 10,000 or more (Table 1e). The minimum recommended Reynolds number will vary from 10,000 to 15,000 for 2 in. (50 mm) through 4 in. (102 mm) pipe sizes for β ratios up to 0.5, and 20,000 to 45,000 for higher β ratios. The Reynolds number requirement will increase with pipe size and β ratio and may range up to 200,000 for pipes 14 in. (355 mm) and larger. Maximum Reynolds numbers may be 10^6 through 4 in. (102 mm) pipe and 10^7 for larger sizes.

Location of Pressure Taps

For liquid flow measurement, gas or vapor accumulations in the connections between the pipe and the differential pressure measuring device must be prevented. Pressure taps are generally located in the horizontal plane of the center line of horizontal pipe runs. The differential pressure measuring device is either mounted close-coupled to the pressure taps, or connected through downward sloping connecting pipe of sufficient diameter to allow gas bubbles to flow up and back into the line. For gas, similar precautions to prevent accumulation of liquid are required. Taps may be installed in the top of the line, with upward sloping connections, or the differential pressure measuring device may be close-coupled to taps in the side of the line (Figure 14e). For steam and similar vapors which are condensable at ambient temperatures, condensing chambers, or their equivalent, are generally used, usually with down-sloping connections from the side of the pipe to the measuring device. There are five common locations for the differential pressure taps: flange taps, vena contracta taps, radius taps, full flow or pipe taps, and corner taps.

In the United States, flange taps (Figure 14e and 14f) are predominantly used for pipe sizes 2 in. (50 mm) and larger.

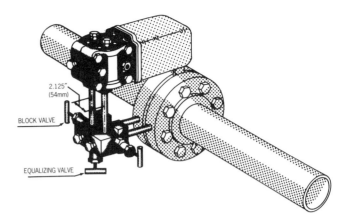

FIG. 14e
Measurement of gas flow with differential pressure transmitter and three-valve manifold.[3]

The manufacturer of the orifice flange set drills the taps so that the centerlines are 1 in. (25 mm) from the orifice plate surface. This location also facilitates inspection and cleanup of burrs, weld metal, and so on, that may result from installation of a particular type of flange. Flange taps are not recommended below 2 in. (50 mm) pipe size and cannot be used below 1½ in. (37.5 mm) pipe size, since the vena contracta may be closer than 1 in. (25 mm) from the orifice plate. Flow for a distance of several pipe diameters beyond the vena contracta tends to be unstable and is not suitable for differential pressure measurement (Figure 14a).

Vena contracta taps use an upstream tap located one pipe diameter upstream of the orifice plate, and a downstream tap located at the point of minimum pressure. Theoretically, this

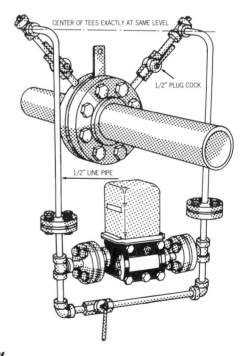

FIG. 14f
Steam flow measurement using standard manifold.[3]

is the optimum location. However, the location of the vena contracta varies with the orifice-to-pipe diameter ratio, and is thus subject to error if the orifice plate is changed. A tap location too far downstream in the unstable area may result in inconsistent measurement. For moderate and small size pipe, the location of the vena contracta is likely to lie at the edge of or under the flange. It is not considered good piping practice to use the hub of the flange to make a pressure tap. For this reason, vena contracta taps are normally limited to pipe sizes 6 in. (152 mm) or larger depending upon the flange rating and dimensions.

Radius taps are similar to vena contracta taps, except that the downstream tap is located at one-half pipe diameter (one radius) from the orifice plate. This practically assures that the tap will not be in the unstable region regardless of orifice diameter. Radius taps today are generally considered superior to the vena contracta tap because they simplify the pressure tap location dimensions and do not vary with changes in orifice β ratio. The same pipe size limitations apply as to the vena contracta tap.

Pipe taps are located $2\frac{1}{2}$ pipe diameters upstream and 8 diameters downstream from the orifice plate. Because of the distance from the orifice, exact location is not critical, but the effects of pipe roughness, dimensional inconsistencies, and so on, are more severe. Uncertainty of measurement is perhaps 50% greater with full flow taps than with taps close to the orifice plate. These taps are not normally used except where it is necessary to install an orifice meter in an existing pipeline and radius or vena contracta taps cannot be used.

Corner taps (Figure 14g) are similar in many respects to flange taps, except that the pressure is measured at the "corner" between the orifice plate and the pipe wall. Corner taps are very common for all pipe sizes in Europe. The relatively small clearances of the passages are a possible source of trouble. Also, some tests have indicated inconsistencies with high β ratio installations, which have been attributed to a region of flow instability at the upstream face of the orifice. For this situation, an upstream tap at one pipe diameter upstream of the orifice plate has been used. Corner taps are used in the United States primarily for pipe diameters of less than 2 in. (50 mm).

ECCENTRIC AND SEGMENTAL ORIFICE PLATES

The use of eccentric and segmental orifices is recommended where horizontal meter runs are required and the fluids contain extraneous matter to a degree that the concentric orifice would plug up. It is preferable to use concentric orifices in a vertical meter tube if at all possible. Flow coefficient data is limited for these orifices and they are likely to be less accurate. In the absence of specific data, concentric orifice data may be applied as long as accuracy is of no major concern.

The eccentric orifice plate, Figure 14h, is like the concentric plate except for the offset hole. The segmental orifice plate, Figure 14i, has a hole which is a segment of a circle. Both types of plates may have the hole bored tangent to the inside wall of the pipe or more commonly tangent to a concentric circle with a diameter no smaller than 98% of the pipe internal diameter. The segmental plate arc is parallel to the pipe wall. Care must be taken so that no portion of the flange or gasket interferes with the hole on either type plate. The equivalent β for a segmental orifice may be expressed as $\beta = \sqrt{a/A}$ where a is the area of the hole segment and A is the internal pipe area.

In general, the minimum line size for these plates is 4 in. (102 mm). However, the eccentric plate can be made in smaller sizes as long as the hole size does not require beveling. Maximum line sizes are unlimited and contingent only on calculation data availability. Beta ratio limits are limited to between 0.3 and 0.8. Lower Reynolds number limit is 2000D (D in inches), but not less than 10,000. For compressible fluids, $\Delta P/P_1 \leq 0.30$ where ΔP and P_1 are in the same units.

Flange taps are recommended for both types of orifices, but vena contracta taps can be used in larger pipe sizes. The taps for the eccentric orifice should be located in the quad-

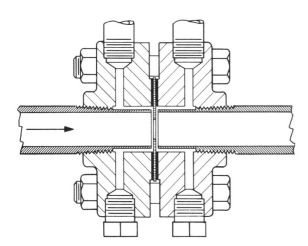

FIG. 14g
Corner tap installation.

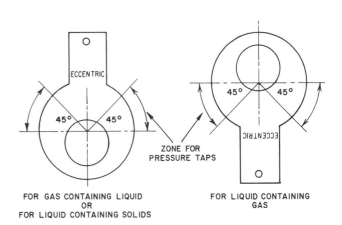

FIG. 14h
Eccentric orifice plate.

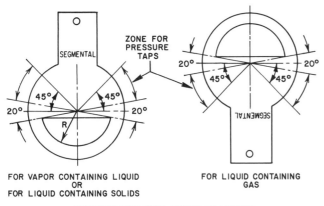

FIG. 14i
Segmental orifice plate.

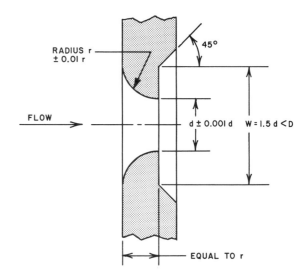

FIG. 14k
Quadrant edge orifice plate.

rants directly opposite the hole. The taps for the segmental orifice should always be in line with the maximum dam height. The straight edge of the dam may be beveled if necessary using the same criteria as for a square edge orifice. In order to avoid confusion after installation, the tabs on these plates should be clearly stamped "eccentric" or "segmental."

QUADRANT EDGE AND CONICAL ENTRANCE ORIFICE PLATES

The use of quadrant edge and conical entrance orifice plates is limited to lower pipe Reynolds numbers where flow coefficients for sharp-edged orifice plates are highly variable, in the range of 500 to 10,000. With these special plates, the stability of the flow coefficient increases by a factor of 10. The minimum allowable Reynolds number is a function of β ratio and the allowable β ratio ranges are limited. Refer to Table 14j for β ratio range and minimum allowable Reynolds number. The maximum allowable pipe Reynolds number ranges from $500,000 \times (\beta - 0.1)$ for the quadrant edge to $200,000 \times (\beta)$ for the conical entrance plate. The conical entrance also has a minimum $D \geq 0.25$ in. (6.35 mm). For compressible fluids, $\Delta P/P_1 \leq 0.25$ where ΔP and P_1 are in the same units. Flange pressure taps are preferred for the quadrant edge but corner and radius taps can also be used with the same flow coefficients. For the conical entrance units reliable data is available for corner taps only. A typical quadrant edge plate is shown in Figure 14k and a typical conical entrance orifice plate is shown in Figure 14l. These plates are thicker and heavier than the normal sharp-edge-type. Because of the critical dimensions and shape, the quadrant edge is difficult to manufacture; it is recommended that it be purchased from skilled commercial fabricators. The conical entrance is much easier to make and could be made by any qualified machine shop. While these special orifice forms are very useful for lower Reynolds numbers, it is recommended that for a pipe $Re > 100,000$ the standard sharp-edge orifice be used. In order to avoid confusion after installation, the tabs on these plates should be clearly stamped "quadrant" or "conical."

TABLE 14j
Minimum Allowable Reynolds Numbers for Conical and Quadrant Edge Orifices

Type						Re Limits					
Conical Entrance	β	0.10	0.11	0.12	0.13	0.14	0.15	0.16	0.17	0.18	0.19
	Re	25	28	30	33	35	38	40	43	45	48
	β	0.20	0.21	0.22	0.23	0.24	0.25	0.26	0.27	0.28	0.29	0.30
	Re	50	53	55	58	60	63	65	68	70	73	75
Quadrant Edge	β	0.25	0.30	0.35	0.40	0.45	0.50	0.55	0.60			
	Re	250	300	400	500	700	1000	1700	3300			

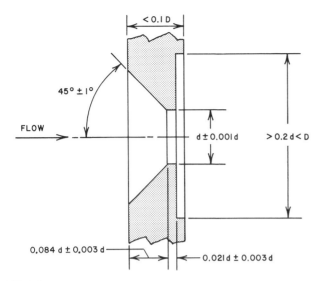

FIG. 14l
Conical entrance orifice plate.

An application summary of the different orifice plates is given in Table 14m. For dirty gas service, the annular orifice plate (Figure 22a) can also be considered.

THE INTEGRAL ORIFICE

Miniature flow restrictors provide a convenient primary element for measurement of small fluid flows. They combine a plate with a small hole to restrict flow, its mounting and connections, and a differential pressure sensor, usually a pneumatic or electronic transmitter. Units of this type are often referred to as "integral orifice" flow meters. Interchangeable flow restrictors are available to cover a wide range of flows. A common minimum standard size is a 0.020 in. (0.5 mm) throat diameter which will measure water flow down to 0.0013 GPM (5 cm^3/min) or air flow at atmospheric pressure down to 0.0048 SCFH (135 cm^3/min) (Figure 14n).

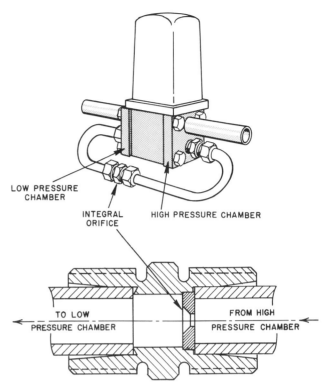

FIG. 14n
Typical integral orifice meter.

Miniature flow restrictors are used in laboratory-scale processes, and pilot plants; to measure additives to major flow streams; and for other small flow measurements. Clean fluid is required, particularly for the smaller sizes, not only to avoid plugging of the small orifice opening, but also because of buildup of even a very thin layer on the surface of the element will cause an error.

There is little published data on the performance of these small restrictors. These are proprietary products with performance data provided by the supplier. Where accuracy is important, direct flow calibration is recommended. Water

TABLE 14m
Selecting the Right Orifice Plate for a Particular Application

Orifice Type	Appropriate Process Fluid	Reynolds Number Range	Normal Pipe Sizes Inches (mm)
Concentric, Square Edge	Clean gas and liquid	Over 2000	0.5 to 60 (13 to 1500)
Concentric, Quadrant, or Conical Edge	Viscous clean liquids	200 to 10,000	1 to 6 (25 to 150)
Eccentric or Segmental Square Edge	Dirty gas or liquid	Over 10,000	4 to 14 (100 to 350)

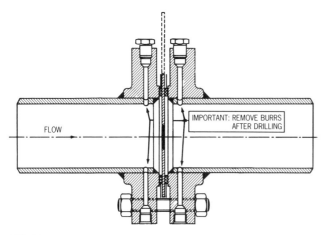

FIG. 14o
Prefabricated meter run with inside surface of the pipe machined for smoothness after welding for a distance of two diameters from each flange face. The mean pipe ID is averaged from four measurements made at different points. They must not differ by more than 0.3%.[3]

flow calibration, using tap water, a stop watch, and a glass graduate (or a pail and scale) to measure total flow, is readily carried out in the instrument shop or laboratory. For viscous liquids, calibration with the working fluid is preferable since viscosity has a substantial effect on most units. Calibration across the working range is recommended since precise conformity to the square law may not exist. Some suppliers are prepared to provide calibrated units for an added fee.

Installation

The orifice is usually mounted between a pair of flanges. Care should be exercised in installing the orifice plate to be sure that the gaskets are trimmed and installed so that they do not protrude across the face of the orifice plate beyond the inside pipe wall (Figure 14o). A variety of special devices are commercially available for mounting orifice plates, including units which allow the orifice plate to be inserted and removed from a flowline without interrupting the flow (Figure 14p). Such manually operated or motorized orifice fittings can also be used to change the flow range by sliding a different orifice opening into the flowing stream.

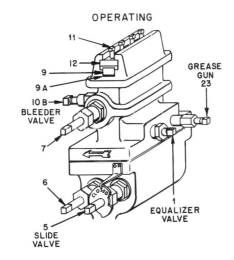

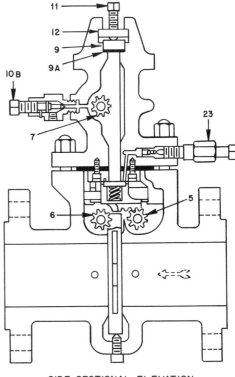

FIG. 14p
Typical orifice fitting. (Courtesy Daniel Flow Products Inc.)

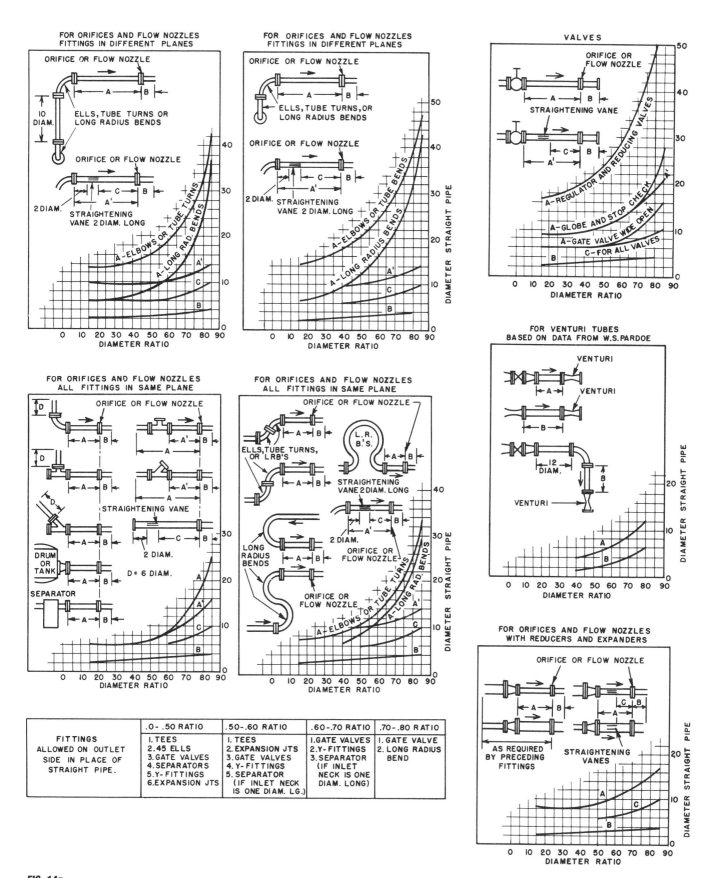

FIG. 14q
Orifice straight-run requirements. (Reprinted courtesy of The American Society of Mechanical Engineers)

In order to avoid errors resulting from disturbance of the flow pattern due to valves, fittings, etc., a straight run of smooth pipe before and after the orifice is recommended. Required length depends on β ratio (ratio of diameter of orifice to inside diameter of pipe) and the severity of the flow disturbance.

For example, an upstream distance to the orifice plate of 45 pipe diameters with 0.75 β ratio is the minimum recommendation for a throttling valve. For a single elbow at the same β, the minimum distance would only be 17 pipe diameters. Figure 14q gives minimum values for a variety of upstream disturbances. Upstream lengths longer than the minimum are recommended. A downstream pipe run of 5 pipe diameters from the orifice plate is recommended in all cases. This straight run should not be interrupted by thermowells or other devices inserted into the pipe.

Where it is not practical to install the orifice in a straight run of the desired length, the use of a straightening vane to eliminate swirls or vortices is recommended. Straightening vanes are manufactured in various configurations (Figure 14r) and are available from commercial meter tube fabricators. They should be installed so that there are at least 2 pipe diameters between the disturbance source and vane entry and at least 6 pipe diameters from the vane exit to the upstream high pressure tap of the orifice.

The installation of the pressure taps is important. Burrs or protrusions at the tap entry point must be removed (Figure 14o). The tap hole should enter the line at a right angle to the inside pipe wall and should be slightly beveled. Considerable error can result from protrusions that react with the flow and generate spurious differential pressure. Careful installation is particularly important when full flow taps are located in areas of full pipe velocity and in positions difficult to inspect.

Limitations

There are certain limitations in the application of the concentric, sharp-edged orifice.

1. The concentric orifice plate is not recommended for slurries and dirty fluids, where solids may accumulate near the orifice plate (Table 14m).
2. The sharp-edged orifice plate is not recommended for strongly erosive or corrosive fluids which tend to round over the sharp edge. Orifice plates made of materials that resist erosion or corrosion are used for conditions which are not too severe.
3. For flows at less than 10,000 Reynolds number (determined in the pipe), the correction factor for Reynolds number may introduce problems in determining the total flow when the flow rate varies considerably (Figure 14b). The quadrant-edged orifice plate is recommended for this application in preference to the sharp-edged plate (Table 14m).
4. For liquids with entrained gas or vapor, a "vent hole" in the plate can be used for horizontal meter runs to prevent accumulation of gas ahead of the orifice plate (Figure 14c). If the diameter of the vent hole is less than 10% of the orifice diameter, then the flow is less than 1% of the total flow. If this error cannot be tolerated, then appropriate correction can be made to the orifice calculation. On dirty service, vent or drain holes are considered of little value because they are subject to plugging, and are not recommended.
5. In a similar fashion, a drain or weep hole can be provided for gas with entrained liquid. However, it is recommended that meters for liquid with entrained gas or gas with entrained liquid services be installed vertically. Normally, the flow direction would be upward for liquids and downward for gases. For severe entrainment situations, eccentric or segmental orifice plates should be used.
6. The basic flow equations are based on flow velocities well below sonic. Orifice measurement is also used for flows approaching sonic velocity but requires a different theoretical and computational approach.
7. For concentric orifice plates, it is recommended that the β ratio be limited to a range of 0.2 to 0.65 for best accuracy. In exceptional cases this range can be extended from 0.15 to 0.75.
8. For large flows, the pressure loss through an orifice can result in significant cost in terms of power requirements. (See Section 1.) Venturi tubes with relatively large pressure recovery will substantially decrease the pressure loss. Lo-Loss Tubes, Dall Tubes, Foster Flow Tubes, and similar proprietary primary elements develop 95% or better pressure recovery. The pressure loss is less than 5% of differential pressure (see Figure 27f). Elbow taps involve no added pressure loss (see Section 5). Pitot tube elements introduce negligible loss. Orifice plates can be sized for full scale differential pressure ranging from 5 in. (127 mm) of water to several hundred inches of water. Most commonly the range is from 20 to 200 in. (508 to 5080 mm) of water. The pressure recovery ratio of an orifice (except for pipe taps) can be estimated by $(1 - \beta^2)$.
9. For compressible fluids, $\Delta P/P_1$ should be ≤ 0.25 where ΔP and P_1 are in the same units. This will minimize errors and corrections required for density changes on flow through the orifice.

FIG. 14r
Straightening vane.

ORIFICE BORE CALCULATIONS

Accurate flow calibration, traceable to recognized standards and using the working fluid under service conditions, is difficult and expensive. For large gas flows, it is nearly impossible. A major advantage of orifice metering lies in the facility with which flow can be accurately determined from a few simple, readily available measurements. In particular, for the concentric, sharp-edged orifice, measurement confidence is supported by a large body of experience and precise, painstaking tests.

Precise flow calculations are quite complex, although the calculation methods and equations have been well standardized. These calculation methods are thoroughly covered in the References 1 through 4 listed at the end of this section. Approximate calculations are practical where moderate accuracy is satisfactory or a firming up of the actual orifice design basis is being done. See Figure 14s for orifice bore determination and Table 14t for maximum flow capacities. The following equations are to be used with Figure 14s for the orifice bore determination.

For liquid flow:

$$Z = \frac{5.663 \, ER \sqrt{h \, G_f}}{GPM \, G_t} \qquad 14(5)$$

*For steam:

$$Z = \frac{358.9 \, ERY}{lbm/hr} \sqrt{\frac{h}{V}} \qquad 14(6)$$

*For gas:

$$Z = \frac{7727 \, ERY}{SCFH} \sqrt{\frac{h \, P_f}{G \, T_f}} \qquad 14(7)$$

where

- E = area factor, determined from curve C on Figure 14s
- R = pipe constant, determined from table on Figure 14s
- G = specific gravity of gas (air = 1.0)
- G_f = specific gravity of liquid at operating temperature
- G_t = specific gravity of liquid at 60°F (15.6°C)
- h = pressure differential across orifice in inches H_2O
- Y = compressibility factor, determined from curve B on Figure 14s
- V = specific volume (ft³/lbm), determined from Steam Tables
- T_f = flowing temperature expressed in °R (°F + 460)
- P_f = flowing pressure in PSIA
- X = pressure loss ratio defined as h/2P

*For steam and gas, "h" expressed in inches H_2O should be equal to or less than "P_f" expressed in PSIA units.

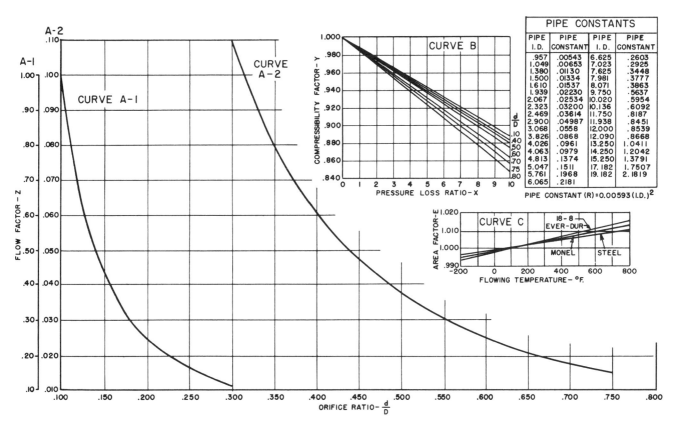

FIG. 14s
Orifice bore determination chart (flange taps). © 1946 by Taylor Instrument Companies (ABB Kent-Taylor Inc.)

TABLE 14t
*Orifice Flowmeter Capacity Table**

				Flange and Vena Contracta Taps			Pipe Taps		
				Liquid	Steam	Gas	Liquid	Steam	Gas
Pipe Size	Actual Inside Diam. (I.D.) Sched. 40	Maximum Orifice Diam.	Meter Range	Water (SG = 1)	100 PSIG Saturated	Air (SG = 1.0) @ 100 PSIG and 60°F	Water (SG = 1)	100 PSIG Saturated	Air (SG = 1.0) @ 100 PSIG and 60°F
Inches	Inches	Inches	Inches of Water	Gal./Min.	Lb./Hr.	Std. Cu. Ft/Min.	Gal./Min.	Lb./Hr.	Std. Cu. Ft./Min.
½	0.622	0.435	200	10.6	338	119	15.7	506	178
			100	7.5	239	84	11.2	358	126
			50	5.3	170	59	7.9	253	89
			20	3.3	107	37	5.0	160	57
			10	2.4	76	27	3.5	113	40
			2.5	1.17	38	13	1.7	56	20
1	1.049	0.734	200	30	963	295	44.8	1 440	507
			100	21.2	682	239	31.7	1 017	358
			50	15.0	482	170	22.4	719	253
			20	9.5	305	108	14.2	455	160
			10	6.7	216	76	10.1	323	113
			2.5	3.35	108	38	5.0	161	56
1½	1.610	1.127	200	70.7	2 270	796	105	3 380	1 190
			100	50.1	1 600	564	75	2 390	844
			50	35.1	1 135	399	52.7	1 690	596
			20	22.4	718	253	33.4	1 070	378
			10	15.8	683	178	23.6	758	267
			2.5	7.9	254	90	11.8	379	133
2	2.067	1.448	200	116	3 740	1 313	174	5 580	1 966
			100	83	2 645	932	123	3 950	1 390
			50	58.5	1 870	658	87	2 790	983
			20	37.0	1 183	417	55	1 768	623
			10	26.1	840	295	39	1 252	440
			2.5	13.1	420	148	19.4	625	220
3	3.068	2.147	200	255	8 240	2 905	383	12 300	4 330
			100	181	5 830	2 080	271	8 700	3 070
			50	128	4 125	1 460	191	6 160	2 175
			20	81.5	2 610	922	121	3 900	1 375
			10	57.5	1 843	653	86	2 760	975
			2.5	28.8	915	325	43	1 366	485
4	4.026	3.02	200	512	16 400	5 780	764	24 500	8 630
			100	362	11 600	4 090	540	17 300	6 100
			50	255	8 170	2 890	382	12 200	4 310
			20	162	5 180	1 830	242	7 730	2 730
			10	115	3 670	1 290	172	5 470	1 930
			2.5	57	1 820	647	85	2 710	965
5	5.047	3.78	200	800	25 600	9 050	1 190	38 200	13 500
			100	557	18 200	6 410	845	27 100	9 560
			50	402	12 900	4 530	598	19 200	6 760
			20	253	8 110	2 870	378	12 100	4 280
			10	180	5 750	2 020	268	8 580	3 020

TABLE 14t Continued

*Orifice Flowmeter Capacity Table**

				Flange and Vena Contracta Taps			Pipe Taps		
				Liquid	Steam	Gas	Liquid	Steam	Gas
Pipe Size	Actual Inside Diam. (I.D.) Sched. 40	Maximum Orifice Diam.	Meter Range	Water (SG = 1)	100 PSIG Saturated	Air (SG = 1.0) @ 100 PSIG and 60°F	Water (SG = 1)	100 PSIG Saturated	Air (SG = 1.0) @ 100 PSIG and 60°F
Inches	Inches	Inches	Inches of Water	Gal./Min.	Lb./Hr.	Std. Cu. Ft/Min.	Gal./Min.	Lb./Hr.	Std. Cu. Ft./Min.
			2.5	90	2 880	1 010	134	4 290	1 510
6	6.065	4.55	200	1 158	37 100	13 100	1 730	55 300	19 500
			100	820	26 300	9 250	1 223	39 200	13 800
			50	580	18 600	6 540	866	27 700	9 760
			20	367	11 700	4 140	547	17 500	6 180
			10	258	8 310	2 930	387	12 400	4 370
			2.5	129	4 150	1 460	193	6 200	2 180
8	7.981	5.9858	200	2 000	64 104	22 511	2 980	95 709	33 692
			100	1 413	45 320	15 952	2 110	67 682	23 853
			50	1 000	32 052	11 285	1 492	47 855	16 846
			20	634	20 275	7 156	943	30 263	10 674
			10	447	14 386	5 054	668	21 468	7 543
			2.5	223	7 186	2 534	333	10 719	3 772
10	10.020	7.5150	200	3 150	101 020	35 475	4 700	150 825	53 094
			100	2 230	71 481	25 138	3 325	106 658	37 589
			50	1 578	50 510	17 785	2 355	75 413	26 547
			20	998	31 950	11 277	1 487	47 691	16 821
			10	706	22 671	7 964	1 052	33 830	11 887
			2.5	352	11 324	3 994	525	16 891	5 944
12	12.000	9.0000	200	4 520	145 000	51 300	6 750	216 000	76 500
			100	3 200	103 000	36 200	4 775	153 000	45 100
			50	2 270	72 400	25 600	3 380	108 000	38 200
			20	1 430	46 000	16 200	2 135	68 600	24 200
			10	1 012	32 400	11 500	1 512	48 300	17 100
			2.5	507	16 200	5 740	757	24 200	8 560
14	13.126	9.8445	200	5 415	173 398	60 891	8 060	258 887	91 135
			100	3 830	122 588	43 148	5 720	183 076	64 520
			50	2 710	86 699	30 526	4 040	129 443	45 567
			20	1 715	54 842	19 356	2 555	81 860	28 873
			10	1 210	38 914	13 670	1 808	58 068	20 404
			2.5	603	19 437	6 855	900	28 994	10 202
16	15.000	11.2500	200	7 065	226 442	79 518	10 520	338 084	119 014
			100	5 000	160 089	56 347	7 460	239 081	84 258
			50	3 535	113 221	39 864	5 275	169 042	59 507
			20	2 240	71 619	25 277	3 335	106 902	37 705
			10	1 580	50 818	17 852	2 360	75 832	26 646
			2.5	788	25 383	8 952	1 175	37 865	13 323
			200	8 920	286 324	100 546	13 320	427 489	150 487
			100	6 330	202 424	71 248	9 270	302 305	106 539
			50	4 475	143 162	50 406	6 675	213 744	75 243

TABLE 14t Continued

Orifice Flowmeter Capacity Table*

Pipe Size	Actual Inside Diam. (I.D.) Sched. 40	Maximum Orifice Diam.	Meter Range	Flange and Vena Contracta Taps			Pipe Taps		
				Liquid	Steam	Gas	Liquid	Steam	Gas
				Water (SG = 1)	100 PSIG Saturated	Air (SG = 1.0) @ 100 PSIG and 60°F	Water (SG = 1)	100 PSIG Saturated	Air (SG = 1.0) @ 100 PSIG and 60°F
Inches	Inches	Inches	Inches of Water	Gal./Min.	Lb./Hr.	Std. Cu. Ft/Min.	Gal./Min.	Lb./Hr.	Std. Cu. Ft./Min.
18	16.876	12.6570	20	2 830	90 558	31 962	4 220	135 172	47 676
			10	1 995	64 256	22 573	2 985	95 885	33 693
			2.5	995	32 095	11 320	1 485	47 876	16 847
20	18.814	14.1105	200	11 100	356 238	125 097	16 550	531 871	187 232
			100	7 870	251 352	88 645	11 720	376 121	132 554
			50	5 565	178 119	62 714	8 310	265 936	93 616
			20	3 520	112 671	39 766	5 250	168 177	59 318
			10	2 485	79 946	28 085	3 715	119 298	41 920
			2.5	1 240	39 932	14 084	1 850	59 566	20 960
24	22.626	16.9695	200	16 060	515 222	180 927	23 950	769 238	270 791
			100	11 375	364 250	128 206	16 960	543 978	191 710
			50	8 035	257 611	90 703	12 000	384 619	135 395
			20	5 090	162 954	57 513	7 585	243 233	85 790
			10	3 590	115 625	40 619	5 375	172 539	60 628
			2.5	1 795	57 753	20 369	2 675	86 150	30 314

*Reproduced by permission of Taylor Instrument Co. (ABB Kent-Taylor).

A useful simplified form of the mass flow equation [Equation 14(3)] is:

$$W = 359 \, Cd^2 \sqrt{\frac{h\rho}{1-\beta^4}} \quad \quad 14(8)$$

where

- W = mass flow in pounds per hour
- d = orifice diameter in inches
- h = differential pressure in inches of water (Water density is assumed to be 62.32 pounds per cubic foot, corresponding to 68°F (20°C).)
- ρ = operating density in pounds per cubic foot
- β = ratio of orifice diameter to pipe diameter in pure number
- C = coefficient of discharge in pure number

This is a modification of the basic equation for mass flow [Equation 14(3)] substituting the $359 \, Cd^2 \sqrt{1-\beta^4}$ for kA. The figure 359 includes a factor for the chosen units of measurement. The coefficient of discharge is involved with the flow pattern established by the orifice, including the vena contracta and its relation to the differential pressure measurement taps. An average value of C = 0.607 can be used for flange and other close up taps, which gives a working equation:

$$W = 218 d^2 \sqrt{\frac{h\rho}{1-\beta^4}} \quad \quad 14(9)$$

For full flow taps, C = 0.715 and the equation becomes

$$W = 275 d^2 \sqrt{\frac{h\rho}{1-\beta^4}} \quad \quad 14(10)$$

These working equations can be used for approximate calculations of flow of liquids, vapors, or gases through any type of sharp-edged orifice. When using orifices for measurement in weight units, errors in determination of ρ must be considered. Accurate determination of density under flowing conditions is difficult, particularly for gases and vapors. In some cases, even liquids are subject to density changes with both temperature *and* pressure (for example, pure water in high pressure boiler feedwater measurement).

For W, d, h, and ρ given in dimensions other than those stated, simple conversion factors apply. Transfer of ρ in Equations 14(8), 14(9), and 14(10) from the numerator to denominator will give volume flow in actual cubic feet per hour at flowing conditions (see Equations 14(2) and 14(3)).

Beta ratio, and hence orifice diameter, can be calculated from a transposed form of the mass flow equation, 14(8).

ORIFICE METERING ACCURACY

If the purpose of flow measurement is not absolute accuracy, but only repeatable performance, then the accuracy in calculating the bore diameter is not critical, and approximate calculations will suffice. On the other hand, if the measurement is going to be the basis for the sale of large quantities of natural gas transported in high-pressure gas lines, for example, absolute accuracy is essential and precision in the bore calculations is critical.

Some engineers feel that instead of individually sizing each orifice plate, bore diameters should be standardized.[6] This approach would make it practical to keep spare orifices on hand in all standard sizes. This approach seems reasonable because the introduction of the microprocessor-based DCS systems means it is no longer important to have round figures for the full scale flow ranges. If this approach to orifice sizing were adopted, the orifice bore diameters and the d/p cell ranges would be standardized round values and the corresponding maximum flow would be an uneven number that corresponds to them.

In the past, in order to increase flow rangeability, the natural gas transport stations used a number of parallel runs (Figure 14u). In these systems the flow rangeability of the individual orifices was minimized by opening up another parallel path if the flow exceeded 90% of full scale flow (of the active paths) or by closing down a path when the flow in the active paths dropped to 80%. By so limiting the rangeability, metering accuracy was kept high, but at the price of a substantial investment in added piping and metering hardware. Another, less expensive choice was to use two transmitters, one for high (10 to 100%) pressure drop, the other for low (1 to 10%) and switch their outputs depending on the actual flow. This doubled the transmitter hardware cost, plus added some logic expense at the receiver, but increased the rangeability of orifice flowmeters to about 10:1.

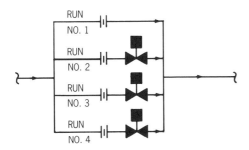

FIG. 14u
Metering accuracy can be maximized by keeping the flow through the active runs between 80% and 90% of full scale.[8]

As smart transmitters come on the market, another relatively inexpensive option is also becoming available: the dual-span transmitter. Some smart d/p transmitters are currently available with 0.1% of span accuracy, and their spans can be automatically switched by the DCS system, based on the value of measurement.[7] Therefore, a 100:1 pressure differential range (10:1 flow range) can be obtained by automatically switching between a high (10 to 100%) and a low (1 to 10%) pressure differential span. As the transmitter accuracy at both the high and the low flow condition is *0.1% of the actual span,* the overall result can be a 1% of actual flow accuracy over a 10:1 flow range.

Where the ultimate in accuracy is required, actual flow calibration of the meter run (the orifice, assembled with the upstream and downstream pipe, including straightening vanes, if any) is recommended. Facilities are available for very accurate weighed water calibrations, in lines up to 24 in. (61 cm) diameter and larger, and with a wide range of Reynolds number. For orifice meters, highly reliable data exists for accurate transfer of coefficient values for liquid, vapor, and gas measurement.

References

1. Spink, L.K., *Principles and Practice of Flow Meter Engineering,* 9th edition, The Foxboro Company, Foxboro, Massachusetts, 1967, Chapter 5 (now published as the *Flow Measurement Engineering Handbook,* R.W. Miller, New York: McGraw-Hill, 1983).
2. ASME, "Fluid Meters, Their Theory and Application," Report of ASME Research Committee on Fluid Meters, American Society of Mechanical Engineers, New York.
3. *Shell Flow Meter Engineering Handbook,* Waltman Publishing Company, Delft, The Netherlands: Royal Dutch/Shell Group, 1968.
4. American Gas Association, *AGA Gas Measurement Manual,* New York: American Gas Association.
5. Miller, O.W., and Kneisel, O., "Experimental Study of the Effects of Orifice Plate Eccentricity on Flow Coefficients," ASME Paper Number 68-WA/FM-1, Page 10, Conclusions 3, 4, 5, New York: American Society of Mechanical Engineers.
6. Ahmad, F., "A Case for Standardizing Orifice-Bore Diameters," *InTech,* January 1987.
7. Rudbäck, S., "Optimization of Orifice Plates, Venturies and Nozzles," *Measurements and Control,* June 1991.
8. Lipták, B.G., "Applying Gas Flow Computers," *Chemical Engineering,* December 1970.

Bibliography

AGA/ASME, "The Flow of Water Through Orifices," Ohio State Univ. Stud. Eng. Ser., Bull. 89, Vol. IV, No. 3.
American Gas Association, Report No. 3, "Orifice Metering of Natural Gas," AGA, 1985.
ANSI/API 2530, "Orifice Metering of Natural Gas," New York: ANSI, 1978.
ANSI/ASME MFC, "Differential Producers Used for the Measurement of Fluid Flow in Pipes (Orifice, Nozzle, Venturi)," New York: ANSI, December 1983.
ASME, "The ASME-OSI Orifice Equation," *Mech. Eng., Vol. 103,* No. 7, 1981.
BBI Standard 1042, "Methods for the Measurement of Fluid Flow in Pipes, Orifice Plates, Nozzles and Venturi Tubes," London: British Standard Institution, 1964.

"Differential Pressure Flowmeters," *Measurement and Control,* September 1991.

Kendall, K., "Orifice Flow," *Instruments and Control Systems,* December 1964.

Sauer, H.J., "Metering Pulsating Flow in Orifice Installations," *InTech,* March 1969.

Shichman, D., "Tap Location for Segmental Orifices," *Instruments and Control Systems,* April 1962.

Starrett, P.S., Nottage, H.B., and Halfpenny, P.F., "Survey of Information Concerning the Effects of Nonstandard Approach Conditions upon Orifice and Venturi Meters," presented at the annual meeting of the ASME, Chicago, November 7–11, 1965.

Stoll, H.W., "Determination of Orifice Throat Diameters," Taylor Technical Data Sheets TDS-4H603.

15 Pitot Tubes, Averaging, and Duct Section Units

W. H. HOWE (1969), **J. O. HOUGHEN** (1982),
B. G. LIPTÁK, M. PTÁČNÍK (1993)

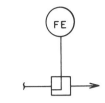

Flow Sheet Symbol

Types:	A. Standard, single-ported B. Multiple-opening, averaging C. Area averaging for ducts
Applications:	Liquids, gases, and steam
Operating Pressure:	Permanently installed carbon or stainless steel units can operate at up to 1400 PSIG (97 bars) at 100°F (38°C) or 800 PSIG (55 bars) at approximately 700°F (371°C); pressure rating of retractable units is a function of isolating valve
Operating Temperature:	Up to 750°F (399°C) in steel, 850°F (454°C) in stainless steel construction, when permanently installed
Flow Ranges:	Generally 2 in. (50 mm) pipes or larger; no upper limit
Materials of Construction:	Brass, steel, stainless steel
Minimum Reynolds Number:	In the range of 20,000 to 50,000
Rangeability:	Same as orifice plates
Straight-Run Requirements:	Downstream of valve or two elbows in different planes, 25–30 pipe diameters upstream and 5 downstream; if straightening vanes are provided, 10 pipe diameters upstream and 5 downstream
Inaccuracy:	For standard industrial units: 0.5 to 5% of full scale. Full-traversing Pitot Venturis under National Bureau of Standards type laboratory conditions can give 0.5% of actual flow error. Industrial Pitot Venturis must be individually calibrated to obtain 1% of range performance. Inaccuracy of individually calibrated multiple-opening averaging pitot tubes is claimed to be 2% of range when Reynolds numbers exceed 50,000. Area-averaging duct units are claimed to be between 0.5 and 2% of span. The error of the d/p cell is additional to the errors listed
Costs:	One-inch dia. averaging pitot tube in stainless steel costs $750 if fixed, $1400 if retractable for hot tap installation. Cost usually doubles if the pitot tube is calibrated. Hastelloy units for smokestack applications can cost $2000 or more. A local pitot indicator cost $400; a d/p transmitter suited for pitot applications with 4 to 20 mA DC output costs about $1000
Partial List of Suppliers:	ABB Kent-Taylor Inc. (A); Air Monitor Corp. (C); Alnor Instrument Co. (A); Andersen Instruments Inc. (A); Blue White Industries (A); Brandt Instruments (C); Davis Instrument Mfg. Co. (A); Dietrich Standard, a Dover Industries Company ("Annubar"—B); Dwyer Instruments Inc. (B); Fischer & Porter Inc. (A); Foxboro Co. ("Pitot Venturi"—A); Land Combustion Inc. (A); Meriam Instrument, a Scott Fetzer Company (B); Mid-West Instrument ("Delta Tube"—B); Preso Industries ("Elliptical"—B); Sirco Industries Ltd. (A); Ultratech Industries Inc. (A); United Electric Controls Co. (A)

In 1732 Henri de Pitot invented the pitot tube for the measurement of the flowing velocities of fluids. Pitot tubes detect the flowing velocity either at one point (standard), at several inlet points into an averaging probe (multiple-ported), or at many points across the cross section of a pipe or duct (area-averaging). Their advantages are low cost, low permanent pressure loss, and capability of inserting the probe-type sensors (wet- or hot-tapping) into existing or operating pipes. Their disadvantages are low accuracy, low rangeability, and limitation to clean liquid, gas, or vapor service, unless purged.

THEORY

The stagnation or impact pressure on a body immersed in a moving fluid is the sum of the static pressure and the dynamic pressure. Thus,

$$P_t = P + P_v \qquad 15(1)$$

where:

P_t = total pressure which can be sensed by a fixed probe provided the fluid stagnates at the sensing point in an isentropic manner
P = the static pressure of the fluid whether in motion or at rest
P_v = the dynamic pressure equivalent to the kinetic energy of the fluid considered as a continuum

With respect to a static state, the energy relation at the isentropic stagnation point of an ideal probe is:

$$\int_P^{P_t} \frac{dp}{\rho} = \int_o^{V_p} \frac{V_p dV}{gc} \qquad 15(2)$$

where:

V_p = the approach velocity at the probe location
ρ = the fluid density
gc = constant

For a liquid of constant density, integration yields, at a point:

$$(P_t - P) = P_v = \frac{P V_p^2}{2gc} \qquad 15(3)$$

For a compressible perfect gas for which $\dfrac{P}{\rho \gamma}$ remains constant during an isentropic change, a similar relation emerges:

$$(P_t - P) = P_v = \rho \frac{(\gamma - 1)}{\gamma} \frac{V_p^2}{2gc} \qquad 15(4)$$

where γ is the ratio of specific heats.

Assuming isentropic stagnation at the sensing point of the probe:

$$\int_P^{P_t} \frac{dp}{\rho} = \int_o^{V_p} \frac{V_{pd}V}{gc} \qquad 15(5)$$

where, using English units

V_p = velocity of approach, ft/s
P = pressure, lbf/ft^2
ρ = fluid density, lbm/ft^3
gc = 32.2 $\dfrac{lbm}{lbf} \dfrac{ft}{s^2}$

If density is constant, integration yields:

$$(P_t - P) = P_v = \frac{\rho(V_p^2)}{2\,gc} \qquad 15(6)$$

For a compressible perfect gas, for which $\dfrac{P}{\rho \gamma}$ remains constant during an isentropic change, a similar relation is obtained:

$$(P_t - P) = P_v = \frac{(\gamma - 1)}{\gamma} \frac{p(V_p^2)}{2\,gc} \qquad 15(7)$$

where γ is the ratio of specific heats.

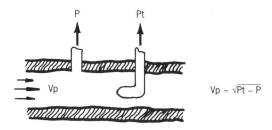

FIG. 15a
The velocity at a point (in the turbulent flow range) is related to the square root of the pressure difference between total and static pressures.

In order to compute point velocities, it is necessary to measure both point values of static pressure (P) and total (P_t) pressures from whence:

$$V_p = C \frac{(P_t - P)^{0.5}}{\rho} \qquad 15(8)$$

where C is a dimensional constant.

STATIC PRESSURE MEASUREMENT

For fluids flowing through conduits, the static pressure is commonly measured in one of three ways: through taps in the wall, from static probes inserted into the fluid stream, or from small apertures properly located on an aerodynamic body immersed in the flowing fluid.

The data of Shaw[1] as presented by Benedict[2] show that errors in the measurement of static pressure are minimal for velocities up to 200 ft/s (60 m/s) if wall tap dimensions conform to those in Figure 15b for a tap diameter, d, of 0.0635 in. and if $D \cong 2d$ and $1.5 < l/d < 6$.

Static pressure errors are also dependent upon fluid viscosity and velocity and whether the fluid is compressible or not. Shaw[1] states that for incompressible fluids flowing in a circular conduit with a pipe Reynolds number of 2×10^5, an error of about 1% of the mean dynamic pressure may occur using a wall tap having a diameter 1/10 that of the pipe. Rayle[3] mentions that a tap 0.03 in. (0.75 mm) in diameter with a conical countersink 0.015 in. (0.34 mm) deep will ensure nearly true static pressure sensing.

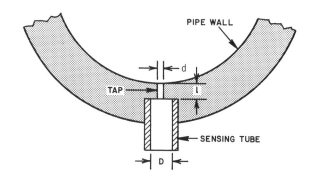

FIG. 15b
Wall tap for static pressure measurement.

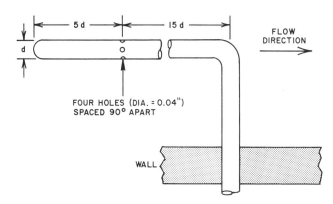

FIG. 15c
Typical static pressure-sensing probe.

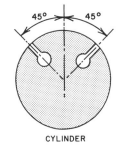 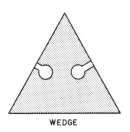

FIG. 15d
Two shapes of aerodynamic probes used to sense static pressure.

Static pressure may also be sensed through a tube inserted into the moving fluid. One configuration is shown in Figure 15c.

Other static probe designs are described in the literature.[2] The aerodynamic probe is a bluff body inserted into the flowing fluid with appropriately located holes on its surface through which pressure signals are obtained. The probe is oriented so that the sensed pressure is a measure of the static pressure. Two configurations taken from Benedict,[2] the cylinder and the wedge, are shown in Figure 15d. The probes are rotated until the pressure sensed from each hole is the same or, alternatively, the two taps may be manifolded to obtain an averaged pressure.

The total pressure develops at the point where the flow is stagnated isentropically, which is assumed to occur at the tip of a pitot tube or at a specific point on a bluff body immersed in the stream. Figure 15e is a typical pitot tube which also shows the taps for sensing static pressure. Another variation is shown in Figure 15f.

SINGLE-PORTED PITOT TUBE

Such probes must be carefully aligned with respect to the direction of fluid flow, which is sometimes uncertain when turbulence exists. The pitot tube is made less sensitive to flow direction if the impact aperture has an internal bevel of about 15 degrees extending about 1.5 diameters into the tube. Benedict[2] shows the behavior of various designs with orientation. Figure 15g shows typical performance data for a pitot tube.

To increase the magnitude of the pressure signals from in-stream velocity sensors, the Pitot Venturi and double venturi have been developed as shown in Figures 15h and 15i. These elements are intended to remain in a fixed position in a duct and hence signals must be correlated with flow

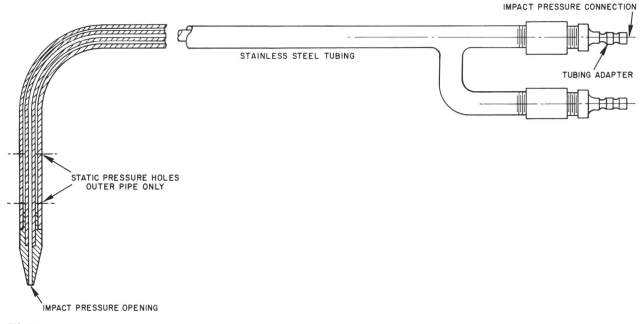

FIG. 15e
Typical pitot tube. (Courtesy Andersen Instruments, Inc.)

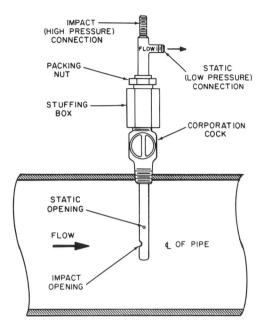

FIG. 15f
Schematic of an industrial device for sensing static and dynamic pressures in a flowing fluid.

rate by calibration that accounts for fluid properties and flow conditions (e.g., Reynolds number). A smooth straight section of conduit of at least 10 to 15 pipe diameters, or the equivalent, is recommended on both sides of the probe.

The determination of flow rate in a conduit using a pitot tube, for example, requires a carefully conducted traverse of the flow conduit to obtain a measure of the true average velocity. For circular pipes such an average is obtained from measurements of $(P_t - P)$ on each side of the cross section at

$$\left(\sqrt{\frac{2n-1}{N}} \right) \times 100\%, \left(n = 1,2,3 \ldots \frac{N}{2} \right) \quad 15(9)$$

of the pipe diameter measured from the center, where N is the number of measurements per traverse. Two measurements normal to each other are recommended.

To enhance the ability to obtain measurements near the walls of pipes in excess of about 6 in. (150 mm), a reduction nozzle inserted in the pipeline as shown in Figure 15j is suggested.

Calibration of Pitot Tubes

In high-precision laboratory tests the pitot tube is traversed across the cross section of the pipe, thereby establishing the velocity profile that exists in the pipe. In industrial applications the pitot tube is fixed and measures the flow velocity only at one point on the velocity profile (Figure 15k). If the velocity (V) measured by this fixed pitot tube is not the average velocity (U), a substantial error will result. This error cannot be easily eliminated, because even if the pitot tube insertion is carefully set to measure the average velocity U under one set of flow conditions, it will still be incorrect as soon as the flow velocity changes. At Reynolds numbers under 1000 (in the fully laminar region) the ratio between the average velocity and the center velocity is 0.5 ($V/U_c = 0.5$ in Figure 15k). In fully developed turbulent flow (Re = 50,000 or more) this same ratio is about 0.81 ($V/U_c = 0.81$).

Unfortunately, the velocity profile is affected not only by the Reynolds number but also by the pipe surface roughness and by upstream valves, elbows, and other fittings. In order to reform the velocity profile it is recommended to provide a straight pipe length of about 25 pipe diameters between the upstream disturbances and the pitot element. If the pitot tube is properly installed and if the pipe surface is smooth, it should be possible to design a microprocessor-based "smart pitot tube" that measures only the center velocity (U_c) and the Reynolds number and, based on these two readings, accurately calculates the flow under all flow velocity conditions.

The (absolute) method of calibrating a pitot tube, as demonstrated by the National Bureau of Standards, is to mount the device on a carriage which can be drawn through stagnant air at a known velocity. Smoke is introduced into the room to verify the absence of turbulence. Such tests have shown that pitot tubes with coefficients very close to unity can be designed.

Devices such as Pitot-Venturis or double Venturis, when used as flow rate meters, require extensive in situ calibration for each specific installation if high-precision measurements (within 1%) are desired.

MULTIPLE-OPENING PITOT TUBES

One approach in attempting to overcome the inherent limitation of the pitot tube—that of being a point velocity sensor—was to measure the velocities at several points and average these readings. It was argued that by averaging the velocities measured at four fixed points, for example (see Figure 15l), changes in the velocity profile will be detected, and therefore the reading of multiple-opening pitot tube will be more accurate than that of the single-point sensor pitot tube. The manufacturers of the averaging pitot usually claim that the flow coefficient (K) will stay within 2% between the Reynolds numbers of 50,000 and 1,000,000. This is probably so, but it might not be attributable to averaging action but rather to the fact that in this highly turbulent region the velocity profile is flat and changes very little.

Critics of this device argue that it offers little improvement over the single-opening pitot tube because it is ineffective at Reynolds numbers under 50,000, which means it is not applicable for a large portion of industrial liquid flows. The other argument the critics mention is that the openings on the averaging pitot tubes are too large. Therefore it is said that these devices are not true averaging chambers and therefore the pressure detected by the sensing port, as shown in Figure 15l, does not reflect the average, but rather is dominated by the effect of the nearest port. It would serve the interest of the instrumentation profession for the indepen-

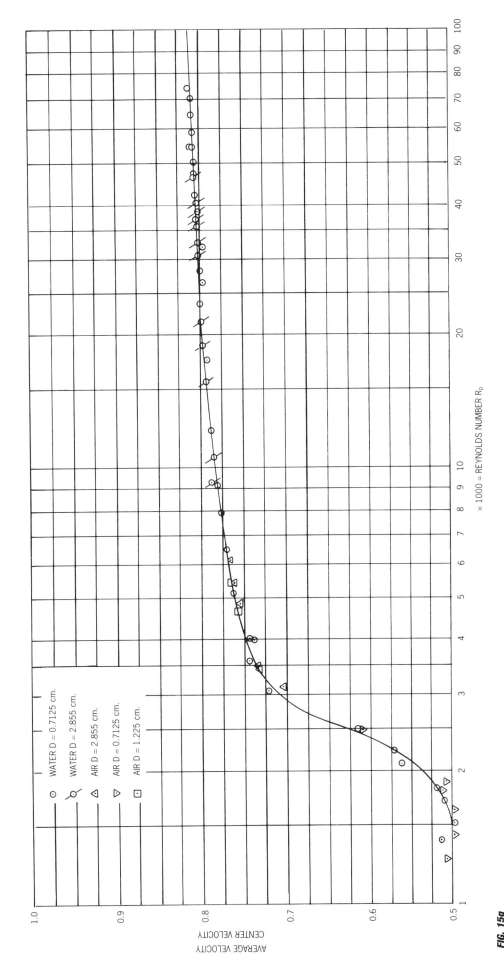

FIG. 15g *Center to average velocity ratios in straight and smooth pipes.*[11]

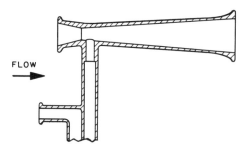

FIG. 15h
A Pitot Venturi produces a higher differential pressure than the standard pitot tube.

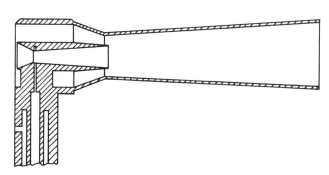

FIG. 15i
The double Venturi produces a higher differential pressure than the standard Pitot tube. (Courtesy of Foxboro Co.)

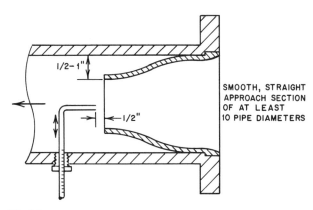

FIG. 15j
Reduction nozzle used to expedite velocity traverses.

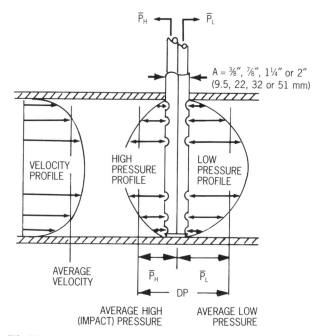

FIG. 15l
The design of a particular averaging pitot tube. (Courtesy of Dietrich Standard, a Dover Industries Company)

dent testing laboratories to publish their findings on these effects.

The reason why the ports are so large in most averaging pitot tubes is to prevent plugging. Some manufacturers of area-averaging pitot tubes do overcome this limitation by purging. These purged units can act as true averaging chambers, because the small port openings are kept clean by the purge gas. Naturally, they can be used only on such processes where the introduction of a purge media is acceptable.

One advantage of the averaging pitot tubes that both its manufacturers and its critics agree on is the ability to be installed on operating, pressurized pipelines. This hot-tapping capability and the ability to remove the sensor without requiring a plant shutdown are important advantages of all probe-type instruments (Figure 15m).

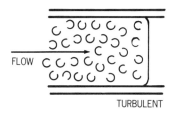 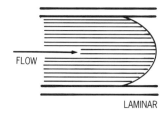 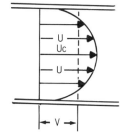

FIG. 15k
The velocity profile becomes flatter as the Reynolds number rises (the flow becomes more turbulent), and the task of the pitot-type flow sensor is to find the insertion depth corresponding to the average velocity (V).

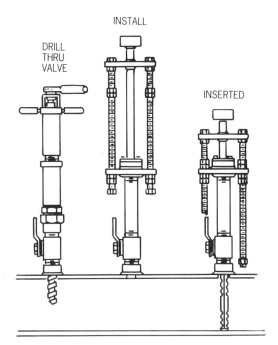

FIG. 15m
The hot-tap installation of an averaging pitot tube involves the same steps used in the "retractable" installation of all probe-type instruments. (Courtesy of Dietrich Standard, a Dover Industries Company)

Pressure Differential Produced

One of the problems with pitot tubes is that they do not generate strong output signals. The minimum span of "smart" d/p cells is 0 to 2 in. of H₂O (0 to 0.5 kPa), and

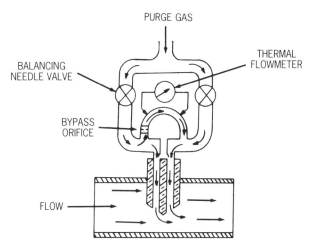

FIG. 15n
Pitot-tube rangeability can be increased by replacing the d/p cell detector with a thermal flowmeter.

these units are accurate up to 0.1% of actual span. For even narrower differentials, down to 0 to 0.1 in. H₂O (0 to 25 Pa), the membrane-type d/p cells can be used. In addition to d/p cells, one can also use elastic element or manometer-type devices, variable-area flowmeters (Figure 25h), and thermal flowmeters (Figure 15n) as detectors on pitot tubes. The thermal detector gives the highest rangeability, but its use is limited to applications where the pitot tube is purged.

In calculating the pressure differential produced by an averaging pitot tube, one might use the equations listed in Table 15o. For the metric equivalents of the units used in this table, refer to the Appendix. The flow coefficient (K) of

---

TABLE 15o
Equations for Calculation of the Pressure Differential Produced by the Averaging Pitot Tube Described in Figure 15l

Liquid, gas, steam (mass rate of flow)

$$h_w = \left(\frac{1}{\rho_f}\right)\left(\frac{lb_m/hr}{358.94 \ K \ D^2}\right)^2$$

Liquid (volume rate of flow)

$$h_w = \left(G_f\right)\left(\frac{GPM}{5.666 \ K \ D^2}\right)^2$$

Gas (standard volumetric flow)

$$h_w = \left(\frac{T_f G}{P_f}\right)\left(\frac{SCFH}{7{,}711 \ K \ D^2}\right)^2$$

Gas (actual volume rate of flow)

$$h_w = \left(\rho_f\right)\left(\frac{ACFH}{358.94 \ K \ D^2}\right)^2$$

h_w = differential pressure, inches of water at 68°F
K = flow coefficient
D = internal pipe diameter, inches
lb_m/hr = pounds mass per hour
GPM = U.S. gallons per minute
$ACFH$ = Actual cubic feet per hour
$SCFH$ = Standard cubic feet per hour (at 14.73 psia and 60°F)
ρ_f = flowing density, lb_m/ft^3 for gas:

$$P_f = \frac{P_f}{14.73} \times \frac{520}{T_f} \times .076487 \times G$$

.076487 lb_m/ft^3 = air density at 14.73 psia and 60°F
G_f = specific gravity of liquid
G = specific gravity of gas (molecular weight of air = 28.9644)
T_f = temperature of flowing gas in degrees Rankine (°R = °F + 460)
P_f = flowing pressure, psia

(Courtesy of Dietrich Standard, a Dover Industries Company)

TABLE 15p
The Flow Coefficient K for the Averaging Pitot Tube shown in Figure 15l Having the "A" Dimension Also Defined in That Figure*

Pipe Size			Flow Coeffeicient-K			
Size/Sch	D-in	D-mm	A = 3/8"	A = 7/8"	A = 1¼"	A = 2"
2" sch 40	2.067	52.50	.5912			
2½" sch 40	2.469	62.71	.6026			
3" sch 40	3.068	77.93	.6134			
3½" sch 40	3.548	90.12	.6192			
4" sch 40	4.026	102.26	.6235			
5" sch 40	5.047	128.19	.6297	.5934		
6" sch 40	6.065	154.05		.6047		
8" sch 40	7.981	202.72		.6173		
10" sch 40	10.020	254.51		.6250		
12" sch std.	12.000	304.80		.6298	.6186	
14" sch std.	13.250	336.55		.6321	.6220	
16" sch std.	15.250	387.35		.6349	.6263	
18" sch std.	17.250	438.15		.6370	.6296	
20" sch std.	19.250	488.95		.6387	.6321	
24" sch std.	23.250	590.55		.6411	.6357	.6247
30" sch std.	29.250	742.95		.6435	.6393	.6308
36" sch std.	35.250	895.35		.6450	.6416	.6346
42" sch std.	41.250	1047.7		.6461	.6432	.6373
48" —	48.00	1219.20			.6445	.6395
60" —	60.00	1524.0			.6461	.6422
72" —	72.00	1828.80			.6472	.6439

*(Courtesy of Dietrich Standard, a Dover Industries Company)

the pitot tube varies with its design. The K values of the averaging pitot tube shown in Figure 15l are listed in Table 15p. The distance "A" given in Table 15p is defined in Figure 15l.

AREA-AVERAGING PITOT STATIONS

For the measurement of large volumes of low-pressure gas flows, such as the detection of combustion air to boilers, air flow to dryers, or in HVAC systems, area-averaging pitot stations have been designed. These units are available with circular or rectangular (Figure 15q) cross sections and can be mounted in large pipes or ducts, including the suction or discharge of fans. These stations are so designed that one total pressure port and one static port is located in each unit area of the duct cross section, and each is connected to its own manifold. The manifolds act as averaging chambers and are purged to protect the individual ports from plugging.

The straight-run requirement of these units is reduced by the addition of a hexagon-cell-type flow straightener and a flow nozzle in front of the area-averaging pitot flow sensor.

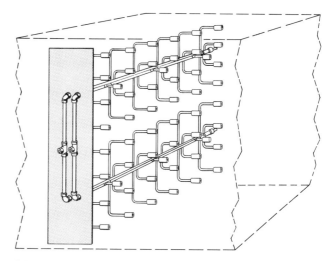

FIG. 15q
Installation in rectangular duct of area-averaging Pitot tube ensembles for metering the flow rate of gases. (Courtesy of Air Monitor Corp.)

The nozzle in addition serves to amplify the differential pressure produced by the unit. The manufacturer claims that with this design (Figure 15r) the straight-run requirement of the installation will range between 0 and 10 diameters. The 10-diameter straight run is recommended when a butterfly valve or a damper is located upstream to the flow meter station.

As the area-averaging pitot stations generate very small pressure differentials, special d/p cells are needed to detect these differentials. One such detector is the membrane-type design which can have as small a span as 0 to 0.01 in. H_2O (to 2.5 Pa). When such extremely small differentials are detected, the pressure drop in the tubing between the d/p cell and the pitot station must be minimized. This is achieved by making the connecting tubes short, and large in diameter. The pressure differential generated by the flow element shown in Figure 15r can be calculated by using the equations in Table 15t. For the equivalent SI units for use in these equations, refer to the Appendix of this book.

SPECIAL PITOT TUBES FOR PULSATING FLOW

Unsteady flow mean velocity measurements are usually inaccurate when using standard pitot probes.[4] Error is typically of about 5 to 30% of mean total pressure value in this case. This problem can be partially solved by means of specially designed probes for unsteady or pulsating flow measurements.

Figure 15s shows a design provided with a low-capacity capillary probe filled with silicon oil. The oil serves to transmit the process pressure to the d/p transducer. This type of probe was developed and is used by Deutsche Forschungs- und Versuchsanstalt für Luft und Raumfahrt in Germany.[5]

Figure 15u shows another sample of a probe designed to measure unsteady flow. This probe was developed in the

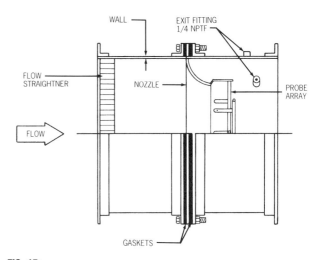

FIG. 15r
The flow straightener and the nozzle serve to reduce the upstream straight pipe-run requirement and increase the pressure differential generated. (Courtesy of Brandt Instruments)

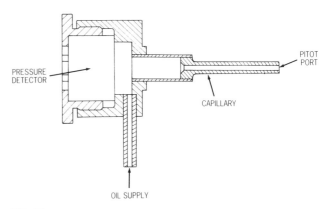

FIG. 15s
On highly pulsating flow measurements a minute flow of silicon oil through a capillary can serve as a pressure-averaging purge.

TABLE 15t
Calculation of Pressure Differentials Generated By Area Averaging Pitot Stations*

Equations for Differential Pressure Calculations	Terms Used
$DP = \left(\dfrac{ACFM}{Area}\right)^2 \times \dfrac{DENS}{(1096.845)^2}$	**Area** = Cross sectional area of duct section in ft²
	ACFM = Actual Cubic Feet per Minute
	DP = Differential Pressure in inches w.c.
$DP = \left(\dfrac{SCFM}{4000.7 \times Area}\right)^2$	**M** = Mass Flow in pounds per hour
	SCFM = Standard Cubic Feet per Minute
	V = Velocity in Feet per Minute
$DP = (V)^2 \times \dfrac{DENS}{(1096.845)^2}$	**PABS** = Absolute pressure in PSIA
	PATM = Atmospheric Pressure in PSI
	Ps = Static Pressure in inches w.c.
	T = Temperature in degrees F
$DP = \left(\dfrac{M}{60 \times Area}\right)^2 \times \dfrac{1}{DENS \times (1096.845)^2}$	**DENS** = Density at actual conditions lbs/ft³
	DENSTD = Density at Standard Conditions lbs/ft³

*Courtesy of Brandt Instruments

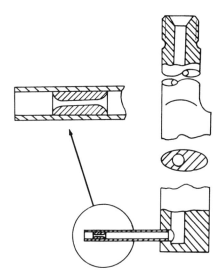

FIG. 15u
Pitot tube designed for pulsating flow averaging using tuned natural frequency.[6]

Aeronautical Research and Test Institute in Czechoslovakia.[6] The main problem in this case is in achieving equal input and output resistance of the probe opening. Natural frequency of the probe must be carefully tuned to avoid resonance during measurement.

References

1. Shaw, R., "The Influence of Orifice Geometry on Static Pressure Measurements," *Fluid Mechanics,* 7, Pt. 4, April 1960, p. 550.
2. Benedict, R.P., *Fundamentals of Temperature, Pressure, and Flow Measurements,* New York: John Wiley & Sons, Inc., 1969, p. 237.
3. Rayle, R.E., "Influence of Orifice Geometry on Static Pressure Measurements," ASME Paper 59-A-234, December 1959.
4. Becker, H.A., "Reaction of Pitot-tube in Turbulent Flow," *Journal of Fluid Mechanics,* Vol. 62, Part I, 1974.
5. Weyer, H., Bestimmung der zeitlichen Druckmittelwerte in stark fluktuirender Strömung, inbesondere in Turbomaschienen, Forschungsbericht 74-34, Deutsche Forschungs- und Versuchsanstalt für Luft- und Raumfahrt-Portz-Wahn, 1974.
6. Neruda, J., Šoch, P., "Measurement System with a Pitot Tube," Czechoslovak Patent No. 218 417.
7. Spink, L.K., "Principles and Practice of Flow Meter Engineering," The Foxboro Co., 1967.

Bibliography

"The Accuracy of the Pitot Tube Traverse Method of Measuring Pipe Flow at Various Distances up to 30 Diameters Downstream of a Smooth Right-Angled Bend," National Engineering Laboratory Flow Measurement Memo, No. 37, 1969.

Andrew, W.G., and Williams, H.B., *Applied Instrumentation in the Process Industries,* Vol. I, 2nd edition, Gulf Publishing Co., 1979.

Beitler, S.R., "Present Status of the Art of Flow Measurement in the Power Industry," ASME Paper No. 68-WA/PTC-7, December 1968.

De Boom, R.J., "Flow Meter Evaluation," 1991 ISA Conference, Paper #91-0509.

Dietrich, P.D., "Primary Flow Meter," *Instruments and Control Systems,* December 1968.

"Flow Meter Survey," *Instruments and Control Systems,* 42 No. 3, pp. 115–130, March 1969, and 42, No. 7, pp. 100–102, July 1970.

Hiser, R., "Increased Functions and Reduced Costs of Differential Pressure Flowmeters," *Measurements and Control,* September 1990.

London, A.V., "Less Traditional Methods of Flow Measurements," *Process Engineering, Plant & Control,* August 1968, pp. 47–50.

Malherbe G., and Silberberg, S., "Device for Measuring the Flow of Pulverized Control," Central Electricity Generating Board, Translation CE 4938 form *Automatisme* 13, No. 3, pp. 114–122, March 1968.

Migliorini, R., "Pitot Sensors," *Measurements and Control,* September 1991.

Ower, E., and Pankhurst, R.D., *The Measurement of Air Flow,* 4th edition, London: Pergamon Press, 1966.

Spencer, E.A., "Flow Measurements at the National Engineering Laboratories," *Process Engineering, Plant and Control,* August 1968, pp. 53–57.

16 Positive Displacement Gas Flowmeters

Flow Sheet Symbol

R. SIEV (1969),   **G. M. CRABTREE** (1982, 1993)

Type of Designs:	A. Positive displacement, B. High precision
Design Pressures:	Low-pressure designs from 5 to 100 PSIG (0.34 to 6.9 bars); high-pressure units are available up to 1440 PSIG (100 bars).
Design Temperatures:	Standard units can be used from -30 to $140°$ F (-34 to $60°$ C)
Materials of Construction:	Aluminum, steel, plastics, and synthetic elastomers
Inaccuracy:	A. 0.5 to 1% of registration B. 0.5% of actual flow over 50:1 range.
Costs:	A household gas meter for 250 SCFH (7 SCMH) capacity costs $130. A 50,000 SCFH (1416 SCMH) capacity, diaphragm-type displacement-type flowmeter in cast aluminum for natural gas service costs $4800. For natural gas service a 70,000 SCFH (1983 SCMH) rotary positive displacement meter in cast aluminum costs about $2800
Partial List of Suppliers:	American Meter (A); Dresser Measurement Div. Dresser Industries Inc. (A); Email-Westinghouse Pty Ltd. (Australia) (A); Equimeter Inc. (formerly Rockwell) (A); Elster-Handel Gmbh (Germany) (A); Instromet Technology Corp. (A); Kimmon Mfg. (Japan) (A); National Meter Co. (A); Perkins & Cowan (England) (A); Pierburg Luftfahrtgerate Union GmbH (B): J.B. Rombach, Gmbh & Co. (Germany) (A); Romet Ltd. (Canada) (A); Schlumberger Industries, Gas Div. (formerly Sprague) (A)

Positive displacement gas meters measure by internally passing isolated volumes of gas that successively fill and empty compartments with a fixed quantity of gas. The filling and emptying process is controlled by suitable valving and is translated into rotary motion to operate a calibrated register or index that indicates the total volume of gas passed through the meter.

The liquid sealed drum meter is the oldest commercial positive displacement gas meter (see Figure 16a). Developed in the early 1800s, it was used for many years during the gaslight era. This type of meter is still available today and remains one of the most accurate of the displacement-type meters. Applications of the liquid sealed drum meter today include laboratory work, appliance testing, pilot plant measurements, and as a calibration standard for other meter types.

Some of the inherent difficulties with the liquid sealed meter, such as changes in liquid level and freezing, were overcome in the 1840s with the development of the diaphragm-type positive displacement meter. The early meters were constructed with sheepskin diaphragms and sheet metal enclosures. Today, meters are made of cast aluminum with synthetic rubber-on-cloth diaphragms. The principle of operation, however, has remained the same for almost 150 years.

THE DIAPHRAGM METER

The operating principle of the four-chamber diaghragm meter is illustrated in Figure 16b. The measurement section consists of four chambers formed by the volumes between the diaphragms and the center partition and between the diaphragms and the meter casing. Differential pressure across the diaphragms extends one diaphragm and contracts

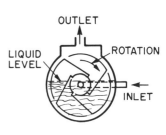

FIG. 16a
The liquid sealed drum meter.

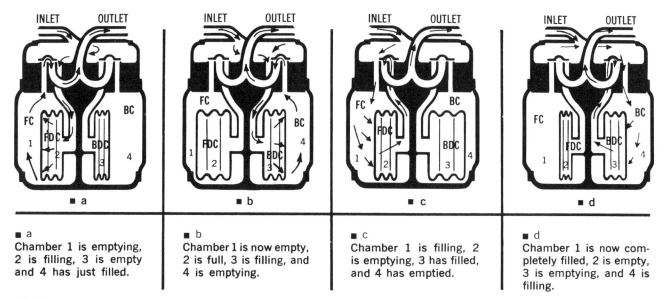

FIG. 16b
The four chamber diaphragm meter; FC = front chamber; BC = back chamber; FDC = front diaphragm; BDC = back diaphragm chamber.

the other, alternately filling and emptying the four compartments. The control for the process is through the "D" slide valves which are synchronized with the diaphragm motion and timed to produce a smooth flow of gas by means of a crank mechanisms. The crank and valve mechanism is designed and adjusted with no "top-dead-center" to prevent the meter from stalling. The rotating crank mechanism is connected through suitable gearing to the index which registers the total volume passed by the meter.

The rating of small diaphragm meters is usually specified in cubic feet per hour (0.03 m³/hr) of 0.6 specific gravity gas which result in a pressure drop of 0.5 in. water column (0.13 kPa). Larger meters are often rated for flow at 2 in. water column (0.5 kPa) differential.

Since most meters are sold to gas utility companies which sell natural gas with a specific gravity of approximately 0.6, it may be necessary to determine the flow rating of a diaphgram for other gases. This is accomplished by:

$$Q_n = Q_c \sqrt{\frac{(SG)_c}{(SG)_n}} \qquad 16(1)$$

where

Q_n = New flow rating (ft³/hr)*
Q_c = Meter rating (ft³/hr)
$(SG)_c$ = Specific gravity for which meter is rated (usually 0.6)
$(SG)_n$ = Specific gravity of new gas

The inaccuracy of diaphragm positive displacement meters is typically ±1% of registration over a range in excess of 200:1. This accuracy is maintained over many years of service. Deterioration of meter accuracy is rare unless unusual conditions of dirt, wear, or moisture in the gas are present.

THE LOBED IMPELLER

The lobed impeller meter (described in Section 17 ["Positive Displacement Liquid Meters and Provers"]) is used for high-volume measurement up to 100,000 ft³/hr (up to 3000 m³/hr). In this meter the close clearance of moving parts requires the use of upstream filters to prevent deterioration of accuracy performance. Typically the inaccuracy of lobed

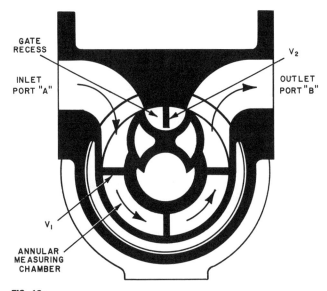

FIG. 16c
The rotating vane meter.

*For SI units refer to Appendix

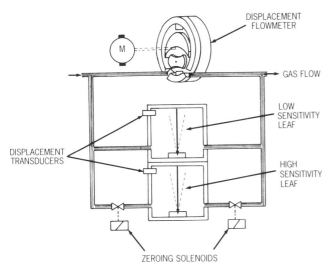

FIG. 16d
High-precision displacement flow meter for gas service. (Courtesy of Pierburg Loftfahrtgerate Union GmbH).

impeller meters is ±1% over a 10:1 flow range at pressure drops of approximately 0.1 psi (0.7 kPa).

An improvement on the lobed impeller meter is the rotating vane meter illustrated in Figure 16c. Here four compartments formed by the vanes rotate in the same direction as a rotating gate. The fixed volumes of gas are swept through the meter by the vanes which are passed from inlet side to outlet side through the gate. The motion of the vanes and gate is synchronized by gears. Typical inaccuracy for the rotating vane meter is ±1% over a 25:1 range at pressure drops of 0.05 in. of water column (0.013 kPa).

HIGH-PRECISION GAS FLOWMETER

For the high-precision measurement of air flows in engine test rigs, positive displacement flowmeters are used. High precision and high rangeability are achieved by eliminating the pressure drop and thereby eliminating the slip or leakage flows. This is achieved by providing a motor drive for the displacement element and using it to introduce only as much driving energy as is needed to keep the pressures at the inlet and outlet of the meter equal (Figure 16d). This flowmeter uses high-sensitivity leaves to detect the pressure differential and displacement transducers to detect the deflection of the leaves. The flowmeter is also provided with automatic rezeroing capability through periodic solenoid isolation of the high-sensitivity leaves.

This flowmeter is claimed to provide a reading with only a 0.25% error over a 50:1 range and a 0.5% error over a 100:1 range. The meter is designed for ambient operating temperatures and 30 PSIG (2 bars) operating pressures. The different models of this flowmeter can detect air or gas flows from 0.3 to 1500 ACFM (0.6 to 2500 ACMH).

APPLICATION NOTES

All displacement gas meters can be used to measure any clean dry gas that is compatible with the meters' construction materials and flow and pressure ratings. Dirt and moisture are the worst enemies of good meter performance; inlet filtering should be used when indicated. Since all gases change volume with pressure and temperature changes, these sources of possible error should be controlled or

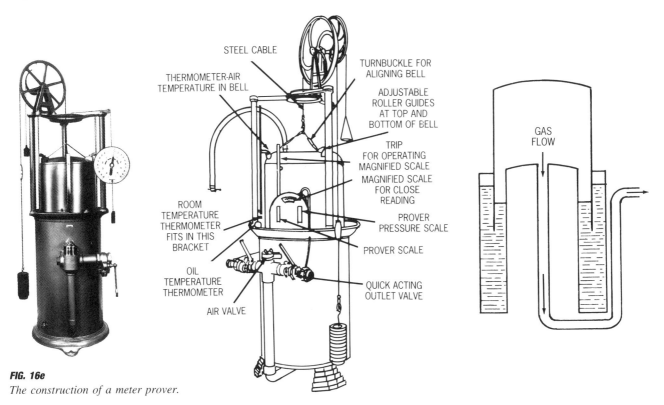

FIG. 16e
The construction of a meter prover.

compensated. The national standard cubic foot of fuel gas is at 14.73 PSIA and 60°F; significant deviation from these values should be accounted for in measuring standard gas volumes. At elevated pressures and lower temperatures a deviation from the ideal gas laws occurs, requiring the application of a compressibility factor to the measured volumes.

TESTING AND CALIBRATION

The testing, or proving as it is called in the gas utility industry, of gas meters is usually done using a special type of gasometer referred to as a prover. The construction of a meter prover is shown in Figure 16e. An accurately calibrated "bell" of cylindrical shape is sealed over a tank by a suitable liquid. The lowering of the bell discharges a known volume of air through the meter under test to compare the volumes indicated. Meter provers are typically supplied to discharge volumes of 2, 5, and 10 cubic feet (0.06, 0.15, and 0.3 m^3) and larger provers of several hundred cubic foot capacity are in use by meter manufacturers and gas utility companies. The volumetric inaccuracy of meter provers is on the order of ±0.1% as determined by physical measurement and comparison with more accurate volumetric standards.

Other standards used to calibrate gas meters are calibrated orifices and critical flow nozzles. These devices compare rates of flow rather than fixed volumes and have inaccuracy ratings typically from ±0.15 to ±0.5%.

Bibliography

Bailey, S.J., "Fit Meter to Stream, Flow Loop for Top Performance," *Control Engineering,* Vol. 29, No. 6, 1982, p. 87.

Beck, H.V., *Displacement Gas Meters,* Singer American Meter Div., Philadelphia, Pennsylvania, 1970.

Berghegger, H.W., "Diaphragm Meter Capacity Ratings," *Gas,* Vol. 44, September 1968, p. 51.

Bernitt, C.C., Holmes, H.H., and Stevenson, J.R., "New Developments in Displacement Metering," *Pipe Line Industry,* Vol. 28, July 1968, pp. 29, 38, 40.

Consodine, D.M., *Encyclopedia of Instrumentation and Control,* New York: McGraw-Hill.

Crabtree, G.M., "Guide to Gas Meters," *Pipeline & Gas Journal,* undated reprint.

Delaney, L.J., "Rotary and Diaphragm Displacement Meters," *Instruments and Control Systems,* November 1962, p. 114.

Evans, H.J., *Turbo-Meters—Theory and Application,* Pittsburgh, Pennsylvania: Rockwell International, 1968.

"Fluid Meters, Their Theory and Application," 6th edition, New York: American Society of Mechanical Engineers, 1971.

Hall, J., "Flow Monitoring Applications Guide," *Instruments and Control Systems,* February 1983, p. 41.

Hall, J., "Solving Tough Flow Monitoring Problems," *Instruments and Control Systems*, February 1980.

Jasek, A.W., "Mechanical Displacement Meter Prover for Gas Meters," *Gas,* Vol. 41, August 1967, p. 52.

Lief, A., *Metering for America,* New York: Appleton-Century-Crofts, 1961.

Lomas, D.J., "Selecting the Right Flowmeter," *Instrumentation Technology,* May 1977.

Miller, J., "High Accuracy transmitters for Custody Transfer of Natural Gas," 1991 ISA Conference, Paper #91-0520.

O'Rourke, E.L., "The Select Compact Gas Meter," American Gas Association, Operating Section Proceedings, Arlington, Virginia, 1991.

Perrine, E.B., "Displacement Gas Meters," *Instruments and Control Systems,* February 1966, p. 127.

Staff, "Rotary Positive Displacement Meter," *Pipeline Engineer,* Vol. 38 (pipeline Handbook), March 1966, p. 121.

Steuernagle, R.L., "Diaphragm Meter Design & Operation," Technical papers, Appalchian Gas Measurement Short Course, Robert Morris College, Coraopolis, Pennsylvania, 1990.

Welch, J.V., "Trends in Low Gas Flow Metering," *InTech,* February 1991.

17 Positive Displacement Liquid Meters and Provers

Flow Sheet Symbol

R. SIEV (1969)   **B. G. LIPTÁK** (1982, 1993)

Reviewed by J. B. Stoddard (1993).

Types of Designs: A. Impeller, propeller, turbine
B. Nutating disc
C. Oval gear
D. Piston
E. Rotating vane
F. Specialized, low-flow, etc.
G. Viscous helix
H. Prover
I. High Precision

Design Pressure: To 1500 PSIG (10.5 MPa)

Design Temperature: To 560°F (293°C)

Strainer Required: Yes

Materials of Construction: Bronze, cast iron, aluminum, steel, stainless steel, Monel, Hastelloy, and plastics

Size Range: ¼ to 16 in. (6 to 406 mm)

Flow Range: 0.01 GPH to 20,000 GPM (0.04 l/hr to 75 m³/m)

Rangeability: 15:1 (some suppliers claim more)

Inaccuracy: ±1/10 to ±2% of actual flow, better in larger sizes

Cost: A 1 in. (25 mm) bronze disk-type, water meter with 2 to 3% error costs about $600.
A 1 in. (25 mm) oval flowmeter for LPG service with ductile iron housing complete with valve, vapor eliminator, and register with printer costs about $3000.
A 2 in. (50 mm) piston meter, in steel construction, having 0.5% error and provided with register, preset valve, and ticket printer costs about $5200.
A 6 in. (150 mm) flanged, birotor meter for fuel oil service with ductile iron preset valve, impulse contactor, large dial register, and ticket printer costs about $13,000.
Prover costs range from $50,000 to $300,000 depending on size

Partial List of Suppliers: Accurate Metering Systems Inc. (D); American Flowmeter Div. if VSI Nitto (D); Badger Meter Inc. (B,D); Brooks Instrument Div. of Rosemount (A,C,D,G,H); Conameter Corp. (F); Daniel Industries Inc. (E,H); Dresser Industries (A); Engineering Measurements Co. (D,G); Flowdata Inc. (A,E); Flow Technology Inc. (H); Fluidyne Instrumentation (C,D,G); Hersey Measurement Co. (A,B,D); ITT Barton (A,E); Kent Meters Inc. (A); Liquid Controls (A,E); Max Machinery Inc. (C,D,G); Meter Equipment Mfg. Inc. (A); Metric Corp. (H); Milton Roy Co. (B); Oilgear Co. (A); Pierburg Luftfahrtgerate Union GmbH (I); Schlumberger Industries (A,B,D); Smith Meter Inc., a Moorco Company (A,C,D,E); Tokheim Corp. (E); Waugh Controls Corp. (A,C)

Positive displacement meters split the flow of liquids into separate known volumes based on the physical dimensions of the meter, and count them or totalize them. They are mechanical meters in that one or more moving parts, located in the flow stream, physically separate the fluid into increments. Energy to drive these parts is extracted from the flow stream and shows up as pressure loss between the inlet and the outlet of the meter. The general accuracy of these meters is dependent upon minimizing clearances between the moving and stationary parts and maximizing the length of this leakage path. For this reason meter accuracy tends to increase as size increases.

NUTATING DISK

This meter, also known as the disk meter, is used extensively for residential water service. The moving assembly, which separates the fluid into increments, consists of an assembly

FIG. 17b
Rotating vane meter.

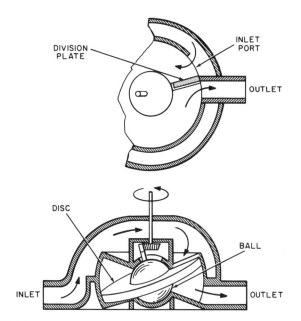

FIG. 17a
Nutating disk meter.

of a radially slotted disk with an integral ball bearing and an axial pin (see Figure 17a). This part fits into and divides the metering chamber into four volumes, two above the disk on the inlet side and two below the disk on the outlet side. As the liquid attempts to flow through the meter, the pressure drop from inlet to outlet causes the disk to wobble, or nutate, and for each cycle to display a volume equal to the volume of the metering chamber minus the volume of the disk assembly. The end of the axial pin, which moves in a circular motion, drives a cam that is connected to a gear train and the totalizing register. This flowmeter has an inaccuracy of about ±1 to 2%. It is built for small pipe sizes. Its temperature range is from −300 to 250°F (−150 to 120°C), and its maximum working pressure is 150 PSIG (1034 kPa). On cold water service the capacity ranges are as follows:

Size	Capacity
½ in. (13 mm)	2–20 GPM (7.5–75 lpm)
1 in. (25 mm)	5–50 GPM (19–190 lpm)
1-½ in. (38 mm)	10–100 GPM (38–380 lpm)
2 in. (51 mm)	16–160 GPM (61–610 lpm)

ROTATING VANE

This flowmeter has spring-loaded vanes that seal increments of liquid (Figure 17b) between the eccentrically mounted rotor and the casing and transport it from the inlet to the outlet where it is discharged due to the decreasing volume. This type of meter is the most widely used in the petroleum industry and is used for such varied service as gasoline and crude oil metering, with ranges from a few GPM of low-viscosity clean liquids to 17,500 GPM (66.1 m³/m, or 25,000 bbl/hr) of viscous particle-ladened crudes. Accuracies of ±0.1% of actual flow are normal, and ±0.05% has been achieved in the larger meters.

This instrument is built from a variety of materials of construction and can be used for fairly high temperature and pressure service. Upper limits are approximately 350°F and 1000 PSIG (177°C and 6.9 MPa).

Another rotary design is illustrated in Figure 17c. Here an abutment rotor operates in timed relation with two displacement rotors and at half their speed.

OSCILLATING PISTON

The moving portion of the oscillating piston meter consists of a slotted cylinder that oscillates about a dividing bridge that separates the inlet port from the outlet port. Spokes connect this cylinder to a pin located on the axis of the cylinder. As the cylinder oscillates about the bridge (Figure 17d) the pin makes one rotation per cycle. This rotation is transmitted to the gear train and registers either directly or magnetically through a diaphragm. This meter, in addition to being in common usage for the measurement of domestic water, has the capability of handling clean viscous and/or corrosive liquids. Inaccuracies are in the range of ±1% of actual flow. This type of flowmeter is normally used in small pipe lines (2 in./50 mm or smaller) to measure low flow rates.

Metering accuracies are increased by reducing the clearance spaces to 0.002 in. (5 microns). Such clearances do necessitate filtering the entering fluid in order to remove larger particulates. The cases are usually made of cast iron,

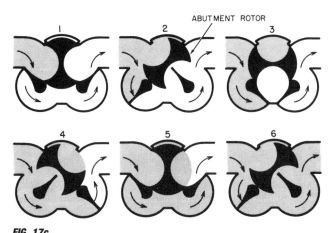

FIG. 17c
Six-phase metering cycle of rotary displacement-type flowmeter.

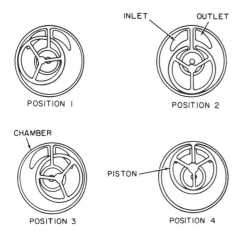

FIG. 17d
Oscillating piston meter.

bronze, or steel, while the chamber and piston materials are usually bronze, aluminum, and Ni-Resist. Iron and bronze meters are good for up to 150 PSIG (1034 kPa) and 200°F (93°C), while steel meters can be used up to 400 PSIG (2760 kPa) and 300°F (149°C).

RECIPROCATING PISTON

The oldest of the positive displacement meters, this meter is available in many forms—multi-piston meters, double-acting piston meters, rotary valves, horizontal slide valves. Figure 17e shows a schematic of one variety of this meter. A crank arm actuated by the reciprocating motion of the pistons drives the register. These meters are widely used in the petroleum industry and can achieve accuracies of ±0.2%.

Another version of this meter is shown in Figure 17f. The liquid enters the cylinder on the left forcing the piston down

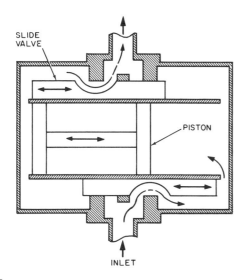

FIG. 17e
Reciprocating piston meter.

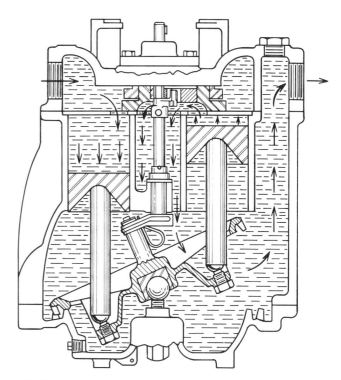

FIG. 17f
Cutaway of reciprocating piston meter with two opposing pistons.

through lever action of the control plate. The piston on the right is forced up, discharging liquid through the port into the inner portion of the valve, down through the center of the meter, and out the meter discharge.

ROTATING LOBE AND IMPELLER

In this type of meter, two lobed impellers, which are geared together to maintain a fixed relative position, rotate in opposite directions within the housing (Figure 17g). A fixed volume of liquid is displaced for each revolution. A register is geared to one of the impellers. They are normally built for service in pipe sizes from 2 to 24 in. (50 to 610 mm) and their maximum capacities range from 8 to 17,500 GPM (30.4 to 66,500 l/m).

The advantages of this design include good repeatability (0.015%) at high flows, the availability of a range of materials of construction, and high operating pressures (1200 PSIG, or 8300 kPa) and temperatures (400°F, or 205°C). The

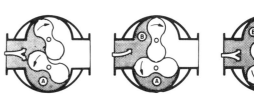

FIG. 17g
Rotating lobe meter.

FIG. 17h
The rotating impeller flowmeter. (Courtesy of Flowdata Inc.)

disadvantages include loss of accuracy at low flows, large size, heavy weight, and high cost.

The rotating impeller design is illustrated in Figure 17h. It has only two moving parts: the two impellers, made out of wear-, abrasion-, and corrosion-resistant thermoplastics. A proximity switch senses the passing magnets implanted in the impeller lobes and transmits the resultant pulses to a counter. Units are available from $1/8$ to 4 in. (3 to 100 mm) sizes with up to 3000 PSIG (21 MPa) pressure and 0 to 400°F (205°C) temperature ratings. The design is suited for high-viscosity operation and the claimed accuracy and rangeability are also high.

OVAL-GEAR FLOWMETERS

A special variety of the rotating lobe flowmeter is the oval-geared metering elements. In this design, as shown in Figure 17i, a precise volume of liquid is captured by the crescent-shaped gap formed between the housing and the gear. This volume is then carried to the outlet, while the gears rotate an output shaft which operate the register. If slippage between the oval-gears and the housing is small, and flow rate and viscosity are high (over 1 GPM and 10 cps, respectively), these flowmeters can provide high accuracies (0.1%). At lower flows the relative proportion of the "slip" leakage increases and therefore accuracy drops to about 0.5%. Viscosity also affects the slip flow. If a meter is calibrated on a fluid with a viscosity of 1 cps, it will read high by 1.2% if the viscosity rises to 100 cps.

These flowmeters are available in sizes from $1/4$ to 16 in. (6 to 406 mm). When the viscosity of the process fluid is between 1.5 and 10 cps they can handle flow ranges from 0.05 to 0.5 up to 250 to 5000 GPM (from 0.2 to 2 up to 950 to 19,000 lpm). Their materials of construction includes brass, carbon steel, 316 stainless steel, and Alloy 20. Operating pressures are available up to 1450 PSIG (10 MPa) and operating temperatures up to 560°F (293°C).

The servo version of this meter has been introduced to completely eliminate slip leakage in smaller sizes (0.2 to 40 GPH, or 0.8 to 150 lph). In this design the servomotor drives the oval-gear elements at a speed which eliminates the pressure drop across the meter and keeps the outlet pressure the same as the inlet. This eliminates the motivating force which causes the slip flow and therefore increases accuracy at low flows or under variable viscosity conditions.

VISCOUS HELIX

The helix flow transducer (Figure 17j) is a positive displacement device utilizing two uniquely nested, radically pitched helical rotors as the measuring elements. Close machining tolerances ensure minimum slippage and thus high accuracy. The design of the sealing surfaces provides the optimum

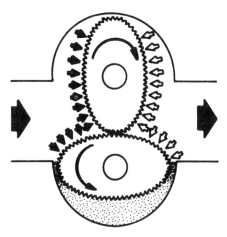

FIG. 17i
Oval-gear flowmeter. (Courtesy of Brooks Instrument, Division of Emerson Electric)

FIG. 17j
Viscous helix flowmeter. (Courtesy of Fluidyne Instrumentation)

ratio of longitudinal to lateral sealing for minimum pressure drop, especially with high-viscosity liquids.

The large inlet size of the progressive cavity allows for the passage of gels, fines, agglomerates, and even undissolved or hydraulically conveyed solids. The meter can measure flow rates from 0.5 to over 4000 GPM (2 to 15,000 lpm). This flow sensor is available in sizes from 1.5 in. up to 10 in. (38 to 250 mm) and can operate at temperatures up to 600°F (315°C) and at pressures up to 3000 PSIG (21 MPa). It is a high pressure drop device requiring at least 10 PSIG (69 kPa) for its operation at full flow. Its turndown can reach 100:1, while its metering inaccuracy is claimed to be under 0.5% of actual flow.

Available design variations include versions that are heated to maintain line temperatures for meltable solids or polymer services. Also available are units with sanitary construction. This meter is suited for high-viscosity (over 1000 centipoises) and slurry services. The pocketless straight-through design serves to simplify cleanout.

It is recommended to filter the process fluid before it enters the flowmeter. U.S. mesh size 30 is sufficient for that purpose.

HIGH-PRECISION DISPLACEMENT FLOWMETER

For the high-precision measurement of fuel and alcohol flows in engine and carburetor test rigs and other applications, positive displacement flowmeters are often used. Their high precision and high rangeability is achieved by eliminating the pressure drop and thereby eliminating the slip or leakage flows. This is achieved by providing a motor drive for the displacement element and using it to introduce as much pumping energy as is needed to keep the pressures at the inlet and outlet of the meter equal to each other (Figure 17k). This flowmeter uses a high-sensitivity piston to detect the pressure differential and photoelectrical sensors to detect the position of the piston. The flowmeter is also provided with a variable-speed controller which adjusts the drive speed whenever the pressure differential is other than zero. Because the response time of the system is less than 0.5 seconds, the flowmeter is able to follow most dynamic flow transients or can be used on short-duration tests.

This flowmeter is claimed to provide a reading with only 0.25% error over a 50:1 range and a 0.5% error over a 100:1 range. The meter is designed for ambient operating temperatures and up to 150 PSIG (10 bars) operating pressures. The different models of this flowmeter can detect diesel, gasoline, or alcohol flows from 0.04 to 40 GPH (0.15 to 150 lph). Because vapor lock is a common problem in fuel flow metering, the unit is provided with a vapor separator.

ACCESSORIES AND INTELLIGENT ELECTRONICS

Standard positive displacement meter accessories include strainers, air release assemblies which remove all the vapors from the flow stream before it enters the meter, automatic batch shutoff valves provided with two-stage closure for full and dribble flow operation, temperature compensators, manual and/or automatic ticket printers, and pulse generators for remote indication, totalization data monitoring, and/or control. Pneumatic pulse generators are also available for interfacing with pneumatic batch controllers. In addition to the totalizer-type digital readout registers, flow rate indication can also be provided. Impulse contactors are also available to actuate predetermining counters, or to serve as electrical interlocks which actuate flow ratio systems, pumps, valves, solenoids, alarms, printers, sampling devices, etc.

The intelligent positive displacement meters are usually provided with solid-state pulse transmitters. These units play a vital role in the metering of household utilities. There is substantial economic justification for substituting a telemetering system, operated either on the telephone lines or by radio, for the current system (human meter readers). It is also feasible to combine the readings of the electric, the water, and the gas meter of a household into a single transmitter and to transmit that information to the appropriate utilities without the need for an individual meter reader to visit the home or apartment. The economic advantage of this method of metering is not only in the saving of labor, but also in the speed and frequency at which the data is available for billing.

PROVERS

All flowmeters that consist of moving and stationary parts that are rubbing against each other (positive displacement or turbine flowmeters) require periodic recalibration. This is necessary because the clearance space and the slip or clearance flow through it increases with wear. Recalibration can be done by removing the flowmeter from the pipeline and sending it to a calibration laboratory, or it can be done inline. The flow provers that allow for inline recalibration without interruption of the process flow are described below.

As shown in Figure 17l, provers consist of a smooth-walled, precalibrated displacement chamber and a barrier piston within it. Usually a follower rod is attached to the back side of the piston and is connected to position sensors.

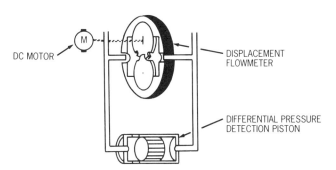

FIG. 17k
High-precision positive displacement flowmeter. (Courtesy of Pierburg Luftfahrtgerate Union GmbH)

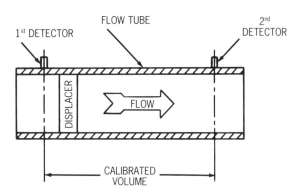

FIG. 17l
Prover operation.

Dividing the volume of the prover by the time it takes to displace its volume gives the calibrated flow rate. This calibrated flow rate is compared to the reading of the flowmeter which is being calibrated.

In order to minimize the disturbance to the process flow, inline ballistic flow provers have been developed. In these units (Figure 17l) the piston is constructed so that it will not disrupt the flow in the line. Therefore, the prover can be permanently installed in an operating pipeline, upstream or downstream of the flowmeter which is being calibrated. The poppet valve within the piston assembly allows for the piston to be withrawn to the start position after a calibration run while the process flow continues undisturbed. The provers can be portable (Figure 17m) or permanently installed, and the calibration cycle can be initiated manually or automatically.

The repeatability of provers is around 0.02% of the actual flow if the seals are tight. It is recommended to periodically check the seals by closing a tight shutoff valve downstream of the prover and applying nitrogen pressure to the upstream face of the piston. If this results in any movement of the piston, the seals need maintenance. Provers are available for up to 3000 PSIG (21 MPa) operating pressure and 165°F (74°C) operating temperature; they can detect flow rates from 0.001 GPM (0.004 lpm) to 20,000 GPM (75,000 lpm). The calibrated displacement volume of provers can range from a fraction of a gallon to several hundred gallons. Large provers can fit on truck beds or trailers (Figure 17m).

POSITIVE DISPLACEMENT FLOWMETER SUMMARY

As a class, liquid positive displacement meters are one of the most widely used instruments for measuring volumetric flow when the fluid is bought and sold on a contract basis. As a result, a large variety of meters, covering a broad spectrum

Skid mounting for truck bed use

Vertical Mount

Skid mounting for ease of portability

Skid/Trailer use
Two Axle

FIG. 17m
Portable prover assemblies. (Courtesy of Brooks Instrument, Division of Emerson Electric)

of requirements, are available. Their good accuracy and ready availability warrant their primary consideration when selecting a volumetric meter.

These flowmeters are especially useful when the fluid to be measured is free of any entrained solids. A typical example is the measurement of water delivered to homes, factories, office buildings, etc.

Wear of parts introduces the major source of error over meter service life. Leakage error increases with lower viscosity fluids but remains relatively constant with time. In the large meters temperature variations and the resulting change in fluid density and viscosity must be taken into consideration.

Positive displacement meters provide good accuracy ($\pm 0.25\%$ of flow) and high rangeability (15:1). They are repeatable to $\pm 0.05\%$ of flow. Some designs are suited for high or variable viscosity services. They require no power supplies and are available with a wide variety of readout devices. Their performance is virtually unaffected by upstream piping configuration. Positive displacement meters are excellent for batch processes, mixing, or blending applications.

These meters are simple and easy to maintain by regular maintenance personnel using standard tools. No specially trained crews or special calibration instruments are needed.

Positive displacement meters require relatively expensive precision-machined parts to achieve the small clearances upon which their accuracy depends. From this it follows that the liquids metered must be clean, for wear rapidly destroys accuracy. Contaminant particle size must be kept below 100 microns, and most of these meters are not adaptable to the metering of slurries. Because of the moving parts, maintenance is required at frequent intervals; where corrosive liquids are metered this may result in high costs. Due to close tolerances the moving components are subject to wear, and therefore the meter requires periodic recalibration and maintenance. Positive displacement flowmeters are expensive in larger sizes or in special materials. They can be damaged by overspeeding and can require high pressure drops. In general, they are not suited for dirty, nonlubricating, or abrasive services.

Bibliography

Barnes, G., "Pipeline Metering with Liquid Positive Displacement," *Advances in Instrumentation, Vol. 39,* ISA, 1984.

Blasso, L., "Flow Measurement Under Any Conditions," *Instruments and Control Systems,* February 1975.

Clark, W.J., "Flow Measurement," Elmsford, New York: Pergamon Press, 1967.

Hall, J., "Solving Tough Flow Monitoring Problems," *Instruments and Control Systems,* February 1980.

Hayward, A.J., "Choose the Flowmeter Right for the Job," *Processing Journal,* 1980.

Hendrix, A.R., "Positive Displacement Flowmeters," *InTech,* December 1982.

Laskaris, E.K., "The Measurement of Flow," *Automation,* 1980.

Lomas, D.J., "Selecting the Right Flowmeter," *Instrumentation Technology,* 1977.

"Positive Displacement Flowmeters," *Measurements and Control,* October 1991.

Spink, L. K., *Principles and Practices of Flow Engineering,* 9th edition, The Foxboro Company.

Watson, G.A., "Flowmeter Types and Their Usage," *Chartered Mechanical Engineering Journal,* 1978.

18 Purge Flow Regulators

E. L. SZONNTAGH (1993)

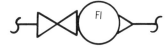

ROTAMETER W/VALVE

PURGE FLOW REGULATOR
Flow Sheet Symbol

Applications:	Low flow regulation for air bubblers, for purge protection of instruments, for purging electrical housings in explosion-proof areas, and for purging the optical windows of smokestack analyzers
Purge Fluids:	Air, nitrogen, and liquids
Operating Pressure:	Up to 450 PSIG (3 MPa)
Operating Temperaure:	For glass tube up to 200°F (93°C)
Ranges:	From 0.01 cc/min for liquids and from 0.5 cc/min and higher for gases. A ¼ in. (6 mm) glass tube rotameter can handle 0.05 to 0.5 GPM (0.2 to 2 lpm) of water or 0.2 to 2 SCFM (0.3 to 3 cmph) of air
Inaccuracy:	Generally 2 to 5% of range (laboratory units are more accurate)
Costs:	A 150 mm glass-tube unit with ⅛ in. (3 mm) threaded connection, 316 stainless steel frame, and 16-turn high-precision valve is $260; the same with aluminum frame and standard valve is $100. Adding a differential pressure regulator of brass or aluminum construction costs about $150 (of stainless steel, about $500). For highly corrosive services, all-Teflon, all-PTFE, all-PFA, and all-CTFA units are available which, when provided with valves, cost $550 with ¼ in. (6 mm) and $1300 with ¾ in. (19 mm) connetions
Partial List of Suppliers:	Aaborg Instruments & Controls Inc.; Blue White Industries; Brooks Instrument, Div. of Rosemount; Fischer & Porter Co.; Fisher Scientific; Flowmetrics Inc.; ICC Federated Inc.; Ketema Inc. Schutte and Koerting Div.; Key Instruments; King Instrument Co.; Krone America Inc.; Matheson Gas Products Inc.; Omega Engineering Inc.; Porter Instrument Co. Inc.; Scott Specialty Gasews; Wallace & Tiernan Inc.

Purge flows are low flow rates of either gases or liquids. They usually serve to protect pressure taps from contacting hot or corrosive process fluids or from plugging. They can also protect electrical devices from becoming ignition sources by maintaining a positive inert gas pressure inside their housings or to protect the cleanliness of the optics of analyzers through purging.

The low flow rates of the purge media can be detected by a variety of devices including capillary, miniature orifice, metering pump, positive displacement, thermal, and variable-area-type sensors. Most of these devices have already been described in other parts of this volume. Capillary flow elements (Section 2.8) are ideal for the measurement of low flow rates and are frequently combined with thermal flowmeters to provide a high-precision, high-rangeability, but also higher cost flow regulator (Figure 12e). Integral orifices (Figure 14n) can be used on both gas and liquid flow measurement, while positive displacement meters are most often used to detect the flow of liquids (Figures 17h to 17k). The purge flow of liquids can be both measured and controlled by metering pumps (Figure 13a). In addition, *Process Control (Instrument Engineers' Handbook,* third edition) includes a complete section devoted to flow regulators.

There is only one type of purge flow regulator that has not been covered in other parts of this reference set: the rotameter-type purge meter. This is the least expensive and most widely used purge meter, and for these reasons it is described in this separate section.

PURGE ROTAMETERS

These are perhaps the most widely used flowmeters and certainly are the most widely used form of the rotameter. These meters take many forms, all of which are inexpensive, and are intended for low flow measurement. Most purge meters are selected to handle inert gases or liquids at low flow rates where these fluids are used as a purge; therefore, accuracy is not critical. Repeatability is normally the required performance characteristic. Purge meters are available with optional needle control valves. Figure 18a shows a typical purge-type rotameter with integral needle control valve.

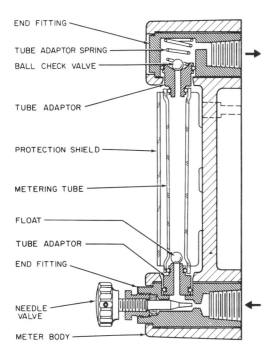

FIG. 18a
Purge rotameter with integral needle valve.

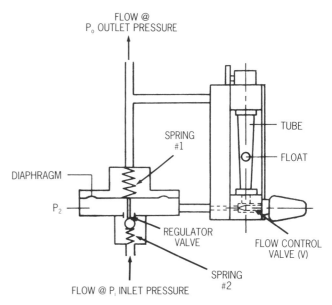

FIG. 18c
Purge flow regulator consisting of a glass tube rotameter, an inlet needle valve, and a differential pressure regulator. (Courtesy of Krone America Inc.)

The metering needle valves are usually multiple-turn units provided with long stems. The opening around their needle-shaped plugs can approach capillary dimensions. The flow rate through these devices is a function of not only the opening of the valve and the pressure differential across it, but also of both the density and the viscosity of the purge media. Table 8b provides information on the density and viscosity of a number of purge gases. Figures 18b shows a high-precision needle valve which is provided with a vernier-type scale that allows a more accurate setting of the valve opening. The dual scale increases the precision and reproducibility of setting by subdividing the smallest reading of the first scale onto the second.

When the purge flowmeter is combined with a differential pressure regulator, it becomes a self-contained flow controller (Figure 18c). By adjusting springs #1 and #2 for a constant pressure difference of about 60 to 80 in. of water (150 to 200 cm of water) this constant pressure drop ($P_2 - P_o$) is maintained across the flow control valve (V) and the purge flow is thereby fixed. Figure 18c describes a configuration in which the outlet pressure (P_o) is constant and the inlet pressure P_i is variable. Units are also available for bubbler and purge applications where the inlet pressure P_i is constant and the outlet P_o is variable. In that case the constant pressure drop across the valve (V) is maintained to equal ($P_i - P_2$). Purge flow controllers on gas service are usually provided with a range of 0.2 to 2 SCFH (6 to 60 slph), an accuracy of 5% of full scale over a range of 10:1, a pressure rating of 150 to 300 PSIG (1 to 2 MPa), and a maximum temperature limit of 212 to 572°F (100 to 300°C).

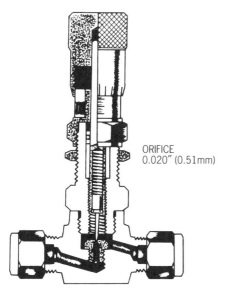

FIG. 18b
Fine-adjustment needle valve with vernier scale. (Courtesy of Swagelok Co.)

Bibliography

Blasso, L., "Flow Measurement Under Any Conditions," *Instruments and Control Systems,* February 1975.

Cheremisinoff, N.P., *Fluid Flow,* Ann Arbor, MI: Ann Arbor Science Publishers, 1982.

Cross, D.E., "Rotameter Calibration Nomograph for Gases," *Instrumentation Technology,* April 1969, pp. 53–56.

Des Marais, P.O., "Variable-Area Meter for Viscous Service," *Instruments and Control Systems,* August 1961.

Hall, J., "Solving Tough Flow Monitoring Problems," *Instruments and Control Systems,* February 1980.

Instrument Society of America, Recommended Practices RP16.1, RP16.2, RP16.3, RP16.4, RP16.5, and RP16.6. [These documents deal with the terminology, dimensions, installation, operation, maintenance, and calibration of rotameters.]

Lomas, D.J., "Selecting the Right Flowmeter," *Instrumentation Technology,* May 1977.

Polentz, L.M., "Theory and Operation of Rotameters," *Instruments and Control Systems,* June 1961.

"Rotameters/Variable-Area Flowmeters," *Measurements and Control,* September 1991.

Sydenham, P.H., et al., *Introduction to Measurement Science and Engineering,* Chichester, England: John Wiley & Sons, 1989.

19 Segmental Wedge Flowmeter

B. G. LIPTÁK (1993)

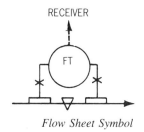

Flow Sheet Symbol

Applications:	Clean, viscous liquids or slurries, and fluids with solids
Sizes:	1 to 12 in. (25.4 to 305 mm) diameter pipes
Designs:	For smaller sizes (1 and 1.5 in.) the wedge can be integral; for larger pipes remote seal wedges are used with calibrated elements
Wedge Opening Height:	From 0.2 to 0.5 of pipe inside diameter
Pressure drops:	25 to 200 in. H_2O (6.2 to 49.8 kPa)
Materials of Construction:	Carbon or stainless steel element; stainless or Hastelloy C seal; special wedge materials like tungsten carbide are available
Design Pressure:	300 to 1500 PSIG (20.7 to 103 bars) with remote seals
Design Temperature:	−40 to 700°F (−40 to 370°C) but have been used in high-temperature processes up to 850°F (454°C).
Inaccuracy:	Elements are individually calibrated; d/p cell error contribution to total measurement inaccuracy is 0.25% of full scale. The error over a 3:1 flow range is usually not more than 3% of actual flow
Cost:	A 3 in. (75 mm) calibrated stainless steel element with two stainless steel chemical tees and with an electronic d/p transmitter provided with remote seals is about $3500
Partial List of Suppliers:	ABB Kent-Taylor Inc.

The segmental wedge flow element provides a flow opening similar to that of a segmental orifice, but the obstruction to flow is less abrupt (more gradual), and its sloping entrance makes the design similar to the flow tube family. It has been designed primarily for use on slurries. Its main advantage is its ability to operate at low Reynolds numbers. While in the cases of sharp-edged orifices, venturis, or flow nozzles the square root relationship between flow and pressure drop necessitates a Reynolds number well above 10,000 (Figure 14b), for segmental wedge flowmeters, a Reynolds number of 500 or 1000 is enough.[1] For this reason it can measure flows at low flow velocities and when the process fluid is viscous. In that respect it is similar to conical or quadrant edge orifices.

For pipe sizes under 2 in. (50 mm) the segmental wedge flow element is made by cutting a V-notch into the pipe and welding a solid wedge accurately in place (Figure 19a). In sizes over 2 in., the wedge is fabricated from two flat plates that are welded together before insertion into the spool piece (Figure 19a). On clean services regular pressure taps are used that are equidistant from the wedge (Figure 19a), while on applications where the process fluid contains solids in suspension, "chemical tees" are added upstream and downstream of the wedge flow element (Figure 19b). The chemical seal element is flush with the pipe, thereby eliminating pockets and making the assembly self-cleaning. The seals are made of corrosion-resistant materials and are also suited for high-temperature services. Some users have reported applications on processes at 3000 PSIG (210 bars) and 850°F (454°C).[2]

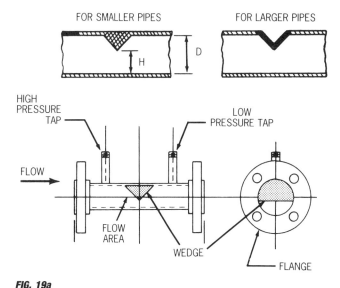

FIG. 19a
The segmental wedge flowmeter designed for clean fluid service.[1,2]

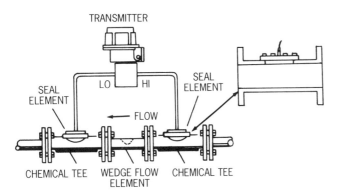

FIG. 19b
Segmental wedge flowmeter designed for corrosive or slurry service.[1]

The segmental wedge flowmeters are calibrated on water. The pressure drop detected by the d/p transmitter is interpreted based on these calibration curves, and users report actual readings within 2.5 and 3.5% in these calibration curves of slurry services.[2] The flow capacities of these sensors are given in Table 19c.

The segmental wedge flowmeter fills the need for corrosion-resistant slurry flowmeters operating at higher process temperatures when the accuracy and rangeability expectations are not high.

TABLE 19c
Segmental Wedge Flowmeter Capacities in GPM Units* (courtesy of ABB Kent-Taylor)

Pipe Size	H/D†	Approximate Differential Pressure Inches H_2O					
		20 in.	40 in.	60 in.	100 in.	120 in.	160 in.
1 in.	0.2	3.43	4.80	5.90	7.66	8.40	9.70
	0.3	5.75	8.14	9.95	12.9	14.1	16.3
	0.4	8.30	11.7	14.4	18.6	20.4	23.4
	0.5	11.0	15.6	19.1	24.6	27.0	31.2
1-1/2 in.	0.2	6.80	9.70	11.9	15.8	16.8	19.4
	0.3	12.8	18.2	22.2	28.7	31.4	36.4
	0.4	22.7	32.1	39.3	50.7	55.5	64.2
	0.5	32.2	45.5	55.8	72.0	79.0	91.0
2 in.	0.2	12.2	17.2	21.1	27.2	29.9	34.6
	0.3	20.3	28.7	35.2	45.4	49.7	57.5
	0.4	34.0	48.0	58.9	76.0	83.4	96.3
	0.5	50.9	72.0	88.0	114.0	125.0	144.0
3 in.	0.2	26.4	37.4	45.6	59.0	64.6	74.6
	0.3	44.5	62.5	77.0	99.5	109.0	126.0
	0.4	75.5	107.0	131.0	169.0	185.0	214.0
	0.5	113.0	160.0	196.0	252.0	277.0	320.0
4 in.	0.2	49.5	70.0	86.0	111.0	121.0	140.0
	0.3	76.1	108.0	132.0	170.0	187.0	216.0
	0.4	127.0	180.0	220.0	284.0	311.0	360.0
	0.5	192.0	272.0	332.0	430.0	470.0	544.0
6 in.	0.2	86.4	122.0	150.0	193.0	212.0	244.0
	0.3	185.0	262.0	320.0	414.0	454.0	524.0
	0.4	294.0	416.0	509.0	657.0	720.0	831.0
	0.5	444.0	628.0	768.0	994.0	1089.0	1255.0
8 in.	0.2	173.0	244.0	300.0	388.0	425.0	490.0
	0.3	311.0	440.0	539.0	695.0	761.0	880.0
	0.4	475.0	671.0	824.0	1060.0	1165.0	1340.0
	0.5	659.0	930.0	1140.0	1470.0	1610.0	1860.0

*The units in the table can be converted as follows: 1.0 in. H_2O = 249 Pa, 1.0 GPM = 3.785 lpm, 1.0 inch = 25.4 mm.
†The H/D values shown above represent ratios between the segmental opening height and the pipe diameter (Fig. 19a).

References

1. Owen, R.E., "Segmental Wedge Flow Element," An ABB Kent-Taylor Publication.
2. Malone, D.P., "Slurry Flow Measurement: A Case History," *InTech*, November 1985

Bibliography

De Boom, R.J., "Flow Meter Evaluation," 1991 ISA Conference, Paper #91-0509.

"Differential Pressure Flowmeters," *Measurements and Control*, September 1991.

Hall, J., "Flow Monitoring Application Guide," *Instruments and Control Systems*, February 1983.

Husain, Z.D., "Flowmeter Calibration and Performance Evaluation," 1991 ISA Conference, Paper #91-0508.

Krigman, A., "Flow Measurement: Some Recent Progress," *InTech*, April 1983.

Krigman, A., "Guide to Selecting Non-Intrusive Flowmeters," *InTech*, December 1982.

Rusnak, J., "The Fundamentals of Flowmeter Selection," *InTech*, April 1989.

20 Sight Flow Indicators

D. S. KAYSER (1982) **B. G. LIPTÁK** (1993)

Flow Sheet Symbol

Design Pressure:	To 600 PSIG (4 MPa) standard
Design Temperature:	To 450°F (230°C) standard
Materials of Construction:	Glass, plastics, bronze, iron, carbon steel, stainless steel, Monel, Hastelloy, Carpenter 20 body materials; Buna N, Vitou A, neoprene, polyethylene, polypropylene, or Teflon seals; and tempered, annealed, acrylic, quartz or polycarbonate windows
Sizes:	¼ to 16 in. (12 to 406 mm)
Cost:	$100 to $200 per inch diameter for bronze or ductile iron, $200 to $400 for steel and stainless steel
Partial List of Suppliers:	W.E. Anderson Inc. Div. of Dwyer Instr.; Brooks Instrument, Div. of Rosemount; Dover Corp., OPW Div.; ERDCO Engineering Corp.; Ketema Schutte and Koerting Div.; Krone America Inc.; J.G. Papailias Co.; Penberthy Inc.; Schutte and Koerting Div. of KETEMA; Tokheim Corp.

Sight flow indicators are used when a visual inspection of the process is necessary. Figure 20a shows three designs. The flapper design is used in transparent or slightly opaque solutions and in gas services. Flow direction is vertically up or horizontal. Some indication of flow variations may be made by observing the position of the flapper (Figure 20b). Bidirectional flappers are also available. In these designs the flapper is hinged in the center of the pipe (Figure 20c). The drip-tube design is used where the flow is vertically down and is particularly suited for intermittent flows. The paddle design is used on dark process fluids since the motion of the paddle can be easily detected. Flow through the paddle design can be vertical or horizontal.

The paddle or propeller design is limited to higher flows, because low flows will not turn the propeller. At high flows the pressure drop of a paddle-type sight flow glass is much higher than that of a drip to be or flapper unit (Figure 20d).

Figure 20e shows the cross section of a sight flow indicator that has flanged connections. The assembly consists of the chamber, glasses, gaskets, covers, and bolts. It is similar in many respects to the transparent level gauge discussed in Chapter 3. The glass is normally a borosilicate which can be rated to 450°F (230°C) for sight glass applications and has good resistance to mechanical and thermal shock. The glasses can also be made of silica glass or quartz, thus enabling temperature ratings in excess of 1000°F (535°C).

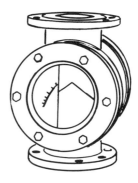

FIG. 20b
Flapper-type sight glass provided with a scale for approximate flow indication. (Courtesy of Turbo Instruments Inc.)

FIG. 20c
Bidirectional flapper. (Courtesy of Dover Corp., OPW Div.)

FLAPPER   DRIP   PADDLE

FIG. 20a
Various designs of sight flow indicators.

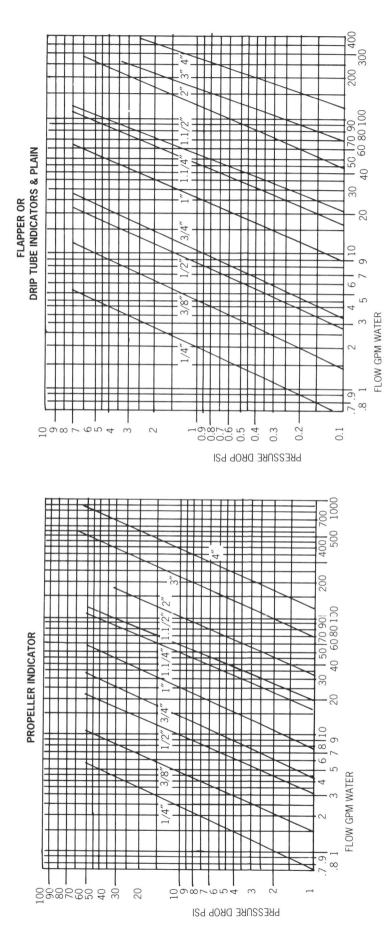

FIG. 20d Sight flow indicator pressure drops on water applications. (Courtesy of Dover Corp., OPW Div.)

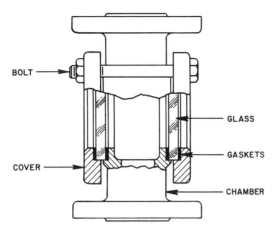

FIG. 20e
Cross-section of sight flow indicator.

Tempered glass is not recommended for fluorine, hydrofluoric acid, or phosphoric anhydride service, because the erosion causes uneven stresses and eventual failure. Annealed glass is a better choice because it signals the approach of failure by turning cloudy. Quartz and polycarbonate can be used instead of glass. Acrylics can also be used if high impact strength and abuse-resistance is required, but they scratch easily.

The flow glass chamber can be obtained in a wide variety of materials and can also be furnished with a number of different plastic liners. This enables a properly specified sight flow indicator to be used in almost all corrosive services. The bolts and covers are normally steel and iron, respectively, but these too may be obtained in different materials depending on corrosion and temperature requirements.

Figure 20f shows the cross-section of a double window assembly. This assembly improves the safety of a sight flow indicator in several ways. In high-temperature services the gradient across each glass is reduced. The outer glass protects the inner glass from thermal shock caused by splashes of cold water. If the outer or inner glass breaks there is a chance that the remaining glass will contain the process until the assembly can be repaired. On steam service mica shields, on hazardous or toxic fluids, abuse and vandalism protection shields are recommended.

Special designs are available that can be used in sanitary services such as food processing. Accessories such as jacketed indicators, illuminators, and spray rings for cleaning the glass in place are also available.

Sight flow indicators offer an inexpensive means of viewing the process to assure that the stream is flowing or to note process characteristics such as color, tubidity, or other properties that might indicate process deterioration or equipment malfunction. In fact, several types of analyzers require flow glasses so that the operator can see the probe or other sensing element in the process. They are also used in secondary services, such as condensate pot installations. Nonetheless, their use is limited in the industrial process area. It is difficult to estimate flow rate through a sight glass and a hazard is created if the glass breaks. They are used more commonly on utility services outside of the industrial processing area.

Bibliography

Green, C.R., "Tank Sight Glasses," *Chemical Engineering,* September 25, 1978.

Sunderhous, C.A., "Sight Indicators Allows Positive Flow Check," *Machine Design,* October 24, 1985.

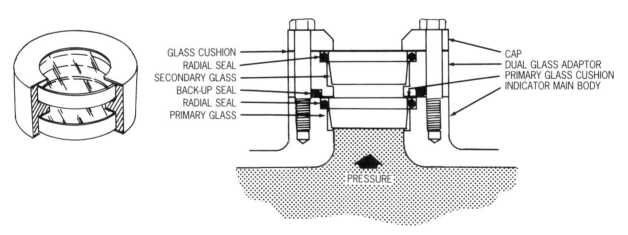

FIG. 20f
Cross-sections of double window assembly. (Courtesy of Dover Corp., OPW Div.)

21 Solids Flowmeters and Feeders

R. SIEV (1969)   **D. C. MAIR** (1972)   **B. G. LIPTÁK** (1982,1993)

BELT TYPE

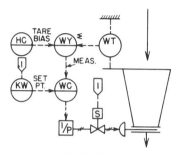

LOSS IN WEIGHT

Flow Sheet Symbol

Types of Designs:	A. Accelerator
	B. Belt-type gravimetric
	C. Volumetric, capacitance
	D. Impulse or impact
	E. Loss-in-weight
	F. Switch (Section 6)
	G. Dual chamber
	H. Cross-correlation (Section 4)
	I. Nuclear
Capacities:	A. 1000 to 80,000 lbm/hr (450 to 36,000 kgm/hr)
	B. Up to 180,000 lbm/hr (80,000 kgm/hr) or up to 3600 ft^3/hr (100 m^3/hr)
	C. Up to 3600 ft^3/hr (100 m^3/hr)
	D. 3000 to 3,000,000 lbm/hr (1400 to 1,400,000 kgm/hr)
	E. Determined by hopper or duct size
	F. Unlimited on-off
	G. 1000 to 300,000 lbm/hr (450 to 140,000 kgm/hr)
	H. Unlimited
	I. Same as B
Costs:	$600 to $1200 (F)
	Around $3000 (C)
	$3000 to $5000 (A,D)
	$5000 to $15,000 (B,H)
	Over $15,000 (E,G,I)
Inaccuracy:	±½% of rate over 10:1 range (B—digital, G)
	±½ to ±1% of full scale (I)
	±1% of rate over 10:1 range (E)
	±1 to ±2% of full scale (D)
	±2 to ±3% of full scale (A,F)
	±2 to 4% of full scale (C)
Partial List of Suppliers:	ABB Kent Inc. (C); Acrison Inc. (B,E); AM Sensors Inc. (F); Auburn International Inc. (F); Babbitt International Inc. (D); Bailey Controls Co. (D); Cutler-Hammer Products, Thayer Scale Div. (B,D,E); Dezurik, a Unit of General Signal (A); Endress + Hauser Inc. (B,C,D,F,H); Kay-Ray Inc. (I); Kistler-Morse (B); K-Tron Corp. (B); Leeds and Northrup, a Unit of General Signal (E); Merrick Scale Mfg. Co. (B,D); Milltronics Inc. (B,D); Monitor Mfg. (F); Ohmart Corp. (I); Schenck Weighing Systems (E); Technicon Industrial Systems (G); TN Cannon Bear (F); Wallace & Tiernan Inc. (B,G)

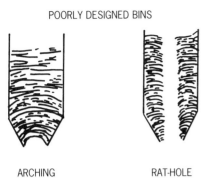

FIG. 21a
Good bin design is a critical element in a successful solids metering installation.

Many types of solids flowmeters are currently available. The majority use some method of weighing, while others utilize a variety of other phenomena, from radiation to impact force determination and from electrical properties to centrifugal force. This section will cover them all except Triboflow (surface charge) switches (see Section 6) and cross-correlation flowmeters (see Section 4).

First to be discussed are the gravimetric devices utilizing some form of weighing; within that category, the belt-type units will be covered first. Before the belt-type gravimetric mass flowmeters are discussed, however, the feeders that bring the solids from the storage vessel onto the belt will be described. Because the keeping of solids in motion and the prevention of arching and rat-holing in the supply bins (Figure 21a) is a serious problem, the description of feeders will be preceded by yet another topic, feeder accessories.

The bin, the feeder, and the solids flowmeter should be designed in an integrated manner, taking into account the characteristics (density, particle size, moisture content, temperature, or hazardous properties) of the solids. For example, the bed depth on a belt must be less than the height of the skirts (to avoid spillage), but it must be at least three times the maximum lump size in order to guarantee good solids flow. Coarse materials (+60 mesh) or wet ores are likely to bridge or rat-hole in the bin and require vibrators and special feeders. Similarly aerated, dry and fine solids (−200 mesh) are likely to either free flow or be compacted and thereby plug the standard rotary vane or screw feeders. Changing the pitch or inserting additional flights can alleviate flushing. Vibrators usually also help, although in some cases they might worsen the situation by packing the solids. In general, the addition of high-amplitude and low-frequency vibrators or air pads and the use of mass flow bins (steep walls at 10 to 30 degrees from the vertical) tend to improve material flow.

HOPPERS AND THEIR ACCESSORIES

Figure 21b illustrates a surge hopper which, when located between the storage hopper and feeder inlet, provides a means of deaerating material and thus conditioning it so that it can be fed by a gate-controlled belt feeder without flooding. The material supply to the surge hopper is controlled by bin switches to maintain a level zone in the hopper by on-off control of the hopper supply device. The supply control device may be a rotary vane feeder, screw conveyor, or a knife gate with suitable actuator. If process requirements are such that the required feed rate is constant or nearly so, the bin switches are located to provide a hopper capacity equivalent to about 2 minutes retention time when operating at the design feed rate. If feed rates must be varied, it can be seen that the retention time will vary if the supply control device is directly actuated by the high and low bin switches. Excess retention time may be undesirable in some cases in which the material may compact in the hopper and interrupt the supply to the feeders. In situations where the feed rate will be varied, an adjustable timer is incorporated in the level control circuit to control the running time of the hopper supply device. This timer is controlled by the upper bin switch and shuts off the hopper supply when the material contacts the probe. The shutoff condition is maintained for a predetermined and adjustable period, after which the timer starts the hopper supply system, which continues to feed material to the hopper until the upper bin switch level is reached. In this arrangement, the lower bin switch serves as a low-level alarm actuator and is usually also utilized to shut down the feeder. Shutdown is usually desirable to prevent loss of the plug of material ahead of the belt feeder. With some materials, particularly those that aerate easily, loss of the plug of deaerated material can cause production to be delayed while steps are taken to re-establish the supply of deaerated material. Deaeration of some materials takes place

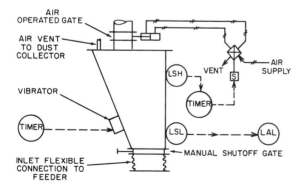

FIG. 21b
Deaerating surge hopper.

in the surge hopper without the need of vibration. Other materials require some assistance through vibration, and hoppers can be furnished with suitable electric or pneumatic vibrators. The frequency and duration of vibration required will vary with the characteristics of the material and for this reason vibrators are supplied with accessory devices to permit adjustment of these variables.

A number of common materials, of which sulfur is an example, will compact unless kept in almost continuous motion, and in some instances will compact even while in motion under the pressure of a relatively low head of material. In these applications, small surge hoppers with bin switches are utilized to maintain a low head of material on the feeder belt. Retention time is in the order of seconds, and external vibration is rarely if ever desirable.

The discharge flow pattern of a belt feeder varies with belt speed and material characteristics. A granular free-flowing material such as sugar will flow smoothly off the belt even at low belt speeds. Other materials having a high angle of repose coupled with a tendency to compact will drop off the end of the belt in lumps, especially at low belt speeds. The feed rate as viewed by the process is thus erratic and will result in short-term blend error in a multifeeder system. The discharge flow pattern can be markedly improved by equipping the feeder with a material distributor. This device consists of a blade located across the full width of the belt at the discharge end of the feeder and vibrated by an electric or pneumatic vibrator. The blade is located so that it almost touches the belt and the material is directed across it. The vibration imparted to the material ribbon causes it to smoothly stream off the belt.

Virtually all manufacturers base their guarantee of performance on sample weight versus setpoint in the case of feeders or sample weight versus totalizer reading in the case of the meters. Sample valves can be provided which facilitate performance checkout procedures. These are available either as a sample tray which can be inserted into the feeder discharge stream for a predetermined period and then weighed, or as a flap valve which diverts the discharge stream from the process duct into a sampling container. The flap-type is generally preferred since the tray-type is only suitable for low feed rates; sampling normally involves the taking of 10 consecutive 1-minute samples and comparing the average sample weight to the setpoint. The flap-type is also faster acting than the tray-type and sample weights are thus more accurate.

Many installations involve feeding directly into processes which may be under low pressure or which may discharge corrosive vapors back through the feeder discharge ducting. If pressures are very low, the feeder can be purged with inert gas or a rotary valve can be installed in the ducting. The rotary valve body should be vented to remove process vapors from the valve pockets prior to their reaching the inlet or feeder discharge side of the valve. If the valve is not vented, blowback resulting from the release of pressure in the rotor pockets can cause discharge flow pattern disturbances and, in extreme cases, affect the feeder weigh section. The valve is vented into a dust or vapor collecting system via a vent port in the side of the valve rotor housing.

Each feeder or meter is usually supplied with a test weight or drag chain which may be used to check the calibration of the device without actually running material. The value of the weight is such that when applied to the weight-sensing mechanism, a full scale output signal will be generated. The test weight is also useful in aligning the control setpoints in multifeeder master-slave systems prior to running any material. With the test weight applied to the weighing section of the master feeder, the resultant output signal is applied to the ratio stations of the slave feeders. The ratio stations can thus be preset to provide the desired slave feeder setpoints.

All manufacturers recommend that a feeder or meter be isolated from sources of vibration, and some include shock mounts with each machine. Inlet and discharge flexible connections to isolate the equipment from vibration and pipe strain in the material inlet and outlet ducting are also recommended.

THROTTLING THE FLOW TO THE FEEDERS

A gravimetric feeder consists of a weight-rate measuring mechanism coupled with a volumetric control device. The vertical gate volumetric regulator, which is perhaps the most popular, is not always suitable for controlling materials having large particle size, which are fibrous or irregularly shaped, or which tend to flow like a fluid due to fine particle size or because they become aerated. Unlike fluids, which exhibit predictable flow behavior, solids flow characteristics are extremely difficult to evaluate on any basis other than an actual trial. For this reason, most manufacturers maintain a test and demonstration facility in which samples of a potential customer's material are fed by test feeders which may be equipped with various volumetric feed sections. Recognizing that a wealth of experience with commonly used materials can very often permit a feed section recommendation without the necessity of testing, it should be noted also that even a minor change in the properties of a material can drastically change its feeding characteristics. These changes might be in the form of a particle size or shape variation, entrainment of air during pneumatic conveying prior to entry to the feeder, or the addition of a relatively minor percentage of an additive to produce a preblend.

Vertical Gate

The vertical gate gravimetric feeder is available in a variety of sizes to produce typical material ribbon widths of 2 to 18 in. (50 to 457 mm) and to regulate up to 6 in. (152 mm) depth of material on the weigh belt. Gate actuators may be electromechanical or pneumatic, or in the form of electric servomotors or stepping motors. Manually adjustable gates are also available. The vertical gate has a typical depth control range of 10:1, and is generally suitable for materials which are not fluidized and which have a particle size not larger than about ⅛ in. (3.175 mm). Larger particles will not flow smoothly under the lip of the gate, thus resulting in an irregular belt load. This may necessitate the excessive dampening of the belt load transmitter output, which will have an undesirable effect on both control accuracy and sensitivity. In addition to producing undesirable control characteristics, rangeability will be decreased as particle size increases. As a rule of thumb, the minimum gate opening should be approximately three times the maximum particle size for materials which have irregularly shaped particles of random size. This factor may be reduced somewhat if the material is homogeneous and particles do not tend to interlock and tumble while in motion (typically, if particle shape approaches that of a sphere).

Rotary Vane

Figure 21c shows a rotary vane feeder with pneumatically controlled variable-speed drive and a belt-type gravimetric meter and controller. The rotary vane feeder is used as a volumetric feed section in instances where the material is aerated or has a low bulk density. Rotary feeders are generally not suitable for handling materials which have large particle sizes, and in some instances are undesirable if process requirements dictate that the material must not be abraded by the feeding device. Like the vertical gate, the rotary vane feeder is not suitable for handling fibrous or stringy materials, because sticky or hygroscopic materials tend to clog the pockets of the rotor. Pocket shape and depth selection are based on required volumetric flow rate and material characteristics. Care must be taken in determining a maximum practical rotor speed. Volumetric capacity of a given arrangement is regulated by rotor speed, but if the speed is too high, rotor pockets won't completely fill as they pass under the material inlet opening and volumetric output may decrease above a certain optimum rotor speed. Essentially, then, the rotary vane feeder is limited to use in connection with free-flowing powders or materials having small particle size but, unlike the vertical gate, it can control low density or aerated materials. The gravimetric meter and rotary feeder should be separately mounted and interconnected by means of a flexible connection to prevent transmittal of vibration from the rotary feeder to the weight-sensing mechanism. Figure 21c also shows a cutoff gate installed in the meter inlet ahead of the weighing section. This device drags across the irregular feed pattern created by a rotary feeder, leveling the ribbon of material and producing a more consistent feed to the weigh section and eventually to the process. The shutoff gate at the feeder inlet permits isolation of the feeder from the material supply when inspecting or servicing the feeder.

It is possible to operate several feeders in parallel or in cascade from the same setpoint, and this mode of operation can be used effectively for blending solids.

Screw Feeder

The feeder element in this device is a screw whose rotary motion delivers a fixed volume of material per revolution (Figure 21d). The screw is located at the bottom of a hopper so that the feed element is always flooded with material. Rotation of the screw discharges material, at one or both ends of the screw, into the receiving vessel. Screws grooved in one direction discharge material at one end only. Screws grooved in opposite directions from the middle deliver material at both ends.

A gravimetric feeding system may also utilize a variable-speed screw feeder to control low-density or aerated materials. The screw section is made as long as is necessary to prevent material from flooding through it. Screw feeders

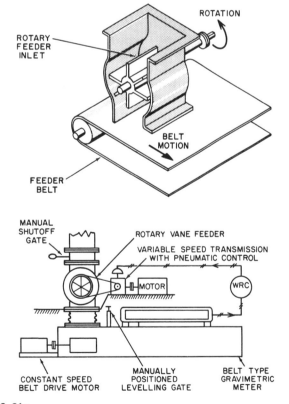

FIG. 21c
Gravimetric feeding system utilizing a rotary valve volumetric feeder controlled by a belt-type gravimetric meter.

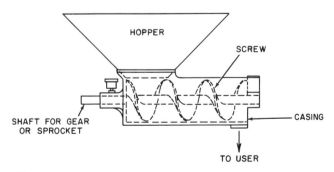

FIG. 21d
Screw feeder.

have also been used successfully to feed fibrous materials and powdered materials which tend to cake. The major advantage of the screw feeder, compared to a rotary vane feeder, is that custom-built screw feeders can be provided with extremely large inlet openings to facilitate entry of fibers and coarse lumps to the conveying screw.

For materials with a tendency to cake or clog the feed screw, the double-ended screw can be provided with a lateral oscillating motion which imparts a cleaning action. In this case material is alternately fed from one end or the other, depending on the direction of lateral motion.

In order to assure an accurate feed, the hopper on the inlet side of the feeder must be designed to provide a uniform supply of material at the feed screw. Vibrators can be added to the hopper to keep the solids agitated and to prevent caking or bridging.

Feeder drives are usually electric motors. With a constant-speed drive, operating feed is adjustable over a 20:1 range by means of a mechanical clutch that varies the operating time per cycle. At 75% feed rate setting, the screw will be operating over 75% of a clutch revolution.

Addition of a variable-speed drive can extend the operating range to 200:1. The variable-speed drive can be electric or mechanical. The electric type will accept any standard milliampere signal; the mechanical type will operate on a 3 to 15 PSIG (0.2 to 1.0 bar) signal.

Vibratory Feeder

Vibratory feeders have been used in gravimetric feeding systems to handle materials consisting of particles which are too large to be handled by screw, rotary vane, or vertical gate feed sections, or in instances where the physical characteristics of the material particles may be adversely affected by passage through other types of volumetric feeding devices. The discharge flow pattern of a vibrating feeder is extremely smooth and thus is ideal for continuous weighing.

The vibratory feeder (Figure 21e) consists of a feed chute (which may be an open pan or closed tube) which is moved back and forth by the oscillating armature of an electromagnetic driver. Material transfer rate to the treatment process can be controlled by adjusting the current input to the electromagnetic driver, which controls the pull of the electromagnet and the length of the stroke. This feature also permits flow rate adjustment or control from a distant control panel.

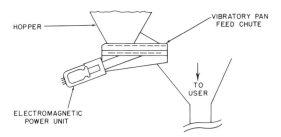

FIG. 21e
Vibratory feeder.

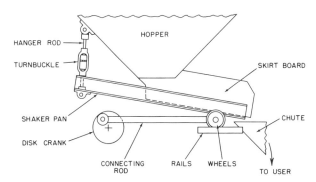

FIG. 21f
Shaker feeder.

The feed chute can be jacketed for heating or cooling, and the tubular chutes can be made dust-tight by flexible connections at both ends. The vibratory feeders can resist flooding (liquid-like flow) and are available for a wide capacity range (from ounces to tons per hour).

Shaker Feeder

The shaker feeder design (Figure 21f) consists of a shaker pan mounted beneath a hopper. The back end of the shaker is mounted on hanger rods. The front end is carried on wheels, and can be moved by a crank. As the pan oscillates, the material is moved forward and dropped into the feed chute.

In most units the number of strokes is kept constant, while the length of the stroke is varied. The angle of inclination of the pan is chosen for the particular material to be fed and varies from about 8 degrees for freely flowing materials to about 20 degrees for sticky materials. If arching is expected in the hopper, special agitator plates are installed in the hopper to break up the arches. The shaker feeder is rugged and self-cleaning, and it can handle most types of material regardless of particle size or condition.

Roll Feeder

Roll feeders are low-capacity devices for handling dry granules and powders (Figure 21g). The feeder consists of a feed

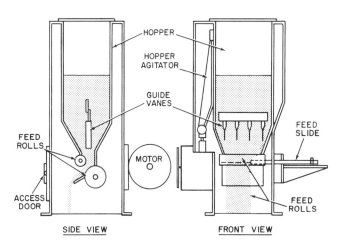

FIG. 21g
Roll feeder.

hopper, two feed rolls, and a drive unit. Guide vanes in the hopper distribute the material and provide agitation by oscillating. The feed rolls form the material into a uniform ribbon, and feed rate is controlled either by means of a slide that varies the width of the ribbon or by means of a variable-speed drive. Operating ranges are typically 6:1 with the feed slide and 10:1 with mechanical or electric variable-speed drive. For materials that tend to cake or bridge in the hopper, agitators can be provided to maintain the material in a free-flowing state.

Revolving Plate Feeder

Revolving plate feeders (Figure 21h) consist of a rotating disk or table located beneath the hopper outlet. This (usually horizontal) table is driven by gears from above or below. As the table rotates, material is drawn from the hopper and is scraped off by the skirt board. Feed rate is controlled by adjusting the height of the adjustable gate or the position of the skirt board.

Revolving plate feeders handle both coarse and fine materials. Sticky materials are also handled satisfactorily because the skirt board is able to push them into the chute. This type of unit cannot handle materials that tend to flood. A variation of the revolving plate feeder utilizes rotating fingers to draw feed material from the bin. Revolving plate feeders can be equipped with archbreaker agitators in the conical throat section of the hopper.

BELT FEEDERS

Belt feeders are compact factory-assembled devices utilizing belts to transport the material across a weight-sensing mechanism. In the case of meters, an uncontrolled solids flow passes across a constant speed belt and the belt load signal is thus a function of gravimetric flowrate.

The feeder in its most basic form consists of a meter to which a controller and volumetric solids flow regulator is added. The flow regulator is normally a simple gate, but may be in the form of a rotary gate, screw, or other volumetric control device capable of being fitted with a suitable actuator. Other methods of control are based on varying belt speed or both belt speed and belt load.

Figure 21i illustrates a simple feeder which is perhaps the

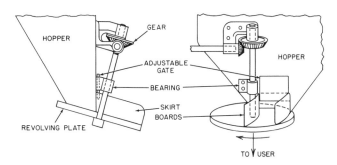

FIG. 21h
Revolving plate feeder.

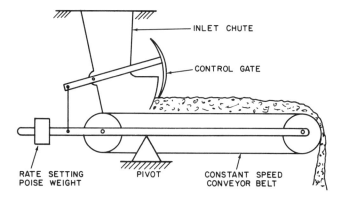

FIG. 21i
Early belt-type mechanical gravimetric feeder.

forerunner of most modern belt feeders. It incorporates a constant speed belt coupled with a gate to modulate the solids flow rate such that belt load is balanced by an adjustable poise weight. The feeder, which is still used in some industrial applications today, is unique in its simplicity but includes a number of disadvantages relative to more modern designs as follows:

1. The entire feeder is weighed rather than only a portion of the belt; consequently, the low ratio of live load to tare coupled with mechanical friction in the linkage pivots results in relatively low sensitivity in the belt load detection system.
2. The position of the gate control element is proportional to the belt load error. In the same manner that a float-operated level control valve cannot maintain level at setpoint if valve supply pressure or tank drawoff vary, this feeder cannot maintain set gravimetric rate if the bulk density of the solids varies.

It should be noted that the basic principle involving the weighing of the entire feeder has been applied in modern designs. Successful operation of these versions has been achieved by adding belt load error detecting instrumentation and by actuating the control gate from an external power source. A controller with reset function eliminates the setpoint error.

Figure 21j describes the basic construction of the electromechanical gravimetric feeder. Here the belt load is balanced by a mechanical beam and poise weight system which energizes one or the other of two clutches via a pair of mercury switches energized by a magnet attached to the beam. These clutches actuate and establish the direction of travel of the gate-positioning mechanism. The gate modulates as required to maintain the desired belt load as established by the position of the poise weight on the balance beam. It can be seen that this feeder will maintain belt load regardless of changes in material density, subject, of course, to the volumetric control limits of the gate. Belt load setpoint is indicated by a mechanical counter geared to the beam poise weight drive. A second counter geared to the belt drive totalizes feet of belt. By varying drive gears, these counters

Solids Flowmeters and Feeders   133

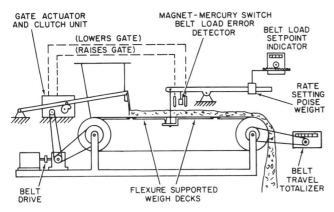

FIG. 21j
Belt-type electromechanical gravimetric feeder.

can be provided to read direct in terms of pounds per foot of belt and belt travel in terms of total feet. Total weight fed can thus be calculated by multiplying the readings of the two counters. Remote belt load setpoint and readout functions are available as well as a belt travel contact switch which may be used to operate a remote counter or to shut down the feeder via a predetermining counter after the desired total weight of material has been fed. Adjustable microswitches actuated by gate position may be utilized to activate alarms indicating either a stoppage of the material supply to the feeder or overtravel of the control gate resulting from abnormally low material density. These feeders are designed to meet Class II, Group G electrical area requirements for hazardous locations when equipped with similarly rated drive motor

Figure 21k illustrates the gravimetric belt meter which is available with either pneumatic or electronic weight detection and transmission system. In the case of the pneumatic version shown, the preliminary calibration procedure involves adjustment of the tare weight with the beam in center position, and location of the nozzle relative to the flapper. This establishes a condition such that balance is achieved when balancing piston pressure is 3 PSIG (0.2 bar). When material crosses the belt, beam movement throttles the nozzle. Nozzle backpressure is imposed on the pneumatic relay, which in turn increases its output pressure until the balancing piston rebalances the beam. The balancing pressure is thus porportional to belt load, and, since the belt speed is constant, balancing pressure is proportional to measured weight-rate. Also shown is an optional ball and disc integrator. The disc is driven by the front belt roll of the feeder and the ball is positioned by a pneumatic positioner. Unlike cam-operated sampling type integrators, the ball and disc type integrates continuously. It is especially recommended for use with gravimetric meters in applications involving the measurement of rapidly varying instantaneous flow rates. The integrator is supplied with a digital totalizer and can be furnished with a pulse transmitting switch to operate a remote counter.

Feed Rate Control

The feed rate of all belt-type gravimetric feeders is a function of the belt speed and the belt load. Belt speed is normally expressed in terms of feet per minute, while belt load is defined as pounds per foot of belt.

$$\text{rate} = \text{belt speed} \times \text{belt load} = \text{lbm/minute} \quad \mathbf{21(1)}$$

In the case of the constant speed belt feeders previously discussed, rate is directly proportional to belt load. Rate setpoint is thus in terms of belt load, and the belt load signal generated by the device can be read out as rate.

Another method of rate adjustment utilizes belt speed variation with belt load a controlled constant. Still another involves variation of both belt speed and belt load wherein the rate signal is the multiplicant of the belt speed and belt load measurement signals generated by the feeder.

Figure 21l illustrates a standard constant-speed belt feeder with pneumatic gate actuator. The length of the weight section and distance from end of weigh section to end of belt have been idealized to some extend but are approximately the same as those in an actual feeder. The response

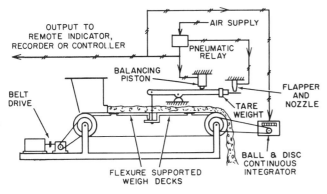

FIG. 21k
Belt-type gravimetric meter.

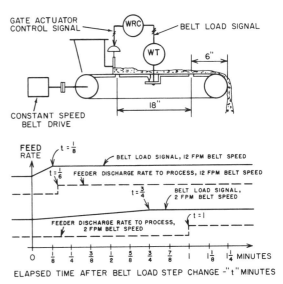

FIG. 21l
Open loop response to belt load step change.

data illustrated in the curves have been simplified somewhat by disregarding the controller response. It is thus not precise since it assumes instantaneous gate response and does not include controller lags, but this is minor relative to the effect of belt transportation lag which is the major source of concern in the application of constant-speed belt feeders.

The uppermost curve represents the belt load signal response of a feeder having a 12 fpm belt speed after a step change in belt loading. The curve immediately below it—as represented by a dashed line—represents the instantaneous feeder discharge rate leaving the end of the feeder belt; i.e., the feed rate as viewed by the downstream process. The effect of the change in belt loading is sensed almost immediately after the step change since the control gate is located at the upstream edge of the weigh section. At 12 fpm belt speed the weigh section has been completely covered by the material at the new rate in $\frac{1}{8}$ minute after the step change. At this instant, however, the feeder is still discharging at the initial rate and an additional $\frac{1}{24}$ minute is required to transport the material at the new rate from the downstream edge of the weight section to the end of the belt, a distance of 6 inches.

The pair of curves immediately below the pair discussed above describe the response of the same feeder when arranged for a belt speed of 2 fpm. Note that in this example the process does not feel the results of the step change until 1 minute after it is made.

Many processes, particularly those involving a single feeder, can tolerate relatively long response times. In continuous blending operations, however, maintenance of instantaneous blend ratio is normally the major performance criterion, and it is in these situations that selection and application of constant-speed feeders require careful study. If two feeders having belt speeds of 12 fpm and 2 fpm were controlled from a common belt load signal, the curves in Figure 211 indicate that the previously established ratio of materials entering the process would be disrupted for a period beginning 10 seconds after the belt load setpoint change and continuing for a period of 50 seconds, at which time the original ratio would be restored.

The curves illustrate the basic fact that two or more constant speed gate feeders cannot maintain blend ratio after a common step change in belt load unless the belt speeds of all feeders are equal. Because of the normal feed rate variations encountered in blending applications, it is rarely possible to size a number of feeders for different rates so that they all have the same belt speed. It is true that if material flow characteristics permit, a narrow feed section can be utilized to decrease the width of the material ribbon on the belt and thus permit an increase in belt speed, but in most applications this does not satisfactorily solve the problem.

Inspection of the belt load signal response curves in Figure 211 suggests the probability that maintenance of blend ratio can be improved by arranging the two feeders in a master-slave relationship wherein the belt load step change is applied to the gate actuator of the master feeder, and its belt load signal is applied to control the gate actuator of the slave feeder. The belt load signal and discharge rate response curves indicate that slaving the low-speed feeder to the high-speed feeder will only increase the duration of blend error. Because of transport time lag, the 2 fpm slave feeder will deliver material to the process at the initial feed rate for a period of 1 minute after the step change is applied to the 12 fpm feeder. Since the gate of the slave is controlled by the belt load signal of the master, the gate of the 2 fmp slave unit will not be at its new setting until $\frac{1}{8}$ minute after the step change. The slave feeder will thus not deliver the corrected rate to the process until $\frac{1}{8} + 1$ or $1\frac{1}{8}$ minutes after the setpoint change is applied to the master. The process "sees" the correction of the 2 fpm slave feeder as a ramp change beginning 1 minute after the master feeder setpoint change and completed $\frac{1}{8}$ minute later. Results of a computer analysis of the response of two constant belt feeders in a master-slave situation have been published in the literature, and this study indicates that optimum maintenance of blend ratio will be obtained when the belt speed of the slave feeder is $1\frac{1}{2}$ times that of the master. The selection and control configuration criteria for constant belt speed feeders in applications involving a single feeder, or in those utilizing two or more feeders for continuous blending, can be summarized as follows:

1. In single-feeder applications, optimum response is obtained by utilizing the maximum possible belt speed commensurate with the characteristics of the material being fed and subject to the belt load limits established by the feeder manufacturer.

2. In applications involving two or more feeders for continuous blending when belt speeds of the feeders can be made equal, blend ratio will be maintained during a total throughput stepchange when the feeders are controlled in parallel from a common system rate-setting signal. If the belt speeds are equal, the feeders should not be operated in a master-slave arrangement.

3. In blending applications utilizing constant-speed belt feeders having different belt speeds, blend ratio cannot be exactly maintained during a total throughput change. Blend ratio error can be minimized, however, by arranging the individual feeders in a master-slave control relationship rather than in parallel and with the feeder having the lowest belt speed as the master. Minimum blend ratio disturbance can be obtained by proper belt speed selection such that the ratio of slave to master feeder belt speed is 1.5.

Belt Speed Control

The advantage of belt speed control over belt load control is that a feed rate change as viewed by the process is almost simultaneous with a change in belt speed setpoint. The use of speed control in multifeeder blending applications eliminates blend error due to differential transport lag which is typical in blending systems utilizing constant-speed feeders. These systems are controlled via a common speed signal that is applied in parallel to all feeders and which increases or decreases the total throughput to the process. Ingredient ratio can be adjusted by varying either the belt loads of the individual feeders or by adjustment of individual speed signal ratio stations. The latter method is preferred in blending situations requiring ratio change while the system is operating; changing belt loading during operation will cause a blend error because of the transport lag between control gate and process. The use of a continuous integrator will provide an accurate totalization of the material fed regardless of changes in belt load or belt speed.

While the variable speed arrangement provides high rate response to setpoint change and eliminates error in blending situations, it includes some undesirable features as far as the operator is concerned:

1. Unlike the constant-speed feeder, the variable-speed arrangement does not provide a readout of feed rate. Feed rate must be calculated by multiplying the speed setpoint signal by the measured belt loading. In addition, the feed rate thus calculated is inferential since the device does not generate a signal proportional to actual belt speed.
2. If this configuration is used in a multifeeder blending system, a blend ratio change will require a change in belt loading or speed ratio setpoint to one or more of the feeders. This will change the total throughput to the process unless a master speed adjustment is made to compensate.

What is needed, therefore, is an arrangement that can accept a single setpoint proportional to feed rate, generate an output that maintains feed rate, and at the same time controls the basis of speed change.

Figure 21m illustrates an arrangement that fulfills the above requirements. In the pneumatic version shown, the speed-changing mechanism has a typical range of 10:1; electronic versions utilizing silicone-controlled rectifier (SCR) drives provide a speed variation of at least 20:1. In the example shown, the feeder is equipped with a fixed gate for use in applications in which the material has a relatively constant density such that the speed actuator can accommodate variations in both density and gravimetric rate while operating within its working range. If the material density varies or if the feeder is to be used on a variety of materials having different bulk densities, such that the total range calculated by multiplying the required gravimetric range times density variations exceeds the range of the speed actuator, internal closed loop belt load control must be added.

As mentioned previously, weighing accuracy is optimized by utilizing the highest possible belt loading in which the ratio of live to dead load is maximum. The arrangement described here is usually sized by the manufacturer to handle the maximum gravimetric feed rate with the belt drive near maximum speed, the design belt loading at about 90% of maximum, and based on the minimum expected material density. In order to assist the operator in setting the manual gate position, a belt load indicator is desirable. This is indicated in Figure 21m, combined with a pair of pressure switches in a commercially available unit. A belt load alarm is desirable to warn the operator in the event that unforeseen density variations dictate a readjustment of the feeder gate or that material flow to the belt has been interrupted.

Belt loading can also be measured by detecting the radiation absorption of a discrete length of material. A source of radiation is located under the belt and a pickup cell located above the material measures the radiant energy which has passed through the material. The difference between the energy transmitted by the source and that measured by the cell is a function of the mass of the material on the belt. The pickup cell converts the measured energy into an electric analog signal proportional to belt load. The balance of the feeder and control accessories is essentially the same as shown in Figure 21m.

Digital Control

A study of Figure 21m indicates that if the continuous integrator was provided with a feed rate transmitter, the belt speed transmitter and feed rate computer could be eliminated, and the feed rate signal from the integrator could be fed directly to the feed rate controller.

Figure 21n describes this arrangement, which has been developed for use with commercial digital control systems. The control system is available in two basic arrangements: one for batching systems, the other for continuous systems.

In the batching version, the master oscillator in conjunction with a timer delivers a total number of pulses proportional to the total weight of material desired. The pulse frequency is adjusted to vary the duration of the batch preparation period. The pulses are applied as the setpoint to the feed rate controllers via ratio setting stations for ingredient ratio. Pulses proportional to material feed rates are generated by photoelectric pulse generators driven by the feeder integrators and applied to the feed rate controllers via scaling and totalizing units. The controllers compare set and measured pulse frequencies and adjust feed rates as required by varying belt drive speed. The batch controller includes a memory feature by which the feeder is kept running until it

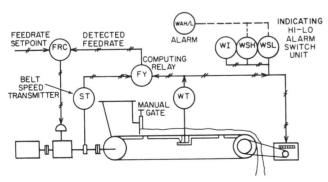

FIG. 21m
Speed-controlled belt feeder with feed rate setpoint and readout.

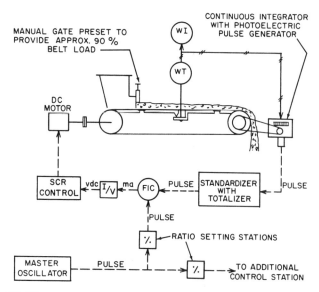

FIG. 21n
Belt-type gravimetric feeder with digital control.

has generated a total number of pulses equal to the total applied as a setpoint to the controller via its ratio station. In a multifeeder batching system, this may result in feeders shutting down at different times, but the batch blend ratio will be correct.

Continuous systems utilize another version of controller which includes a pacing feature. If one of the feeders experiences a decrease in feed rate which cannot be corrected by the controller, this controller gates the output of the master oscillator and thus paces down the feed rates of the other feeders to maintain blend ratio. If the faulty feeder corrects, the entire system is returned to control from the master oscillator. If the condition persists for a predetermined period, an alarm will be activated.

The digital control system is theoretically without error; the pulses generated by the master oscillator must be equalled by those generated by the integrator transmitter on the feeder. Laboratory evaluations confirmed by field tests have indicated that the feeder accuracy based on weighed samples versus total integrator pulses is better than 0.5% of feed rate over a 10:1 feed rate range.

Digital systems are superior to analog ones because each pulse represents a specific increment of weight. Therefore, a pulse rate of 100 pulses per minute, for example, with a pulse value of 2 pounds, signals a solids flow of 200 lb/min. The pulses are totalized on both the measurement and the setpoint side, and therefore errors due to temporary starvation that are common in analog systems cannot occur in digital ones. Another advantage of the digital systems is the flexibility of the microprocessor, which can easily and quickly be reprogrammed from operating like a mass flowmeter to being part of a blending system, for instance.

The microprocessors also provide the capability for automatic recalibration and the retaining of the corrections applied at each test for future reference. The microprocessor-operated units are also capable of functioning in several modes, such as start-up, predetermined fixed flow, or flow-ratio. They can have a variety of ratio or cascade configurations, logic interlocks, input and output signals (BCD, serial, analog), displays, printers, and memory units; they can be tied to other systems for their setpoints or for stop/start signals; and they can operate as batching units with remote resets and can operate as PID loops with dead time compensation utilizing such algorithms as "sample" and "hold."

Digitally controlled gravimetric feeders are utilized in situations involving a number of materials which must be blended in a wide variety of formulations that may be frequently changed. High accuracy, high speed, ease of formula change, and centralized control characterize the digital control system. Although the cost of the feeder and its associated digital control is perhaps twice the cost of a feeder with conventional analog control, digital control systems are gaining acceptance in continuous blending systems, particularly in the food industry.

VERTICAL GRAVIMETRIC FEEDER

A vertical gravimetric feeder is illustrated in Figure 21o. An agitator rotor within the supply bin guarantees a "live" bin bottom. The process material enters through a hole in the top cover of the pre-feeder and is swept through a 180 degree rotational travel by the rotor vanes, where it is dropped out through the discharge pipe. The solids are weighed together with the rotary weight feeder as it transports the solids to the outlet.

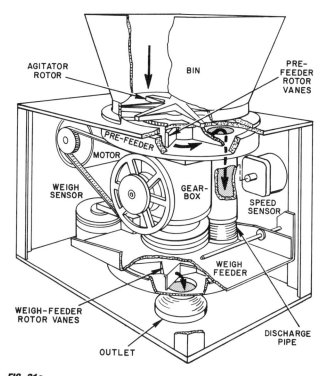

FIG. 21o
Vertical gravimetric feeder.

The advantages of this feeder include its convenient inlet-outlet configuration; its sealed, dust-tight design; and its self-contained nature wherein all associated control instruments are also furnished. After calibration, a ±0.5% of full scale inaccuracy can be expected if a 5:1 rangeability is sufficient. At a 20:1 rangeability, the calibrated inaccuracy is ±1% of full scale.

The main disadvantages of this design are that it has a limited capacity and it can handle only dry and free-flowing powders with particle sizes under 0.1 in. (2.5 mm) in diameter. Large foreign objects cannot be tolerated in the process material, nor can this feeder handle damp, sticky materials that might cake or refuse to flow freely.

NUCLEAR BELT SCALES

Nuclear belt scales are similar to gravimetric belt scales except that the load cells are replaced by nuclear densitometers. They have been first applied on belt conveyors, but later have also been used on screw, drag chain, and vibrating feeders. The radiation source can be cesium 137, cobalt 60, or americium 241. The radiation source is usually placed above the belt and is supported on either side by a C- or A-frame (Figure 21p). The radiation detector is located below the belt and receives a radiation intensity that is inversely proportional to the mass of solids on the conveyor.

Nuclear belt scales are suited for such difficult-to-handle services as hot, abrasive, dusty, or corrosive materials and can provide 0.5% of full scale accuracy when the belt load is high (70 to 100% of full scale) and the moisture content, bulk density, and particle sizes are all constant. When dissimilar solids are measured by the same scale, the differences in radiation absorption characteristics can result in substantial errors. Similarly, these units are not recommended for belt runs that are shorter than 10 minutes or for belt loadings that are below 10% of full scale. The minimum required belt loading is about 2.5 lb/ft^2 (12 kg/m^2).

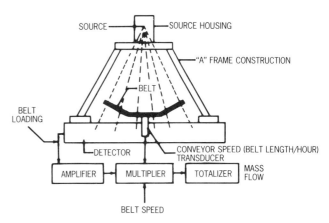

FIG. 21p
Nuclear belt scale supported by A-frame. (Courtesy of Kay-Ray Inc.)

LOSS-IN-WEIGHT FLOWMETERS

Another feeder type is the continuous loss-in-weight feeder (Figure 21q). In one design the weight of material in the hopper is counterbalanced by a retracting poise on a scale beam. The controller adjusts the speed of the rotary feeder to maintain a constant rate of poise retraction. This system can employ rotary screw or vibratory feeders as the modulated control elements. The poise is retracted by a lead screw along the scale beam at a preset rate, and any unbalance of the scale beam is counteracted by the controller by changing the rate of discharge.

In loss-in-weight systems, the weight-sensing section of the system is a tank or silo rather than a horizontal belt surface open on all sides, and thus loss-in-weight feeding systems are suitable for handling liquids and slurries as well as solids. Manufacturers of such systems claim that their systems provide greater accuracy of delivery over short time increments than other continuous feeder systems because the weight-sensing function is located ahead of the material

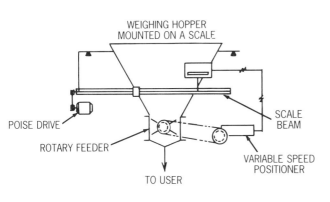

FIG. 21q
Continuous loss-in-weight feeder.

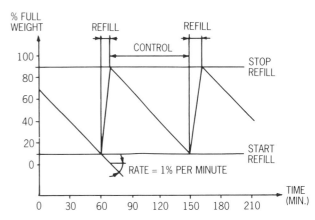

discharge regulating device; that is, if a rate error exists, it is corrected before the material enters the process. These systems are available with either electronic or pneumatic weight-sensing and control functions.

A material hopper or supply tank is supported by or suspended from one or more load cells. In order to eliminate side effect errors due to nonsymmetrical hopper loading, tension-type cells are preferred. The weight signal generated by the load cell system is the measurement signal to the weight rate controller. The setpoint of this controller is set by the operator, by another controller (cascade), or by a programmer. The difference between the weight of material in the hopper and the programmed weight is continually sensed, and the flow rate of the material exiting from the hopper is regulated to counteract any deviations. In other words, the programmer generates a signal corresponding to a fixed reduction rate of the total weight in the hopper, and this signal becomes the setpoint for the controller.

The supply of material in the hopper must be replenished from time to time, and the filling cycle is initiated before the hopper is completely empty. A supply of material referred to as the "heel" always remains in the hopper and serves to minimize the effects of filling shock on the weight-sensing cells and the material exit flow regulator. The filling operation is controlled by a separate subsystem which consists of a differential gap controller and a material supply valve, gate, or feeder. When the weight of material in the hopper reaches a predetermined low value equal to the desired weight of "heel," the differential gap controller starts the filling cycle and at the same time locks the weight controller output signal into the actuator of the material exit flow regulating device. When hopper weight reaches a filled condition, the differential gap controller stops the filling cycle and starts the feeding cycle by returning control of the material exit flow regulator to the loss-in-weight control system. It should be noted that the feeding system is operating on a volumetric rather than gravimetric basis during the filling cycle; hence filling is accomplished as rapidly as possible. It is desirable to design the system such that the filling cycle is a small portion of the total cycle time.

Equipment

The loss-in-weight programmer and weight controller are normally incorporated in a single package suitable for panel mounting. Accessories include loss-in-weight setpoint indicator, set/measured deviation indicator, optional manual controls to regulate filling and feeding cycles, and alarm circuits to detect discrepancy between programmed and actual loss-in-weight, as well as malfunction of the automatic refilling cycle.

Load cells are sealed to withstand dust and corrosion, and include temperature compensation. Hermetically sealed cells also include compensation for barometric pressure changes. Because of the probability of shock loading, cells are normally designed to withstand overloads to 150% of rating or more. Since they are voltage-sensitive, strain gauge load cell power supplies should be closely regulated and compensated for supply voltage variations. As mentioned previously, compression-type strain gauge load cells are generally sensitive to side load forces generated by nonsymmetrical hopper loading or thermal expansion of the hopper and load transfer structure. For these reasons, tension-type cells are normally preferred.

Weigh hoppers can be built in a variety of materials of construction and are often supplied by the user rather than by the feeding system vendor. Maximum sensitivity of the weighing system will be obtained when the ratio of live load to tare is maximum. For this reason, the hopper or tank must be designed not only on the criteria of capacity and structural strength but also for minimum weight.

The material exit flow regulator may be a control valve if the material is a liquid or slurry. Solids may be controlled by a rotary vane feeder, belt feeder, vibrating feeder, or a knife gate valve with positioner. The selection of the specific regulator to be utilized is based on required feed rate as well as on the physical characteristics of the material in question.

System Sizing

Design of these systems involves selection of the proper exit flow regulating device, but perhaps the most critical item is the hopper or tank. Theoretically, the hopper should be as large as possible commensurate with the space available to accommodate it. The larger the hopper, the longer the running cycle and the less frequent the filling cycle. It is obvious, however, that the required load sensitivity of the weighing mechanism must be increased as the hopper weight increases. For a given feed rate (loss-in-weight rate), the system accuracy will decrease as the weight of the material hopper and its contents increase. The hopper should be sized to hold approximately 15 times the maximum pounds-per-minute flow rate. The differential gap controller is set to provide a feeding supply equivalent to $\frac{2}{3}$ of the total supply, thus leaving $\frac{1}{3}$ of the total hopper capacity as the "heel." The refilling cycle is preferably accomplished in one minute or in less than 10% of the total cycle time.

Because of their cost relative to other continuous gravimetric feeding systems, loss-in-weight systems are not used to control the weight rate of easy-to-handle, free-flowing solids.

The loss-in-weight systems are not truly continuous weight rate control systems because the gravimetric rate control is interrupted during the refill cycle. Because of this, it also follows that accurate totalization is not possible, although optional counters are available to indicate the number of times that the hopper supply has been replenished.

Loss-in-weight systems are considered to be most useful for services involving hard-to-handle liquids and slurries. Flow rates of liquids which are highly viscous, nonconductive, corrosive, or contain abrasive solids cannot in many instances be measured or controlled by proportioning pumps or conventional flow control systems. In these instances, gravimetric control on the basis of loss-in-weight has been utilized, and many highly satisfactory systems have been installed.

DUAL-CHAMBER GRAVIMETRIC FEEDER

The gravimetric feeder illustrated in Figure 21r consists of two independently weighed hoppers. While the solids are being discharged into the process from weigh hopper A, the incoming solids are diverted to fill hopper B. When chamber B is filled to its target weight (with the weight of hopper A tared off), the feed is switched to hopper A and hopper B is weighed accurately prior to being charged to the process. This weight is added to the total weight that has previously been supplied to the process. Once chamber B has been accurately weighed, its contents are discharged into the process.

The weighing cycle shown in Figure 21r is monitored and controlled by a microprocessor. The only moving parts are the diverter at the top and the two discharge gates at the bottom of the chambers. Because the hoppers are relatively small, their contents can be weighed at high accuracy. The overall accuracy of the system is about 0.5% of actual flow.

Because the chambers are filled and emptied rather quickly (cycle periods are around a minute) to the process, the solids flow appears to be almost continuous. In addition to accuracy, the small size and vertical flow pattern of the equipment are also advantages in plants where space is limited. This dual-chamber gravimetric feeder is suited for the measurement of free-flowing bulk solids. It can be utilized as a continuous solids flowmeter or as a batch recipe executor operating on many ingredients and weighing the minor and the major ingredients at the same accuracy.

DYNAMIC SOLIDS FLOWMETERS

While the previously discussed devices evaluate the flow rate of solids while they are in a stationary layer on a belt or in a hopper, the devices discussed below look at falling or moving solid streams. These units detect either the forces required to accelerate these solid streams or the forces resulting from the impact of the falling solids.

Impulse-type Solids Flowmeter

When a stream of solids is impacting at an angle on a plate or on a cylindrical surface, the resulting horizontal force com-

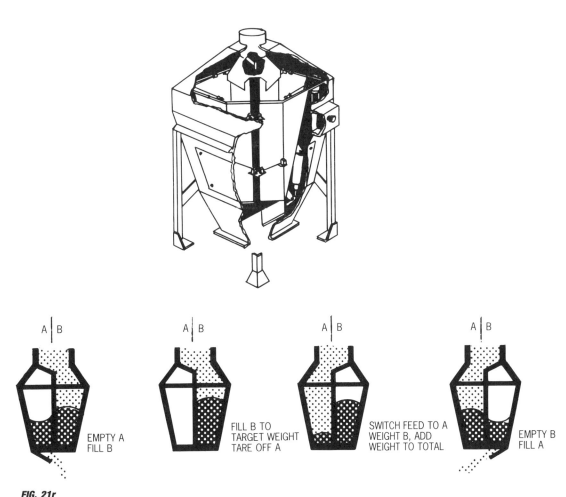

FIG. 21r
Dual-chamber gravimetric feeder. (Courtesy of Technicon Industrial Systems)

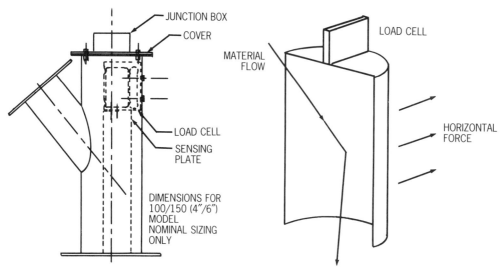

FIG. 21s
Cylindrical impulse flow element. (Courtesy of Milltronics Inc.)

ponent relates to its mass flow rate. The flowmeter illustrated in Figure 21s utilizes this principle. These units are manufactured from steel or stainless steel with the sensing plate being made out of stainless steel. The units can handle free-flowing powders or granular and pelletized materials up to 0.5 in. size. The smallest unit is claimed to have a range of 300 to 30,000 lb/hr (130 to 13,000 kg/hr), while the largest unit can handle flows up to 650,000 lb/hr (300,000 kg/hr). The standard units can be operated at 140°F (60°C) temperature, while special units are available for operation at up to 450°F (232°C). Metering accuracy is claimed to be 1% of full scale. Microprocessor-operated controls are available to integrate this flowmeter into batching or other automated material handling systems.

The principles of impulse and momentum have long been used in liquid flowmeters such as the target or drag-body flowmeters and the angular momentum flowmeter. They are based on Newton's second law of motion and the concept of conservation of momentum. These principles have also been successfully applied to solids flow measurement also. In the design discussed below, solid particles are allowed to fall by

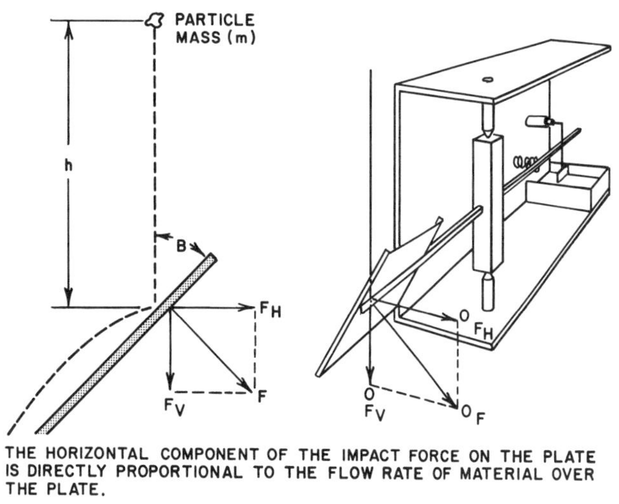

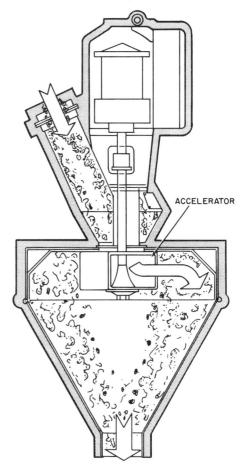

THE HORIZONTAL COMPONENT OF THE IMPACT FORCE ON THE PLATE IS DIRECTLY PROPORTIONAL TO THE FLOW RATE OF MATERIAL OVER THE PLATE.

FIG. 21t
Impulse flowmeter. (Courtesy of Endress & Hauser Inc.)

FIG. 21u
Accelerator-type solids flowmeter.

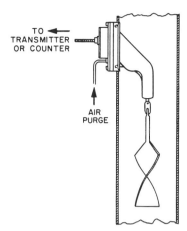

FIG. 21v
Volumetric solids flow detector.

gravity on a calibrated spring-loaded resistance, the displacement of which, caused by the force of the falling particles, is a function of the mass flow rate of the solids. It is measured with a position transducer or transmitter. Figure 21t shows this type of instrument schematically.

These solids flow transmitters have a variety of applications such as continuous weighing; flow control or recording, or both; batching; and alarming. Almost all conceivable types of solids can be measured, such as sugar, salts, cement, and ores.

Accelerator-type Flowmeter

As illustrated in Figure 21u, the solids stream enters the "accelerator" section of the meter by gravity. The accelerator is driven by an electric motor at constant speed. As the flow stream is accelerated, it causes a torque on the motor. The change in torque is sensed by a torque transducer. The amplified pneumatic transmission signal is thus directly proportional to the mass flow rate of solids and can be used as the input to any pneumatic receiver instrument.

This flow sensor error is ±½% and can detect flow rate over a range of 25:1.

The unit is designed for use on a wide range of materials, including powders, granules, pellets, and irregular solids, as well as liquid slurries.

Volumetric Flowmeters

Positive displacement screw impellers provide volumetric flow measurement, but they have to be used on uniform-size solids, such as lead shot, if any reasonable accuracy is to be obtained. This type of instrument is similar to the turbine- or propeller-type flowmeter used for volumetric liquid flow measurement. In the solids flowmeter, a helical vane is used instead of the turbine. As the vane rotates, driven by the flow of granular material, a flexible cable transmits the rotation to a counting mechanism mounted outside of the pipe or duct (see Figure 21v). This counter can be a mechanical counter mounted directly against the piping, or it can be an electrical or pneumatic transmitter so that flow rate and/or total flow can be recorded at a remote location. In the transmitter, a rotary motion produced by a synchronous motor opposes the motion brought in by the flexible cable. This balancing of motions is utilized to position a slotted cam which determines the transmitter output signal.

The vane is normally installed in a vertical position and its bearing surfaces are protected with an air purge. In order to obtain accurate solids flow measurement, the instrument must be calibrated with the material for which its final use is intended.

This unit is capable of detecting volumetric solids flows with ±3% full scale accuracy if the flow rate is between 10 and 100% of design rate.

Bibliography

AWWA Standard for Quicklime and Hydrated Lime, New York: American Water Works Association, 1965.

Beck, M.S., and Plaskowski, A., "Measurement of the Mass Flow Rate of Powdered and Granular Materials in Pneumatic Conveyors Using the Inherent Flow Noise," *Instrument Review,* November 1967.

Colijn, H., and Chase, P.W., "How to Install Belt Scales to Minimize Weighing Errors," *Instrumentation Technology,* June 1967.

Cross, C.D., "Problems of Belt Scale Weighting," *ISA Journal,* February 1964.

"Digitally Controlled Coal Weigh Feeder," *Power Engineering,* 1978.

The Flowmeter Industry, 3rd edition, Venture Development Corp., 1991.

Grader, J.E., "Controlling the Flow Rate of Dry Solids," *Control Engineering,* March 1968.

Jenicke, A.W., "Storage and Flow of Solids," Bulletin 123, Utah Engineering Experiment Station, University of Utah, Salt Lake City, Utah, 1964.

Johanson, J.R., and Colijn, H., "New Design Criteria For Hoppers and Bins," *Iron and Steel Engineering,* October 1964.

Linn, J.K., and Sample, D.G., "Mass Flow Measurement of Solids/Gas Streams Using Radiometric Techniques," Report SAND-82-0228C, Dept. of Energy, Washington, D.C., 1982.

"Mass, Force, Load Cells," *Measurements and Control,* October 1991.

McEvoy, L.D., "Control Systems for Belt Feeders," *InTech,* February 1968.

Miller, R.W., *Flow Measurement Engineering Handbook,* New York: McGraw-Hill, 1983.

Nolte, C.B., "Solids Flow Meter," *Instruments and Control Systems,* May 1970.

"Solids Flowmeter Works Without Obstructing Flow," *Chemical Engineering,* September 18, 1972.

Spitzenberger, R.M., "Long-term Accuracy of Digital Weigh Feeders," *Chemical Processing,* April 1974.

Stepanoff, A.J., *Gravity Flow of Bulk Solids and Transportation of Solids in Suspension,* New York: John Wiley & Sons, Inc., 1969.

Van den Berge, H., "Weighing On-the-Fly Keeps the Process Moving," *Control Engineering,* Vol. 23, no. 9, p. 52.

Vines, G.L., "Digital Weigh Feeders Automate Refractory Production," *Brick & Clay Record,* June 1974.

Zanetti, R.R., "Continuous Proportioning for the Food Industry," *Instrumentation Technology,* March 1971.

22 Target Meters

W. H. HOWE (1969, 1982)   **B. G. LIPTÁK** (1993)

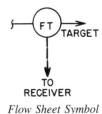

Flow Sheet Symbol

Design Pressure:	Up to 1500 PSIG (10.35 MPa), higher with strain gauge-type
Design Temperature:	Up to 750°F (400°C) with force balance and to 300°F (150°C) through 600°F (315°C) with strain gauge-type; special units up to 1200°F (649°C)
Sizes:	½ to 8 in. (12.5 to 203 mm) with standard and up to 48 in. (1.2 m) pipes with the probe design
Fluids:	Liquids, gases, and steam
Flow Range:	From 1 GPM (3.785 l/m), 1 SCFM (28 l/m), 3 lb/hr (1 kg/hr) to practically any value using the probe design
Inaccuracy:	±½% of full scale for standard and to ±5% of full scale for probe-type
Materials of Construction:	Usually either carbon or stainless steel
Cost:	$1500 to $10,000 as a function of size, design, and materials of construction
List of Suppliers:	Foxboro Co.; Mead Instruments Corp.; Ramapo Products, Hersey Measurement Co.; J.W. Sweet Co.; Turbo Instruments Inc.

Material buildup in front of orifice plates can cause both measurement errors and plugging when the process stream is a liquid slurry or a gas carrying wet solids. The annular orifice (Figure 22a) was introduced to solve this problem by providing an annular opening for the solids to pass through. Target flowmeters (Figure 22b) are similar in design except that the pressure taps have also been eliminated and the detection of differential pressures been replaced by force measurement. Both of these designs are suited for dirty or low-turbulence flowmetering applications, if high precision is not required.

The target meter combines in a single unit an annular orifice and a force-balance transducer. Output is either an electric or pneumatic signal proportional to the square of the flow. Target meters are available in sizes from ½ to 8 in. (12.5 to 203 mm) pipe diameter. The annular orifice is formed by a circular disk supported in the center of a tubular section having the same diameter as schedule 80 pipe of the same nominal size. Flow through the open ring between disk and tube develops a force on the disk proportional to velocity head (the square of the flow). The disk is mounted on a rod passing out through a flexible seal. The force on the disk is measured from the rod outside the seal, using a standard force-balance transducer integrally mounted on the flow tube.

A similar operating principle applies to the drag-body

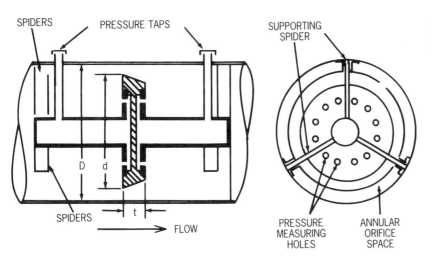

FIG. 22a

Annular orifice plate installation. (Courtesy of the Foxboro Co.)

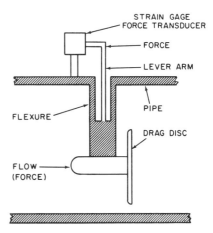

FIG. 22c
The drag-body flowmeter.

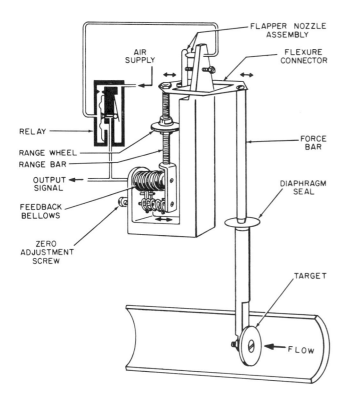

FIG. 22b
Target meter.

flowmeter (Figure 22c), which detects the impact forces by a strain gauge circuitry. This unit is also available in retractable probe designs used in larger pipe sizes where it is desirable to withdraw the sensor periodically for cleaning, without opening the process line.

The target meter is applied in a number of fields for measurement of liquids, vapors, and gases. It allows unimpeded flow of condensates and extraneous material along the bottom of a pipe and at the same time allows unimpeded flow of gas or vapor along the top of the pipe. It has given consistent dependable service on "difficult" measurements such as hot, tarry, sediment-bearing fuels to a pipe still where no other head-type meter has proved successful. There are no differential pressure connections to "freeze." This is useful in steam flow measurement in exposed locations, and for liquids that congeal at ambient temperature in pressure connections. Units are available for service up to 700°F (371°C), which can be used on steam service up to 200 PSIG (14 bars) pressure.

Targets with diameters of 0.6, 0.7, and 0.8 times tube diameter are available. Combined with wide range force measurement transducers, a wide selection of full scale flow rates is provided.

In addition to providing an open flow both at the top and at the bottom of the pipe, the annular orifice has the advantage of being less sensitive to change in Reynolds number and to various upstream piping configurations than a concentric sharp-edged orifice. This is apparently because the main flow is directed outward toward the boundary layer. Greatest improvement occurs at 0.8 diameter ratio.

The advantage of having a single unit combining primary element and transducer also imposes a limitation. The transducer can be zeroed only under zero-flow (nonflowing) condition. If zeroing under flowing conditions is desired, a bypass pipe and suitable valves are required. Modern force-balance transducers are sufficiently stable that zero-checking is seldom demanded under flowing conditions.

Calibration data are provided by the manufacturers. The flow range through a particular size meter can be varied by changing the target size and by replacing or readjusting the transducer.

Repeatability of output is good. Calibration accuracy includes not only the uncertainty of the primary element but also the characteristics of the transducer and the precision of the transducer adjustment. As is the case with some other proprietary devices, test data is unavailable for determination of flow coefficients from physical dimensions for different process fluids and operating conditions. On the other hand target meters with accurate water flow calibration over almost any range of Reynolds numbers can be obtained. Transfer characteristics to other fluids based on Reynolds number are reliable. Because the transducer and the primary element are calibrated as a unit, overall accuracy of calibrated target meters is better than that of orifice-type systems.

Bibliography

Blasso, L., "Flow Measurement Under Any Conditions," *Instruments and Control Systems*, February 1975.

The Flowmeter Industry, 2nd edition, Venture Development Corp., 1985–1990.

Hall, J., "Solving Tough Flow Monitoring Problems," *Instruments and Control Systems*, February 1980.

Miller, R.W., *Flow Measurement Engineering Handbook,* New York: McGraw-Hill, 1983.

Spink, L.K., *Principles and Practices of Flow Engineering,* 9th edition, The Foxboro Company.

Stapler, M., "Drag-Body Flowmeter," *Instruments and Control Systems,* November 1962.

23 Turbine and Other Rotary Element Flowmeters

J. G. KOPP (1969)   **D. J. LOMAS** (1982)   **B. G. LIPTÁK** (1993)

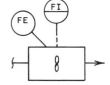

Flow Sheet Symbol

Types:	A. Turbine flowmeters
	B. Propeller, impeller, and shunt-flow types
	C. Insert, probe, or paddlewheel designs
Services:	Relatively clean liquids, gases, and vapors.
Sizes:	A. 3/16 to 24 in (5 to 610 mm) in flow-through designs.
	B. Impeller designs are available from 3 to 72 in. (75 mm to 1.8 m)
	C. Paddlewheel units are available for up to 12 in. (305 mm) pipes. Insertion turbine probes are not limited by pipe size and can also be used in open channels
Outputs:	Generally linear frequency outputs are provided, but 4 to 20 mAdc can also be obtained through conversion
Operating Pressure:	A. 1500 PSIG (10.3 MPa) in standard and 5000 PSIG (34.5 MPa) in special designs
	B. Impeller designs are usually designed for 150 PSIG (1 MPa)
	C. The plastic paddlewheel units can operate up to 200 PSIG (1.4 MPa) at ambient temperatures
Pressure Drops:	A. Usually one velocity head or about 3 to 5 PSIG (20 to 35 kPa)
	B. Usually less than 1 PSID (7 kPa) for the impeller types
	C. Negligible
Operating Temperature:	A. −58 to 300°F (−50 to 150°C) in standard and −328 to 840°F (−200 to 450°C) in extended pickup designs
	B. Up to 160°F (71°C) for the impeller design
	C. The plastic paddlewheel units can operate at up to 220°F (105°C) if the operating pressure is under 25 PSIG (172 kPa)
Materials of Construction:	A. Normally stainless steel housing and rotor with tungsten carbide sleeve bearings, but Hastelloy C or other housing materials and ceramic or PTFE bearings are also available
	B. The impeller-type unit is provided with a plastic impeller and with aluminum, epoxy-coated carbon steel, or stainless steel housing
	C. The plastic paddlewheel units are made of polypropylene, PVDF, ryton, and metallic parts
Error or Inaccuracy:	A. Linearity is 0.25% of actual flow for turbine meters larger than 3/4 in. (19 mm) and 0.5% for smaller units. The repeatability (after calibration) is 0.02% of actual flow. This performance assumes constant viscosity (within 0.3 and 3 cps) and density, proper installation including flow straighteners, a 10- to 15-diameter straight pipe run, and the use of a DC power supply and a preamplifier located at the meter
	B. Shunt flowmeters are accurate to 2% of actual flow over a range of 10:1. The impeller-type units are also claimed to have a 2% of actual flow accuracy if operated at velocities exceeding 1 fps (0.3 mps)
	C. Linearity is 1% relative to actual velocity at point of insertion. Accuracy similar to pitot tubes, or 2 to 5%
Rangeability:	A. 10:1 unless limited by use of line-size units or by high process fluid viscosity
	B. 10:1 for the shuntflow design
	C. The optical designs provide flow rangeabilities in excess of 20:1
Cost:	A. A turbine flowmeter with a preamp (but without readout electronics) and with 150carbon steel flanges can be estimated as follows (1 in. = 25.4 mm): 1/2 to 1 1/2 in.—$2000; 2 to 3 in.—$2500; 4 in.—$3,300; 6 in.—$4,600; 8 in.—$8000; 10 in.—$12,000; 12 in.—$15,000; 16 in.—$27,000; 18 in.—$30,000; 20 in.—$47,000; 24 in.—$75,000. Electronic readout devices might include: auxiliary, explosion-proof power supply—$1000; remote register drive—$3,000; frequency-to-analog converter with digital display—$1000; locally mounted, explosion-proof totalizer/flow indicator—$1000. Accessories include flow straighteners, strainers, batch control units, and two-stage shutoff valves

C. The flow element of the plastic paddlewheel units for sizes between ½ and 12 in. (13 to 305 mm) costs between $200 and $450. Flow elements can be provided with analog indicators ($300), digital readouts ($400), recorders ($750), or batch totalizers ($500)

Partial List of Suppliers: ABB Kent-Taylor Inc. (A); Accuflo Measurement, Inc. (A); Accurate Metering Systems, Inc. (A); AMETEK Inc.; Schutte & Koerting Div. (Cox); Aqua Matic Inc.; Atlantex Industries Inc. (A); Badger Meter Inc. (A); Ball Products, The Oilgear Co.; BIF Unit of General Signal; Brooks Instrument Div. of Rosemount (A); Colorado Engineering Exp. Station; Daniel Industries Inc. (A); Dantec Electronics Inc.; Data Industrial Corp. (C), Digital Precision Flowmeter (C); EFM Inc. (A,C); EG&G Flow Technology Inc. (A,C); Engineering Measurements Co. (A); Fischer & Porter Co. (A); Flow Measurement System Inc.; Flowmetrics, Inc. (A,C); Floscan Instrument Co.; Flo-tech Inc. (A,C); Flow Research Corp.; Flow Technology, Inc.; Foxboro Co. (A); Grindley Systems Inc.; Grinnell Corp.; Hays Div./Romac Ind. Inc. (C); Hersey Measurement Co. (A); Hersey Products Inc. (A,B); Hoffer Flow Controls, Inc. (A); Hydril Co. (B); Invalco Inc. (A); Ion Science Ltd.; Ista Energy Systems Corp. (B); ITT Barton Instruments Co. (A); Kent Meters, Inc. (A); Ketema Inc.; Schutte & Koerting (A); McCrometer Flowmeters (B); McMillan Co. (A); Mead Instruments Corp. (C); Mentor Industries Inc.; Meter Equipment Mfg. Inc.; Miniflow Systems Inc. (A); Neptune Measurement Co. (A); Nitto American Flowmeter Div. (A); Nixon Instrumentation Ltd.; Oilgear/Ball Products; Omega Engineering; Onicon Inc.; Presto-Tek Corp. (C); Rockwell International, Municipal & Utility Div.; J.B. Rombach Inc. (A); Schlumberger Industries Inc. Measurement Div. (A); Signet Scientific Co. (C); Smith Meter Inc.; Spirax Sarco Inc. (A); Sponsler Co. (A); Tejas Controls Inc.; Tokheim Corp.; Tulsa Fluid Measurement Inc. (A); Vortab Corp.; Waugh Controls Corp. (A); Webster Instruments (A); XO Technologies Inc. (A)

Turbine meters are available for liquid, gas, and very low flow rates in both full bore and insertion designs. The most widely used type is the full bore meter for liquid service.

LIQUID TURBINE METERS

A turbine meter consists of a multibladed rotor suspended in the fluid stream on a free-running bearing (see Figure 23a). The axis of rotation of the rotor is perpendicular to the flow direction and the rotor blades sweep out virtually to the full bore of the meter. The fluid impinging on the rotor blades causes the rotor to revolve. Within the linear flow range of the meter, the angular speed of rotation is directly proportional to the volumetric flow rate. The speed of rotation is monitored by an electromagnetic pickup coil, which is fitted to the outside of the meter housing. Two types of pickup coil are used primarily: reluctance and inductance. Both operate on the principle of a magnetic field moving through a coil.

In the reluctance pickup coil system, the permanent magnet is the coil. The field produced is concentrated to a small point by the cone (see Figure 23b). The turbine rotor blades are made of a paramagnetic material, i.e., a material that is attracted by a magnet. As a blade approaches the cone point, its magnetic properties deflect the magnetic field. This deflection causes a voltage to be generated in the coil. As the blade passes under the cone point, the voltage decays, only to be built back up in the opposite polarity as the departing blade deflects the magnetic field in the opposite direction. Thus, each blade produces a separate and distinct voltage pulse as it passes the cone. Since each blade sweeps a discrete volume of fluid, each electrical impulse represents the same discrete volume of fluid.

With the inductance pickup coil system (see Figure 23b), the permanent magnet is embedded in the rotor. As the magnet rotates past the pickup coil position, it generates a voltage pulse for every complete revolution of the rotor.

The typical operating temperature range for standard pickup coils is −58 to 300°F (−50 to 150°C). Specially modified pickup coils are available, however, to cover operation at temperatures ranging from −328 to 840°F (−200 to 450°C). If the meter is located in a hazardous area, the pickup coil can be mounted in a flameproof or explosion-proof conduit box, or alternatively an intrinsically safe pickup coil can be used in conjunction with zener barrier to provide an inherently safe system.

Electronic Display Units

The output signal from the turbine meter is a continuous sine wave voltage pulse train with each pulse representing a

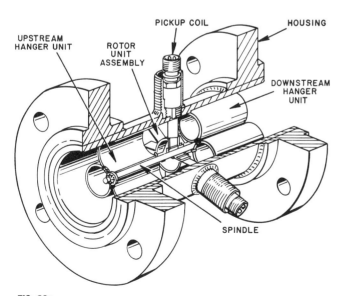

FIG. 23a
Cutaway view of a typical turbine meter.

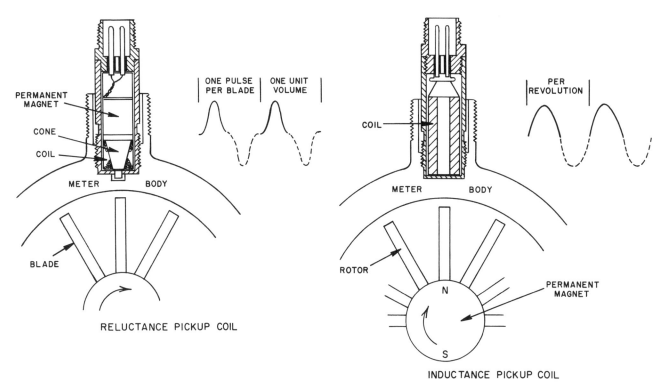

FIG. 23b
Alternative signal generation systems.

small, discrete volume of fluid. Associated electronic units display total volumetric flow or flow rate and perform preset batching, control, automatic temperature correction, and other functions.

Most turbine meter systems incorporate a totalizer unit with a factorizing and scaling function. The pulse output from the turbine meter is not in direct engineering units. For example, each pulse might represent 0.001231 gallons. The factorizer is set to this value and the incoming pulses are multiplied by 0.001231. The display presented is then in gallons.

Alternatively, the totalizer can be a preset batch unit for automatically dispensing predetermined quantities of liquid. The required value is preset and the totalizer then counts down to zero and provides an output—that is, contact closure—to operate a valve and terminate the batch. In order to provide better system repeatability and avoid hydraulic shock, the preset batch unit can be fitted with an advance warning contact, or it can incorporate a ramp function. In the former case, an output is provided, typically 2 to 5% before batch completion. This output partially closes the valve and the batch is "topped off" at a low flow rate up to the final preset quantity. The latter system includes a ramp function in the preset batch unit providing an analog output signal at the start of the batch to open the valve at a predetermined rate. As the batch nears completion, the valve is progressively closed down to a low flow rate. The final valve closure signal is then given at the preset batch size.

Turbine meters measure volume flow at actual operating conditions. Consequently, if high accuracy is required and the fluid temperature is subject to variation, automatic temperature correction is necessary. This involves measuring the liquid temperature with a platinum resistance thermometer and providing an analog control signal proportional to temperature. The temperature/volume relationship for the metered liquid is built into the automatic temperature correction (ATC) unit. Depending upon the measured temperature, the ATC unit modifies the totalizer volume reading in accordance with the preset temperature coefficient of the liquid to give volume readout at the required reference temperature.

To safeguard against interference or lost pulses during signal transmission, a pulse comparator is often used on high-accuracy systems. This involves using two pickup coils (A and B) and taking two separate signal leads to the electronics. The pulse comparator unit monitors the two signals for integrity. If any pulses are lost or picked up on either line, the correct pulse sequence (A, B, A, B, A, B, etc.) will be interrupted. Any such false pulses are logged and the associated totalizer reading corrected accordingly.

Most turbine meter systems require flow rate indication or an analog control signal. These options can generally be provided from the basic totalizer unit.

Linearity and Repeatability

The nominal K factor (the number of pulses per unit volume) is primarily determined by the size and type of turbine meter. In practice, the actual K factor varies slightly between

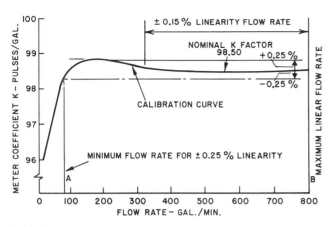

FIG. 23c
Typical calibration curve for a turbine meter.

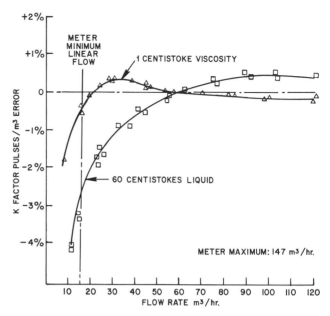

FIG. 23d
Calibration curves illustrating the effect of high viscosity on meter performance.

apparently identical meters, due to manufacturing tolerances. Consequently, it is essential to calibrate each meter to establish its own specific K factor. A typical turbine meter calibration is shown in Figure 23c.

The graph is a plot of K factor against flow rate. It will be noted that over the flow range A to B GPM, the K factor is a constant within the linearity tolerance band. The linearity tolerance band is typically ±0.25% of point over a 10:1 flow range for meters ¾ in. (20 mm) and larger and ±0.5% of point over a 5:1 or 8:1 flow range on meters smaller than ¾ in. (20 mm). It is important to note that the linearity is specified "of point" or "of actual reading" and is not "of full scale deflection."

The calibration in Figure 23c has a typical turbine meter hump in the low flow region (the lower 30% of the flow range). If this region is avoided the turbine meter linearity can be improved to ±0.15% on the larger meters and ±0.25% on the smaller meters.

The repeatability of the turbine meter is typically ±0.02% of point at any flow rate within the linear range of the meter.

Viscosity and Density Effects

The principal fluid parameter which affects a turbine meter is viscosity. High viscosities change the nominal K factor and cause the calibration curve to fall away at a higher minimum flow rate (see Figure 23d). This causes a deterioration in the linearity tolerance over the full flow range or, alternatively, a shorter usable flow range at the standard linearity tolerance.

The effect of viscosity cannot be easily quantified because it is dependent upon the size and type of turbine meter. In general, larger meters are less affected by viscosity than are smaller sizes. This does not imply that an oversize meter should be used on a viscous application. In fact, quite the reverse is true. On a high-viscosity application it is advisable to size the meter so that its maximum permitted flow rate is as close as possible to the application flow rate. Thus, by tending to undersize the meter, the nonlinear portion of the calibration is avoided and the best possible flow range is achieved.

The above comments about viscosity are applicable to the linearity of the meter. Turbine meter repeatability will not be affected in this way, and the standard repeatability tolerance will still be maintained at high viscosities. Consequently, a turbine meter can be used for such duties as on-off control on very viscous products. The control points can be determined impartially and the meter will then repeat these readings even though its calibration may be completely nonlinear. To achieve reliable repeatability, the operating conditions must be constant.

Density has a small effect on the turbine meter's performance. On low-density liquids, the meter's minimum flow rate is increased due to the lower driving torque, but the change in density has a minimal effect on the meter's calibration.

Meter Sizing

Turbine meters are sized by volumetric flow rate. Each meter size has a specified minimum and maximum linear flow figure and the meter should not normally be used outside these values. Typical flow capacities for a range of turbine meters from ¾ in. (19 mm) to 20 in. (508 mm) are shown in Table 23e.

When sizing the meter, it is recommended that the maximum flow rate of the application should fall at approximately 70 to 80% of the maximum flow rate of the meter. This results in a good flow rangeability (about 8:1) and yet there is still approximately 25% spare capacity to allow for future expansion in production or increased metering requirements. Exceptions to this rule of thumb are applications

TABLE 23e
Typical Flow Capacity for a Range of Turbine Meters

Nominal Diameter		Minimum Linear Flow		Maximum Linear Flow	
Inches	mm	GPM	m³/h	GPM	m³/hr
¾	20	2.5	0.68	25	6.8
1	25	3.3	0.90	50	13.6
1½	40	7.2	1.96	108	29.5
2	50	20	5.45	160	43.6
3	75	60	16.3	400	109
4	100	180	27.2	1000	272
6	150	250	68.1	2000	545
8	200	415	113	4150	1130
10	250	715	195	6400	1750
12	300	1025	280	9160	2500
14	350	1210	330	10800	2950
16	400	1830	500	14650	4000
18	450	2310	630	18500	5050
20	500	2930	800	24000	6540

which demand maximum rangeability, and high-viscosity applications.

In order to achieve optimum performance and flow range, most turbine meters are designed for a maximum velocity of 30 ft/s (9.14 m/s). This velocity is higher than the velocities that exist in typical process pipelines, which are typically 7 to 10 ft/s (2.13 to 3.05 m/s). Consequently, if the turbine meter is the same size as the pipeline, the meter flow range will be limited to approximately 2:1 or 3:1. Hence, it is important to size the turbine flowmeter on the basis of volumetric flow rate and not on the basis of pipe diameter. If the turbine meter is sized on volumetric flow rate, it will end up to be smaller than the pipe size. This is a perfectly acceptable and normal practice, provided the meter is installed with the appropriate upstream and downstream straight pipe lengths and cone type reducers (see Figure 23f).

Another aspect which must be considered when sizing the meter is available line pressure. Turbine meters have a typical pressure loss of 3 to 5 PSIG (20.7 to 34.5 kPa) at maximum meter flow rate. The pressure loss reduces with the square of flow rate. Consequently, if the meter is operating at 50% of maximum capacity, the pressure loss is 25% of that at maximum flow rate.

A typical pressure distribution through a turbine meter is shown in Figure 23g. As will be noted, the minimum pressure point occurs in the region of the rotor, with a subtantial pressure recovery occurring immediately thereafter. It is essential to provide sufficient line pressure to prevent liquid cavitation or gassing in the rotor region. To ensure that cavitation does not occur, the downstream line pressure must be at least 2 times the net meter pressure loss plus 1.25 times the vapor pressure of the flowing fluid at its maximum operating temperature. When the backpressure on the meter is not sufficient to meet this requirement, either the back pressure should be increased, or a larger meter operating in a lower region of its flow range (with a resultant lower pressure loss) should be considered. The meter flow range will be reduced by this approach.

If cavitation occurs, it will cause an error in the meter output, and the meter will read high. If severe cavitation is

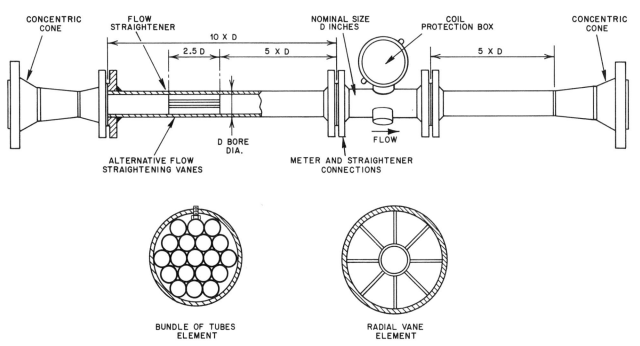

FIG. 23f
Recommended turbine meter installation pipework.

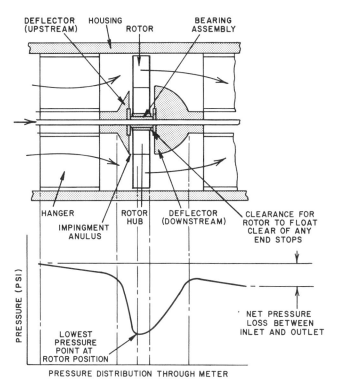

FIG. 23g
Typical pressure distribution through a turbine meter.

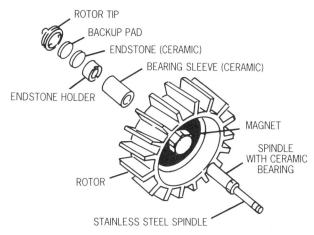

FIG. 23h
Ceramic bearings. (Courtesy of Badger Meter Inc.)

present, it will destroy some of the metallic parts and will cause serious overspeeding of the rotor, resulting in possible mechanical damage to the rotor and bearing.

Pelton Wheel Meters

It is not practical to make turbine meters for very low flow rates below 0.25 GPM (1.58×10^{-5} m³/s). Pelton wheel meters have been developed for these very low flow rates. The meter has a small orifice that projects the liquid onto a small Pelton wheel. The velocity of rotation is then measured electromagnetically and a frequency output signal produced. By varying the diameter of the orifice, a range of flow rates can be covered from 0.001 GPM through to 2 GPM (6.3×10^{-8} to 1.26×10^{-4} m³/s). Flow range varies with meter type but is generally between 10 and 20:1. The meters offer good repeatability ($\pm 0.1\%$) but are generally nonlinear and have a high pressure loss, typically 15 to 20 PSIG (103 to 138 kPa). Typical applications for this type of meter are metering internal combustion engine fuel flows in test rigs and additive dosing.

Meter Characteristics and Features

The wetted materials of a turbine meter are generally stainless steel throughout except for the bearing. The most widely used bearings at present are tungsten carbide, or ceramic sleeve bearings which offers exceptional reliability and immunity to wear. These materials provide good corrosion-resistance capability on a wide range of process liquids (Figure 23h). Where these materials are not suitable, other, more expensive possibilities, such as Hastelloy C with PTFE bearings, are feasible. On clean liquids, some meter designs use ball race bearings in order to achieve greater rangeability.

Turbine flowmeters have also been manufactured without bearings (Figure 23i). In this design the hydraulic forces of the flowing fluid kept the dual turbine in a suspended, "hovering" state. This meter is no longer being manufactured but is mentioned here because of the interesting concept behind its operation.

In very small sizes (under 1 in., or 25 mm) a single turbine can also be rotated without having any physical contact to the meter body. In the design shown in Figure 23j, the process fluid enters as a tangential jet and spins and stabilizes the turbine as it exits through the center of the rotor. The speed of rotation is detected optically by a photodetector. In the 8 cm³/m to the 8 GPM (330 lpm) flow range, up to 30:1 rangeability is claimed.

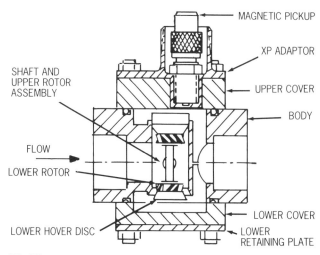

FIG. 23i
Bearing(s) turbine flowmeter. (Discontinued)

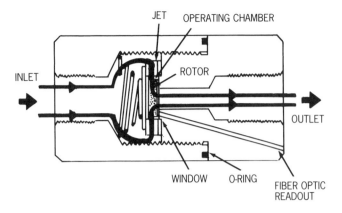

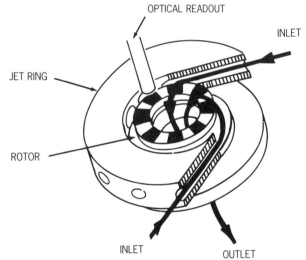

FIG. 23j
Unsupported single rotor with optical readout. (Courtesy of Miniflow Systems Inc.)

Turbine meters are suitable for extremes of temperature. When appropriate pickup coils and bearings are selected, turbine meters can operate at temperatures varying from $-328°F$ ($-200°C$) to $840°F$ ($450°C$). The turbine meter housing is a very good pressure vessel since there are no tappings or protrusions into the meter bore. Consequently, most small turbine meters are suitable for operating pressures up to 5000 PSIG (34.5 MPa), subject to the pressure limitation on the flanges or other end connections.

Another significant feature of the turbine meter is that it has a high throughput for a given size and is small in size and weight, relative to the pipeline. Consequently, turbine meters can handle large volume flow rates with a minimal requirement for space without needing special mounting stands or pads. Other features of the turbine meter include fast response time, suitability for hygienic applications, linear digital output, ease of maintenance, and simple installation.

The main limitations of the turbine flowmeter include its high cost; limitation to clean and nonviscous services; the error caused by viscosity and density changes; the requirement for filtration and for 15 to 20 diameters of straight upstream pipe; the need for periodic recalibration (at operating conditions) and maintenance because the moving components are subject to wear; the potential problems of gassing, cavitation, and overspeeding; the need for relatively high backpressure; and the need for secondary components in providing a readout.

Due to its excellent performance characteristics, the turbine meter is widely used for high-accuracy royalty and custody transfer of crude oil, refined hydrocarbons, and other valuable liquids. Turbine meters are used throughout the petrochemical industry for many other applications, such as process control metering, blending, and pipeline leak detection. Turbine meters are also used in other industries for a broad range of applications, flow rates, and duties. More specialized applications include measurement of cryogenic liquids (liquid oxygen and nitrogen), high-pressure water injection to oil wells, aircraft fuel metering, test rig duty, and road tanker filling. Some of these applications require modified or special meters—for example, aircraft meters are made from aluminum alloy to save weight—but fundamentally the same meter is used in all cases.

Mechanical Installation

The turbine meter's high accuracy can easily be negated by a substandard installation. Upstream disturbances such as bends, valves, or filters may cause swirl and/or a nonuniform velocity profile which, in turn, affects both the linearity of the meter and the nominal K factor. The errors may be positive or negative, depending on the direction of the swirl. If there is sufficient straight pipe between the source of the disturbance and the meter, the fluid shear or internal friction between the liquid and the pipe wall will condition the flow to an acceptable degree. The length of straight pipe required depends upon the upstream disturbance and, in some instances, may have to be as long as 50 times the nominal meter diameter.

To avoid excessively long straight lengths of pipe, an internal flow-straightening element is generally used where good accuracy is required. The flow-straightening element may either be a bundle of thin-wall tubes or a series of radial vanes inserted longitudinally in the upstream section of the straight pipe. The location of the vane is important; the recommended position is shown in Figure 23f. When a flow-straightening element is used, the upstream straight pipe requirement is reduced to 10 times nominal meter diameter. The required downstream length is 5 times nominal meter diameter. It is, nevertheless, good practice not to install the meter downstream of any severe source of disturbance, such as regulating control valves, whenever possible.

If the meter is smaller in diameter than the process pipework, 15-degree inclined angle concentric cones should be fitted at either end of the metering pipework, as shown in Figure 23f. Care should be taken with the internal alignment of all flange joints in the metering section; no gaskets should protrude into the fluid path.

To avoid mechanical damage to the turbine meter and to

TABLE 23k
Typical Strainer Recommendations for Turbine Meter Installations

Turbine Meter Size Inches	Recommended Strainer			
	U.S. Sieve No.	Wire Size (Inches)	Meshes/ Linear Inch	Opening (Inches)
½ and smaller	120	0.0034	120.48	0.0049
¾ to 1½	45	0.0087	44.44	0.0138
2 and larger	18	0.0189	17.16	0.0394

ensure optimum life, a suitable mesh strainer should be fitted upstream of the meter. The recommended mesh size depends on the size and type of turbine meter, but typical guidelines are given in Table 23k. Close attention should be paid to any application where there are fibrous particles in the fluid. Contaminants of this type are frequently not removed by the strainer; the fibrous strands tend to wrap around the rotor and bearing causing the rotor to slow down and the calibration to change.

Electrical Installation

The output frequency from a typical turbine meter pickup coil varies in frequency and amplitude with flow range. At low flows, the signal may be as small as 20 mV peak-to-peak. Consequently, if the turbine meter and electronic readout equipment are not from the same manufacturer, care must be taken to ensure that the two units are compatible with regard to pulse shape (sine wave or square wave), signal frequency, and pulse amplitude and width.

Careful attention should also be given to the cable routing between the turbine meter and the electronics. Areas of electrical noise should be avoided, cable lengths should be kept as short as possible, impedance matching should be verified, and the appropriate shielded cable should be used. When long transmission distances are involved or the area is electrically noisy, a preamplifier should be fitted to the meter (see Figure 23l).

The preamplifier output signal amplitude is independent of flow rate and is typically a 12-volt square wave signal. This high-level signal can be transmitted for great distances, typically 15,000 ft (4,572 m) and is far more immune to electrical interference than an unamplified pickup signal. The limitations of a preamplifier include increased cost and the necessity for a DC power supply at the meter. In some designs, an additional cable is required (a three-wire system as opposed to a two-wire system) and the ambient temperature is typically limited to 212°F (100°C).

GAS TURBINE METERS

The operating principle for the gas turbine meter is the same as that already described for the liquid turbine meter. The

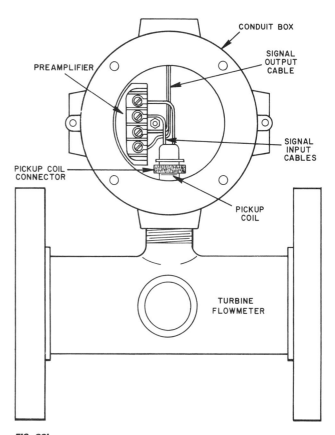

FIG. 23l
Complete turbine flowmeter assembly showing pickup coil and preamplifier.

major difference is that due to the much lower density of the gas, the available fluid driving torque is greatly reduced. Consequently, gas turbine meters feature various design changes to enable the meter to operate at higher fluid velocities and to compensate for the lower driving torque. The principal changes are the use of larger hub diameters to give a smaller ratio of rotor annular area to pipe area (see Figure 23m), lightweight rotors, increased number of blades, modi-

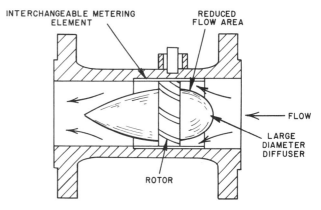

FIG. 23m
Typical gas turbine meter showing low ratio rotor annular-to-pipe area.

fied blade angle, and alternative bearings. Some designs feature local mechanical volume flow indication, achieved by reduction gears in the rotor driving external gears via a magnetic coupling.

Gas turbine meters find application in fuel and other gas measurement applications because of their simplicity and wide rangeability. Figure 23n shows the principle of the axial flow gas turbine meter. A flow diffuser increases the flowing gas velocity and directs it to a multibladed rotor mounted in precision bearings. The calibrated index is driven by the rotor through suitable gearing. Gas turbine meters are available in sizes from 2 to 12 in. pipe diameter (50 to 305 mm) and flow ratings up to 150,000 ft^3/hr (4500 m^3/hr). A desirable characteristic of gas turbine meters is their increase in rangeability at elevated operating gas pressures. Rangeabilities in excess of 100:1 are attainable in the large size meters operating at 1400 PSIG (9.7 MPa).

Due to the lower driving torque of the gas, it is essential to keep bearing frictional resistance to a minimum. The liquid turbine meter journal bearing is usually replaced by a ball race bearing. Any change in the bearing frictional resistance will result in a change in the meter calibration. Meters are frequently used in dust-laden gases and the ball races are frequently of the sealed, self-lubricated type. Some designs, however, use gas bearings.

It is essential to calibrate the gas turbine meter initially, preferably under simulated operating conditions, to establish its own specific K factor. A typical calibration curve is shown in Figure 23o. Linearity is normally ±1% of actual flow over a flow range of 20:1. Gas turbine meters have specific minimum and maximum volumetric flow rate values, and it is essential to select the meter on the basis of these volumetric flow rates and not on the basis of the pipe size. The meter must be sized on the basis of actual volume flow and not on the basis of standard reference units.

The turbine meter output frequency is proportional to the volumetric flow rate at the actual operating pressure and temperature. Pressure and temperature correction are required to convert the meter output into volume flow at reference conditions. If readout in mass units is required, either pressure and temperature correction can be used—although it does not compensate for variations in the composition of the gas—or the meter reading can be multiplied by a density gauge reading to give true mass flow.

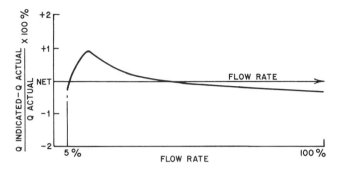

FIG. 23o
Typical gas turbine flowmeter calibration.

In any compensation system, the volume and pressure or density should be measured at the same flow rate. The gas turbine meter has a typical pressure loss of one velocity head ($\frac{1}{2} \frac{\rho V^2}{g}$) and a similar pressure distribution to that of the liquid turbine meter shown in Figure 23g. Consequently, if the pressure or density measurement is not taken at the rotor, a slight correction factor may be necessary to relate the measured value back to that pertaining at the rotor position.

Gas turbine meters are less sensitive to damage by grit and dust particles than are other positive displacement meters. Gas turbine meters can also operate at higher pressures and have a high flow rate capacity for a given meter size. In addition, if the meter fails, the gas flow is not obstructed, ensuring continuity of flow.

Typical upstream pipe requirements are 20 times the nominal meter diameter.

Due to possible variations in the meter bearing characteristics, calibration checks should be made at regular intervals if optimum performance is to be achieved.

IMPELLER AND SHUNT FLOWMETERS

Another flowmeter which is widely used in steam and gas flowmetering and totalizing applications is the shunt flowmeter illustrated in Figure 23p. It consists of an orifice plate in the main flow line and a self-operating rotor assembly in the bypass.

As gas flows through the meter body, a portion of flow is diverted to drive the fan shaft assembly, rotating on a jewel bearing. A second set of blades on the fan shaft, rotating in damping fluid, acts as a damper or governor.

Rotational speed of the shaft is proportional to the rate of flow at all rates within the normal range of the meter.

These flowmeters are available in sizes of 2 in. (50 mm) and larger. Their inaccuracy is around ±2% of the actual flow, and their rangeability is about 10:1.

Impeller- and propeller-type flowmeters are widely used in wastewater and irrigation applications where large flows and line sizes (up to 48 in., or 11.2 m) are required and cost

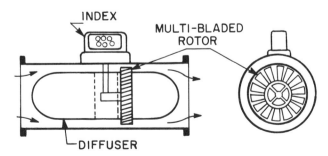

FIG. 23n
The axial flow gas turbine meter.

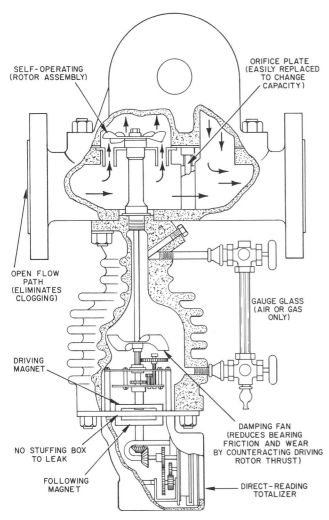

FIG. 23p
Shunt flowmeter.

pling drives an external mechanical register without requiring gears for its operation. The register is sealed from the process and requires no external power for operating a 6-digit totalizer and a flow rate indicator. Easy access and removal of the complete flowmeter is provided through a cover plate. Straightening vanes are provided to improve the flow profile. The materials of construction can be aluminum, epoxy-coated carbon steel, plastic, or stainless steel.

INSERTION-TYPE FLOWMETERS

Both the liquid and gas turbine meters described above are full bore metering devices; all flow passes through the meter. Their cost increases proportionately with pipe diameter. The insertion turbine meter is a set of small turbine meter internals mounted on a probe in a larger diameter pipe (see Figure 23r). The meter operating principles are the same as described previously except that the meter only measures the fluid velocity at a single point on the cross-sectional area of the pipe and does not "see" all the fluid. Total volumetric flow rate for the pipeline can then be inferred if certain assumptions are made about the velocity distribution across the pipe compared with the velocity at measurement point. The velocity distribution can either be established by "profiling" the line—that is, taking a series of measurements across the pipeline and establishing the fluid velocity profile—or by establishing the optimum compromise insertion depth for a range of pipe diameters.

The insertion meter cannot be as accurate as a full bore meter since it is only measuring velocity at one point on the cross-sectional area. It does, however, provide a very low cost metering system for large-diameter gas or liquid pipelines where accuracy is not important.

Insertion meters can be hot-tapped into existing pipelines through a valving system without shutting down the pipeline. A flanged riser, complete with valve, is welded to the pipeline. A hot-tap device is coupled to the valve, the valve is opened, and the pipe is penetrated. The hot-tap unit is withdrawn and the valve is closed. The insertion meter is

is more important than accuracy. Accuracy is claimed to be 2% of reading. As illustrated in Figure 23q, in this meter a corrosion-resistant plastic impeller is connected to a flexible and self-lubricating cable, which through a magnetic cou-

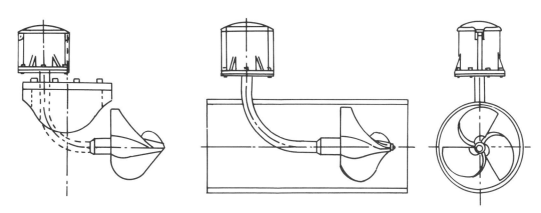

FIG. 23q
Impeller flowmeters are available in the paddle or the flow-through design. (Courtesy of McCroueker Div. of AMETEK).

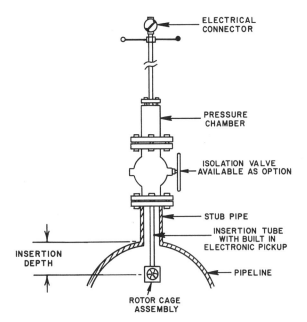

FIG. 23r
Insertion turbine flowmeter installed in large-diameter pipe.

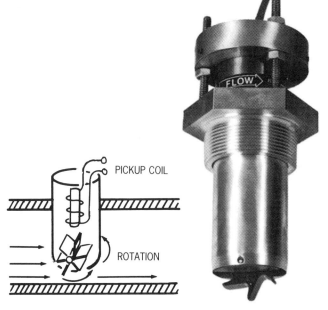

FIG. 23s
Paddlewheel flowmeter. (Courtesy of Data Industrial Corp.)

then installed, the valve is opened, and the meter is screwed in to the appropriate depth.

Insertion meters can be used on pipelines above 4 in. (102 mm) and, due to the small cross-sectional area relative to the pipe area, their pressure loss is very low. Typical linearity and repeatability figures are ±1% and ±0.25% respectively. These are point velocity readings; in overall volumetric accuracy terms, the effects of changes in velocity profile must also be considered.

Optical Flow Sensors

A specialized version of the insertion-type turbine flowmeter is the optical photoflow sensor. The flow transducer consists of a probe supporting a low mass rotating element which interrupts a light ray traveling from a light source to a photo transistor. The result is a pulse train which is converted into a volumetric flow representation.

This flow transducer provides flow ranges as high as 100:1, bidirectional measurement without additional calibration, and extremely low pressure drop. The transmitter has only one moving part, the flow-sensing element. The bearing for the element is not located directly in the flow stream, enabling the transducer to handle severe flow conditions such as heavy surging and pulsating flows.

The installation requirements include the need for 10 or more diameters of upstream straight run and the need to eliminate rotary valves (such as butterflies) at the ends of the measuring run.

Paddlewheel Flowmeters

One of the least expensive ways of measuring liquid flow in larger pipes (up to 12 in., or 305 mm) is to use one of the paddlewheel-type probes, illustrated in Figure 23s. The rotation of the paddlewheel can be detected magnetically or optically, and the different manufacturers offer these probe units in both plastic and metallic materials. Accuracies, pressure ratings, and temperature ratings are low, but rangeability is reasonable, as these units are responsive to velocities as low as 1 ft/s (0.3 m/s) and can handle just about any maximum velocity. The fixed-insertion-length designs tend to be less accurate than the adjustable ones as they cannot be moved as velocity profiles change. Some manufacturers claim these units to be useable on slurry service, but this is likely to require frequent cleaning.

Bibliography

American Petroleum Institute, "Measurement of Liquid Hydrocarbons by Turbine Meter Systems," A.P.I. Standard 2534.

Hall, J., "Flow Monitoring Application Guide," *Instruments and Control Systems,* February 1983.

Instrument Society of America Recommended Practice, RP31.1, "Specification, Installation and Calibration of Turbine Flowmeters," ANSI/ISA-1977.

May, D.L., "Accurate Flow Measurements With Turbine Meters," *Chemical Engineering,* March 8, 1971.

Murphy, H.N., "Flow Measurement by Insertion Turbine Meters," Measurement Technology for the 80's, ISA Symposium, Delaware, 1979.

Nichol, A.J., "An Investigation into the Factors Affecting the Performance of Turbine Meters," Conference on Fluid Flow Measurement in the Mid-1970's, East Kilbride.

Royek, S., "Flowmeters Help Tucson Conserve Water," *Water and Wastewater,* Vol. 2, Issue 5, October 1988.

"Turbine Flowmeters," *Measurements and Control,* February 1991.

Welch, J.V., "Trends in Low Gas Flow Metering," *InTech,* February 1991.

Withers, V.R., Inkley, F.A., and Chesters, D.A., "Flow Characteristics of Turbine Flowmeters," Conference on Modern Developments in Flow Measurement, Harwell.

24 Ultrasonic Flowmeters

R. SIEV (1969)   **L. D. DiNAPOLI** (1982)   **B. G. LIPTÁK** (1993)

Reviewed by L. D. DiNapoli (1993).

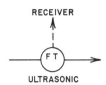

Flow Sheet Symbol

Types:	A. Transmission (contra-propagating transit time) B. Reflection (Doppler frequency shift or multi-pulse time shift) C. Open-channel (Note: A and B can be either "wetted" or "clamp-on." Type A is more often wetted; type B is more often clamp-on. Clamp-on designs cannot be as easily used on concrete or lined metal pipe as on ordinary metal pipe. Type C is usually noncontacting.)
Applications:	A. Clean liquids with little or no solids or bubbles; gases B. Slurries with solids (0.2 to 60% concentration, depending on particle size), liquids that are aerated or contain bubbles, gases with sound-reflecting particles; single-phase turbulent clean liquid C. Open-channel flow measurement based on upstream level in front of flumes or weirs
Flow Velocity Range:	A. Normal is 1 to 50 ft/s (0.3 to 15 m/s); maximum reported is 0.1 to 100 ft/s (0.03 to 30 m/s) B. Minimum velocity for solids to stay in suspension is about 2.5 ft/s (0.75 m/s); bubbles require 6 ft/s (1.8 m/s). Otherwise, 0.2 to 60 ft/s (0.06 to 18 m/s) would be usable C. Limits unavailable in the literature
Process Temperature:	A and B. −300 to 500°F (−184 to 260°C); higher or lower with special sound-transmitting wedges
Design Pressure:	A. Up to 3000 PSIG (207 bars) for wetted; unlimited for clamp-on B. Unlimited for clamp-on C. Usually atmospheric
Materials of Construction:	A. Spools or transducer probes can be steel, stainless steel, or alloys B. Usually clamp-on C. Noncontacting
Sizes:	A. 0.125 to 120 in. (3 mm to 3 m) diameter B. 0.5 to 72 in. (13 mm to 1.8 m) diameter C. Not applicable
Straight Pipe Required:	A and B. 10 to 20 diameters upstream, 5 downstream; very disturbed profiles require even longer straight runs or flow straighteners C. See requirements for weirs or flumes
Inaccuracy:	A. From 1% of actual flow to 2% of full scale. Error can be reduced by careful determination of pipe ID and by increasing number of paths B and C. 2 to 5% of full scale
Costs:	A. Spool designs in steel, not including options or special features and of the single-path design: $4000 for 4 in. (100 mm) $6000 for 10 in. (250 mm) $12,000 for 24 in. (600 mm). For clamp-on design, add $2000 regardless of size; add $1000 per acoustic coupling with thermal expansion chamber B. Clamp-on design, not including options or special features, is about $3000, independent of pipe size; acoustic coupling is additional C. $2000
Partial List of Suppliers:	Badger Meter Inc. (A*); Baird Controls Inc. (B); Bestobell Meter Flow Ltd. (A*,C); Bindicator (A); Bonitron Inc. (C); Caldon Inc. (A*); Controlotron Corp. (A*,B*); Davis Instrument Mfg. (A); Delta Controls Corp. (A); Delavan Inc. (A); Dynasonics Inc. (A,B*); Endress + Hauser Instruments (A,C); ESI-US (B-gas); Ferranti O.R.E. Inc. Accusonic Div. (A); Fischer & Porter Co. (C); Greyline Instruments Inc. (B); Hersey Measurement Co. (B*); ICC Federated (B); Inventron Inc. (C); Isco Inc. (C); Kam Controls Inc. (A); K-Flow (A); Krone America Inc. (A*); Leeds & Northrup (B*); Milltronics Inc. (C); Monitek Technologies Inc. (C); Moore Technologies Inc. (C); Nusonics Inc. (A,B*); Panametrics Inc. (A*,B*); Polysonics Inc. (A,B*); Quality Control Equipment Co. (C); Smith Meter Inc. (A,B*); Sparling Instruments Co. Inc. (A,C); TN Technologies Inc. (A); Tokyo Keiki Co. Ltd.; Transonic Systems Inc. (A*)

*Clamp-on design available.

When ultrasonic flowmeters were first introduced in the 1950s and 1960s the process control industry had high expectations for them. It was anticipated that they could be used for all types of process fluids (transit-time designs for clean fluids, Doppler reflection types for dirty, slurry-type streams); they could be installed without requiring a process shutdown (clamp-on types); their prices would be unaffected by pipe size and therefore would be economical for larger pipes; and they would not generate any pressure drop and would provide wide rangeability in both directions. These expectations generally did not come true until very recently.

The reason the image of ultrasonic flowmeters was rather tarnished for decades is that the complexity of the measurement produced was not realized initially. Factors that affect the measurement include the influence of the pipe, the flow profile, and the many practical obstacles and environment interferences that can occur. For the clamp-on designs it turned out to be difficult to maintain a good acoustic coupling that would not fail due to thermal expansion or drying out of couplant (at higher temperatures for prolonged periods) and would not short-circuit by "ringing around the pipe." These problems have now been solved, or at least well defined.

For wetted designs, the effect of changes in velocity profiles (Figure 24a), the effect of variations in the refractive index, upstream piping configuration, making the acoustic path long enough in smaller sizes, overcoming the effects of turbulence, ambient conditions, fluid property changes, including the change in the velocity of sound with temperature, all of these contributed to some failures. These problems have also been solved or at least better identified.

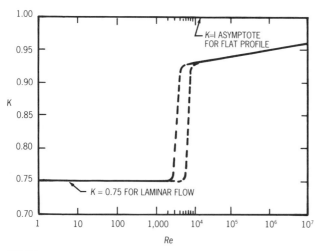

FIG. 24a
The ultrasonic flowmeters operate by averaging the velocity profile. This profile is flat at high Reynolds numbers and elongated at low Reynolds numbers. Consequently, the meter factor (K) also varies with Re, and it is rather unpredictable in the transition zone between laminar and turbulent regions.[1]

TRANSIT-TIME FLOWMETERS

As the name implies, these devices measure flow by measuring the time taken for an ultrasonic energy pulse to traverse a pipe section, both with and against the flow of the liquid within the pipe. Figure 24b is a diagram of a representative transit-time flowmeter.

The time (t_{AB}) for the ultrasonic energy to go from transducer A to transducer B is given by the expression:

$$t_{AB} = L/(C + V \cdot \cos\theta) \qquad 24(1)$$

The time (t_{BA}) to go from B to A is given by:

$$t_{BA} = L/(C - V \cdot \cos\theta) \qquad 24(2)$$

where C is the speed of sound in the fluid, L is the acoustic path length in the fluid, and θ is the angle of the path with respect to the pipe axis. By combining terms and simplifying it can be shown that for V ≪ C:

$$\Delta t = t_{BA} - t_{AB} = 2 \cdot L \cdot V \cdot \cos\theta / C \qquad 24(3)$$

It can also be shown that:

$$V = L \cdot \Delta t / 2 \cdot \cos\theta \cdot t_A^2 = K \cdot \Delta t / t_A^2 \qquad 24(4)$$

where t_A is the average transit time between the transducers.

Since the cross-sectional area of the pipe section or "spool piece" is known, the product of area and velocity will yield volumetric flowrate.

Frequency Difference-type

In sing-around flowmeters the reciprocals of transit times are used. This leads to a frequency difference (Δf) which is proportional to the flow velocity V or to the Mach number V/C. The difference in frequencies is related to the velocity as follows:

$$V = \Delta f \cdot L / 2 \cdot \cos\theta \qquad 24(5)$$

The multi-pulse time shift reflection method uses one or more pulses and times them to determine the change in range per second to an ensemble of scatterers. The change in range per unit time yields the velocity of scatterers.

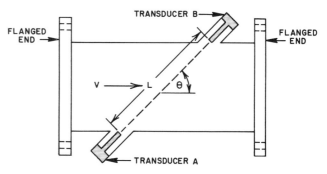

FIG. 24b
Transit-time flowmeter.

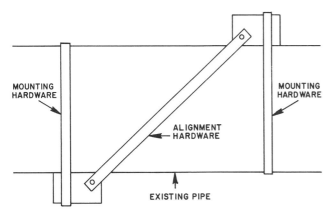

FIG. 24c
External transducers.

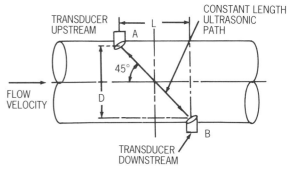

FIG. 24d
Wetted transducers communicate over a path that is fixed and independent of fluid sound speed, unless sound speed is nonuniform. (Courtesy of Panametrics Inc.)

Flowmeter Construction

The flowmeter usually consists of an electronics housing, transducers, and a pipe section. Several options are available as to the construction of the transducers and pipe section. Some designs allow removal of the transducers without interrupting process flow. A spool piece with integral transducers is one of the most common types of construction and is shown in Figure 24b. The manufacturer mounts the transducers to a flanged pipe section (spool piece). Usually the unit is calibrated by the manufacturer to meet the customer's specifications. The spool piece thus becomes an integral part of the hydraulic system so it is not easily retrofitted into an existing system.

Clamp-on transducers are capable of being mounted outside an existing pipe, as shown in Figure 24c. This type of system can be calibrated by the manufacturer only if detailed information on pipe diameter, pipe wall thickness, process fluid, percent of solids concentration, process temperature, variations in process temperature, etc. are also provided by the customer. This type of flowmeter is easily retrofitted onto an existing system, since no pipe section needs be installed.

Some manufacturers provide wetted transducers and mounting hardware that the user installs into an existing pipe. The user drills holes into the existing pipe and attaches the transducer mounting hardware by welding or other suitable means. The transducers are then mounted and aligned. Usually this type of unit may be calibrated by the user after measurements of transducer angle and spacing, and pipe diameter are made.

In case of "wetted" transducers it is important to keep the distance which the ultrasonic pulse travels constant. The best way to achieve this is to keep the transmitting and reflecting surfaces perpendicular to each other (Figure 24d). This eliminates the effect of changes in the angle of refraction, which can result from changes in process temperature or composition. Particularly in smaller pipes, the travel distance of the ultrasonic pulse might not be sufficient to produce accurate flow signals. Figure 24e shows how such distances can be increased.

In the early clamp-on designs one of the difficulties was establishing an acoustically efficient sound-conductive path between the transducer and the process fluid inside the pipe. In case of clamp-on designs the transmitting and reflecting surfaces are the inside surfaces of the pipe. Potential errors caused by "ringing around the pipe"—pulse blockage by a

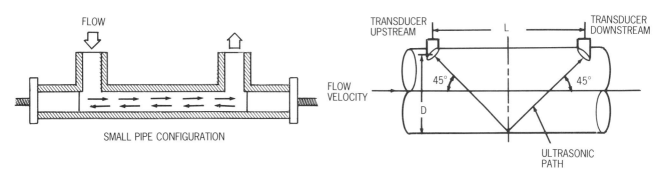

FIG. 24e
In smaller pipelines the time difference can be amplified by locating the transducers at the ends of a straight pipe section or by "bouncing" the ultrasonic pulse (one or more times) as shown on the right. (Courtesy of Panametrics Inc.)

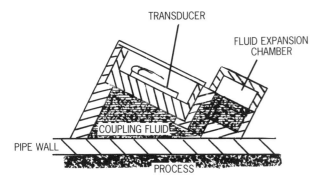

FIG. 24f
Liquid-filled ultrasonic coupling assembly.

molecular layer of air between the pipe and its lining, or variations in travel distance due to changes in the angle of refraction—are more difficult to solve with clamp-on units than with wetted ones. The acoustic coupling between the transducer and the pipe can be improved (Figure 24f) by eliminating the possibility of dry-out and by allowing for thermal expansion and contraction. Unfortunately such couplings are relatively expensive and require periodic refilling of the fluid.

Application and Performance

As with most flowmeters, the spool piece or pipe section must always be full to assure proper operation and volumetric flow indication. Most manufacturers will specify the minimum distance from valves, tees, elbows, pumps, etc. that will ensure accurate flowmeter performance. Typically, 10 to 20 diameters upstream and 5 diameters downstream are required. The flowmeter relies upon an ultrasonic signal traversing across the pipe; therefore, the liquid must be relatively free of solids and air bubbles. Bubbles in the flowstream generally cause more attenuation of the acoustic signals than solids do and therefore can be tolerated less. The flowmeter can tolerate a larger percentage of solids than bubbles.

Depending on the process fluid, proper transducer materials and protection must be chosen to prevent transducer damage due to chemical action. Process temperature limitations must also be considered for proper flowmeter application.

Accuracy is usually specified as a percent of rate. Typically for a single path flowmeter it is around 1 to 2% of rate, depending upon design, velocity, pipe size, and process. This accuracy can be expected only of calibrated flowmeters and only within the range of their calibration.

Repeatability is usually specified as a percent of rate, typically about 0.5% depending upon velocity range and calibration.

To improve performance and accuracy for larger pipe sizes, some suppliers offer flowmeters with two, four, or more pairs of transducers arranged to interrogate multiple acoustic paths. The cost of such units is higher than that of a single path flowmeter. The inaccuracy of multiple-path flowmeters can reach 0.5% of actual reading within a narrower range, if the flow velocity exceeds 1 ft/s (0.3 m/s).

DOPPLER FLOWMETERS

In 1842 Christian Doppler discovered that the wavelength of sound received by a stationary observer from a source that is moving toward the observer appears to be shorter and that the wavelength received when the source is moving away from the observer appears to be longer. The transmitter of a Doppler flowmeter projects an ultrasonic beam at a frequency of about 0.5 MHz into the flowing stream and detects the reflected frequency, which is shifted in proportion to stream velocity. The difference between transmitted and reflected velocities is called the "beat frequency" and its value relates to the velocity of the reflecting surfaces (solid particles and gas bubbles) in the process stream.

As shown in Figure 24g, an ultrasonic wave is projected at an angle through the pipe wall into the liquid by a transmitting crystal in a transducer mounted outside the pipe. Part of the energy is reflected by bubbles or particles in the liquid and is returned through the pipe wall to a receiving crystal. If the reflectors are traveling at the fluid velocity, the frequency of the reflected wave is shifted according to the Doppler principle, in proportion to the flow velocity.

Combining Snell's Law and the classical Doppler equation, the flow velocity can be determined as follows if $V \ll C$:

$$V = \Delta f \cdot C_t / (2 \cdot f_o \cdot \cos\theta) = \Delta f \cdot K \qquad 24(6)$$

where Δf is the difference between transmitted and received frequency, f_o is the frequency of transmission, θ is the angle of the transmitter and receiver crystal with respect to the pipe axis, and C_t is the velocity of sound in the transducer. As shown in Equation 24(6), velocity is a linear function of Δf. Since the inside diameter of the pipe is known, volumetric flow rate can be measured using Equation 24(7):

$$GPM = 2.45 \cdot V \cdot (ID)^2 \qquad 24(7)$$

The single transducer is a popular design. Both the transmitter and receiver crystal are contained in a single trans-

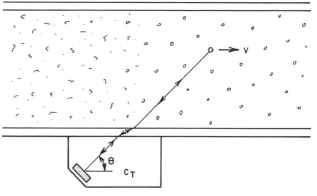

FIG. 24g
Doppler flowmeter principle of operation.

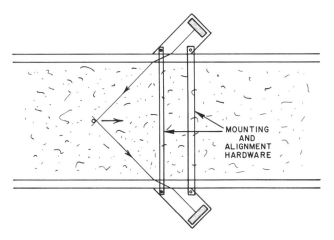

FIG. 24h
Two-transducer approach.

ducer assembly that mounts on the outside of the pipe. Alignment of the crystals is thus controlled by the manufacturer. This approach is shown in Figure 24g.

In one form of the dual transducer design, the transmitter crystal and the receiver crystal are mounted separately on opposite sides of the outside of the pipe. Alignment is maintained by a mounting assembly that maintains the relative positions of the transducers, as shown in Figure 24h.

When the process stream contains large amounts of solids (is sonically highly attenuative)—and bearing in mind that the velocity of the solid particles near the wall is likely to be less than the average, and recognizing that particles near the wall would dominate the readings in a single-transducer installation (causing large errors)—it is recommended that the two-transducer approach (supplemented by range gating) be used. With this approach the reflected ultrasonic radiation is received from a more representative portion or portions of the flow stream.

Each manufacturer provides instructions on how to mount the transducer or transducers to the pipe. The acoustic coupling to the pipe and the relative alignment of the transducers must be maintained in spite of pipe temperature changes and vibration, in order to obtain acceptable performance.

Application and Performance

As with transit-time flowmeters, in order to properly indicate volumetric flow, the pipe must always be full. A Doppler unit will, however, indicate velocity in a partially full pipe as long as the transducer is mounted below the liquid in the pipe.

The minimum straight pipe distance required from valves, elbows, tees, pumps, etc. is typically 10 to 20 diameters upstream and 5 diameters downstream for relatively clean fluids. This requirement can increase with process solids concentration or solids composition.

A Doppler flowmeter relies upon reflectors in the flow stream to reflect the ultrasonic energy. There is a lower limit for the concentration and size of solids or bubbles in the liquid that will give reliable, accurate operation. The flow must also be fast enough to keep the solids or bubbles in suspension, typically 6 ft/s (1.8 m/s) minimum for solids and 2.5 ft/s (0.75 m/s) for small bubbles, according to one manufacturer.

In the past few years some manufacturers have introduced flowmeters that operate at frequencies of 1 MHz or higher. The claim for these high-frequency units is that they will operate on virtually clean liquids because reflections will occur off the swirls and eddys in the flow stream. While this might be so, a high-frequency unit will generally not be suitable if the concentration of bubbles or particles exceeds 0.05% because the penetration depth of the higher frequency energy is much lower. Thus, for proper operation, a high-frequency Doppler flowmeter is limited to low concentration applications.

On horizontal pipes, the best place to locate the transducer around the circumference should be determined on the basis of empirical testing and application experience.

The Doppler flowmeters will operate independent of pipe material provided the pipe is sonically conductive. Such pipes as concrete, clay, and very porous cast iron absorb the ultrasonic energy and are not suited for Doppler-type flowmetering. Similarly, the lining in lined pipes is not bonded well enough to allow the use of this type of clamp-on flowmeter because even just a molecular layer of air is enough to block the transmission of the ultrasonic radiation.

The maximum operating temperatures of some ultrasonic flow transducers is about 212°F (100°C). The inaccuracy or error of the Doppler-type flowmeter is about 3% of full scale or span. The error does vary with flow velocity, pipe size, and flowmeter calibration. The error will also increase as the open flow area in the pipe changes either due to material buildup on the inside of the pipe, or because the inside diameter was incorrectly measured in the first place. Repeatability is usually about 1% of full scale or full span.

DISPLAYS, RECEIVERS, AND INTELLIGENT UNITS

The electronics can be mounted either integrally with the ultrasonic flowmeter or remotely, connected by cable. Remote location can be the choice on high-temperature services, although thermally isolating waveguides (Figure 24i) are also available. Routinely available transmitters can provide 4 to 20 mA DC analog, voltage, pulse train, or digital outputs, while the displays can provide both analog and digital indication in addition to totalization and alarming functions.

"Intelligent" flowmeters are capable of bidirectional flow measurement and of distinguishing true reflector movement from stationary particle or bubble vibration under no-flow conditions. One of the most promising tasks for smart ultrasonic flowmeters is to automatically evaluate the presence or absence of reflector particles or bubbles and to automatically switch from the transit-time to the Doppler

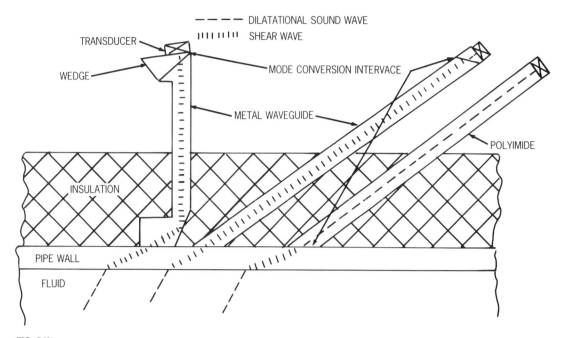

FIG. 24i
Three forms of thermally isolating waveguides.[2]

reflection mode of operation and back. If and when this capability is fully developed, this "all-purpose" ultrasonic flowmeter (Transfection®) will be able to measure the flow of both clean and dirty fluids. This capability, in combination with reduction in the unit costs, would make the ultrasonic flowmeter industry highly competitive.

References

1. Lynnworth, L.C., "Clamp-on Ultrasonic Flowmeters," *Instrumentation Technology,* September 1975.
2. Raptis, A.C., et al., "State of the Art of Acoustic Instrumentation for Coal Conversion Plants," Argonne National Laboratory, ANL/FE-49628-TM04, October 1981.

Bibliography

Addie, G.R., Maffett, J.R., DiNapoli, L.D., and Punis, G., "Doppler Flow Meter Tests at Georgia Iron Works," 1980 ISA Mining and Metallurgy Industries Division Symposium, Phoenix, Arizona.

Bersaas, S.L., "Experience with Ultrasonic Flare Gas Metering Onshore at Kårsto Terminal," Metering Seminar No. 7, K-Lab (Kårsto Metering and Technology Laboratory, a joint venture between Den norsek stats oljeselskap a.s (Operator) and TOTAL Marine Norsk a.s.), January 25, 1989.

Engineer's/User's Guidebook to Doppler Flow in Liquids, second edition, Polysonics Inc., 1989.

Faddick, R., Pouska, G., Connery, J., DiNapoli, L., and Punis, G., "Ultrasonic Velocity Meter," Sixth International Conference on the Hydraulic Transport of Solids in Pipes, BRHA Fluid Engineering (Cranfield UK), September 1979.

Fish, P.J., and Cope, J.A., "Effect of Frequency-Dependent Processes on Pulsed Doppler," *Ultrasonics,* July 1991.

Lipták, B.G., "Ultrasonic Instruments," *Instrumentation Technology,* September 1974.

Lynnworth, L.C., "Selected Alternatives to Conventional Ultrasonic Flowmeter," Ultrasonics International Conference of 1977, held in Guildford, England.

Lynnworth, L.C., "Ultrasonic Flowmeters," *Trans. Inst. M.C.,* Vol. 3, No. 4, Oct.–Dec. 1981.

Lynnworth, L.C., "Ultrasonic Measurements for Process Control," *Academic Press,* 1989.

Miller, R.W., *Flow Measurement Engineering Handbook,* second edition, McGraw-Hill, 1989. *Transit-Time Ultrasonic Flow Measurement in Liquids,* Panametrics application literature.

Munk, W.D., "Ultrasonic Flowmeter Offers New Approach To Large-Volume Gas Measurement," *Oil and Gas Journal,* September 6, 1982.

Mylvaganam, K.S., "High Rangeability Ultrasonic Gas Flowmeter For Monitoring Flare Gas," *IEEE Trans. UFFC,* Vol. 36 (2), 1989.

Mylvaganam, K.S., "Ultraschall—Durchflussmessung strömender Gase unter niedrigem Druck für einen breiten Messbereich," *Technisches Messen,* Heft 3, 1989.

Mylvaganam, K.S., "Ultrasonic Flowmeters Measure Flare Gas in the North Sea," *Oil & Gas Journal,* October 17, 1988.

Mylvaganam, K.S., Folkestad, T., Bo, R., "Fluenta FGM 100 Ultrasonic Gas Flowmeter," Internal Report of Chr. Michelsens Institut, CMI-No. 871413-1, Bergen, Norwegen, 1987.

Oehmke, R.L.T., "Flow Measurement Techniques," *Measurements and Control,* September 1991.

Raptis, A.C., "Acoustic Doppler Flowmeter," Fossil Energy I&C Briefs, March 1980.

Rienzenman, M.J., "Ultrasonic Meters Go with the Flow," *Mechanical Engineering,* September 1989, pp. 74–77.

Rütten, O., Deutsches Patent Nr. 520484, 1928.

Sakariassen, R., "Development of a New Gas Metering System—Ultrasonic Gas Flowmeters for Large Gas Flows," Gas Transport Symposium, January 30–31, Stavanger, 1989.

Schmidt, T.R., "What You Should Know About Clamp-on Ultrasonic Flowmeters," *InTech,* May 1981.

Shane, J.L., "Ultrasonic Flowmeter Basics," *Instrumentation Technology,* July 1971.

Siegbert, E., "Bestimmung von Korrekturfaktoren für ein Ultraschalldurchflussmessverfahren in Rohrleitungen mit Störungen," Dissertation, Fakultät Energietechnik, Universität Stuttgart, Stuttgart, 1982.

Smalling, J.W., Braswell, L.D., Lynnworth, L.C., and Wallace, D.R., "Flare Gas Ultrasonic Flow Meter," Proceedings of the Thirty-ninth Annual Symposium on Instrumentation for the Process Industries, 1984, pp. 27–38.

Ullebust, B., "Experience with Ultrasonic Flare Gas Metering Offshore on Gullfaks B," Metering Seminar No. 7, K-Lab (Kårsto Metering and Technology Laboratory, a joint venture between Den norsek stats oljeselskap a.s (Operator) and TOTAL Marine Norsk a.s.), January 1989.

Ultrasonic Flare Gas Flowmeter FGM 100: Product Information of Fluenta A/S, Bergen-Nesttun, Norway, 1987.

Waller, J.M., "Guidelines for Applying Doppler Acoustic Flowmeters," *InTech,* October 1980.

Zacharias, E.M., "Sound Velocimeters Monitor Process Streams," *Chemical Engineering,* January 22, 1973.

25 Variable-Area, Gap, and Vane Flowmeters

J. G. KOPP (1969, 1982)   **B. G. LIPTÁK** (1993)

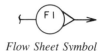
Flow Sheet Symbol

Types:
A. Rotameter (float in tapered tube)
B. Orifice/rotameter combination
C. Open-channel variable gate
D. Spring and vane or piston

Standard Design Pressure:
A. 350 PSIG (2.4 MPa) average maximum for glass metering tubes, dependent on size
Up to 720 PSIG (5 MPa) for metal tubes and special designs to 6000 PSIG (41 MPa)

Standard Design Temperature:
A. Up to 400°F (204°C) for glass tubes and up to 1000°F (538°C) for some models of metal tube meters

End Connections: Female pipe thread or flanged

Fluids: Liquids, gases, and vapors

Flow Range:
A. 0.01 cc/minute to 4000 GPM (920 m^3/hr) of liquid
0.3 cc/minute to 1300 SCFM (2210 m^3/hr) of gas

Inaccuracy:
A. Laboratory rotameters can be accurate to ±½% of actual flow; most industrial rotameters will perform within ±1 to 2% of full scale over a 10:1 range, and purge or bypass meters, ±5 to 10% of full range
B and D. ±2 to ±10% of full range
C. ±7.5% of actual flow

Materials of Construction:
A. TUBE: Borosilicate glass, stainless steel, Hastelloy, Monel, Alloy 20. FLOAT: *Conventional type*—brass, stainless steel, Hastelloy, Monel, Alloy 20, nickel, titanium, or tantalum, and special plastic floats. *Ball type*—glass, stainless steel, tungsten carbide, sapphire, or tantalum, END FITTINGS: Brass, stainless steel, or alloys for corrosive fluids. PACKING: The generally available elastomers are used and O-rings of commercially available materials; Teflon is also available

Cost: A ¼ in. (6 mm) glass tube purge meter starts at $70. A ¼ in. stainless steel meter is about $300. Transmitting rotameters start at about $1000, while with 0.5% of rate accuracy their costs are over $2000. A 3 in. (75 mm) standard bypass rotameter is about $500, while a 3 in. stainless steel tube standard rotameter is about $2000. A 3 in. tapered-plug variable-area meter in aluminum construction is about $1000, while the same unit in spring and vane design is around $750

Partial List of Suppliers: Aaborg Instruments & Controls Inc. (A); Aquamatic Inc. (B); Blue White Industries (A); Brooks Instrument Div. of Rosemount (A); Dwyer Instruments Inc. (A); ERDCO Engineering Corp. (D); ESKO Industries Ltd. (A); Fischer & Porter Co. (A); Flowmetrics Inc. (A); Gilflo Metering & Instrumentation Inc. (D); Gilmont Instruments Div. of Barnant Co. (B); Headland Div. of Racine Federated Inc. (D); ICC Federated Inc. (A); ISCO Environmental Div. (C); Ketema Inc. Schutte and Koerting Div. (A); Key Instruments (A); King Instrument Co. (A); Kobold Instruments Inc.; Krone America Inc. (A); Lake Monitors Inc.; Matheson Gas Products Inc. (A); McMillan Co.; Meter Equipment Mfg. Inc. (D); Metron Technology (A); Omega Engineering Inc. (A); G. A. Planton Ltd. (D); Porter Instrument Co. Inc. (A); Turbo Instruments Inc. (D); Universal Flow Monitors Inc. (D); Wallace & Tiernan Inc. (A); Webster Instruments (D)

The variable-area flowmeter is also a head-type flow sensor, but it does not measure the pressure drop across a fixed orifice; instead, the pressure drop is held relatively constant and the orifice area is varied to match the flow (Figure 25a).

In gravity-type variable-area flowmeters, increasing flow lifts the float, piston, or vane, and it is the weight of these flow elements that have to be balanced by the kinetic energy of the flowing stream. These units can only operate in the

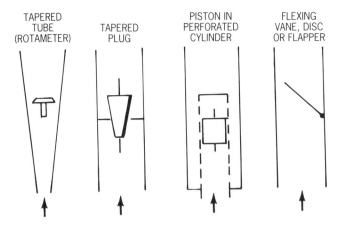

FIG. 25a
The area open to flow is changed by the flow itself in a variable-area flowmeter. Either gravity or spring action can be used to return the float or vane as flow drops.

vertical position. When the lifting of the float, piston, or vane is resisted by a spring instead of gravity, the meter can be installed in any position and can be configured as a pipeline spool piece. This advantage of piping convenience has to be weighed against the fact that a spring in the process stream can be a maintenance problem and may not yield as constant a force as does gravity.

All variable-area flowmeters can be provided with direct local indicators. In addition, most of them can also be furnished with pneumatic, electronic, digital, or fiber-optic transmission or with microprocessors for intelligent and convenient operation. A few of the designs can also be used as self-controlled flowmeters or purge meters. In the discussion below, the different variable-area flowmeters will all be discussed, starting with the tapered tube (rotameter) designs.

ROTAMETERS

Rotameters are popular choices for low flow measurement due to their low cost, simplicity, low pressure drop, relatively wide rangeability, and linear output. They are limited in that they can only be installed vertically, they can only be used on clean fluids, and if glass tubes are used they can represent a safety hazard (and dirt buildup on the glass tube can limit visibility). Glass tubes are also limited in terms of their maximum pressure and temperature ratings. Their safe working pressure drops as the tube diameter increases: For a ¼ in. (6 mm) diameter tube it is 450 PSIG (3 MPa), while for a 4 in. (100 mm) diameter tube it is only 40 PSIG (275 kPa). Because the stainless steel float expands more with temperature than does the glass tube, the operating temperature is limited to 200°F (93°C) for a ¼ in. (6 mm) diameter tube, and this limit drops to 120°F (49°C) for a 4 in. (100 mm) diameter tube. The flow capacity of glass tube rotameters is also limited. A ¼ in. (6 mm) diameter tube will pass 0.05 to 0.5 GPM (0.19 to 1.9 lpm) water or 0.18 to 1.8 SCFM (0.3 to 3 cmph) of air. For a 3 in. (75 mm) diameter

unit the maximum water flow range is 60 to 120 GPM (225 to 450 lpm) and the maximum air flow is 200 to 500 SCFM (350 to 900 cmph). (For a detailed discussion of purge rotameters refer to Section 18.)

The rotameter is a variable-area-type flowmeter. It consists of a tapered metering tube and a float which is free to move up and down within the tube. The metering tube is mounted vertically with the small end at the bottom. The fluid to be measured enters at the bottom of the tube, passes upward around the float, and out at the top. Figure 25b is a representation of a rotameter.

When there is no flow through the rotameter, the float rests at the bottom of the metering tube where the maximum diameter of the float is approximately the same as the bore of the tube. When fluid enters the metering tube, the buoyant effect of the fluid lightens the float, but it has a greater density than the fluid and the buoyant effect is not sufficient to raise it. There is a small annular opening between the float and the tube. The pressure drop across the float increases and raises the float to increase the area between the float and tube until the upward hydraulic forces acting on it are balanced by its weight less the buoyant force. The metering float is "floating" in the fluid stream. The float moves up and down in the tube in proportion to the fluid flow rate and the annular area between the float and the tube. It reaches a stable position in the tube when the forces are in equilibrium. With upward movement of the float toward the larger end of the tapered tube, the annular opening between the tube and the float increases. As the area increases, the pressure differential across the float decreases. The float will assume a position, in dynamic equilibrium, when the pressure differential across the float plus the buoyancy effect balances the weight

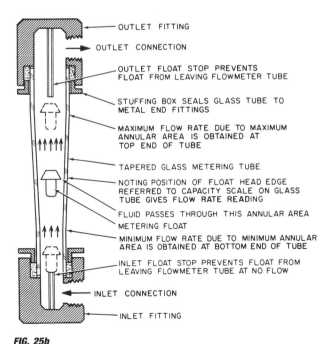

FIG. 25b
The rotameter.

of the float. Any further increase in flow rate causes the float to rise higher in the tube; a decrease in flow causes the float to drop to a lower position. Every float position corresponds to one particular flow rate and no other for a fluid of a given density and viscosity. It is merely necessary to provide a reading or calibration scale on the tube and flow rate can be determined by direct observation of the position of the float in the metering tube.

Metal metering tubes are used in applications where glass is not satisfactory. In this case, the float position must be indirectly determined by either magnetic or electrical techniques. The use of indirect float position sensors also provides functions other than direct visual indication. Rotameters are available which transmit pneumatic, electronic, or time pulse signals, or provide recording, totalizing, or control functions.

Sizing

To size a rotameter, it is customary to convert the actual flow to ''standard flow.'' For liquid flows it is necessary to calculate the GPM (l/m or hr) water equivalent. For gases it is necessary to determine the SCFM (l/m or hr) air equivalent. Capacity tables are based on these ''standard flows'' of GPM or cc/min of water and SCFM or cc/min of air at standard conditions. The tables also are based on using stainless steel floats.

The equations necessary to calculate the water or air equivalent are:

LIQUIDS

Volume Rate
GPM Water Equivalent

$$= \frac{(GPM)(\rho)(2.65)}{\sqrt{(\rho_f - \rho)\rho}} \qquad 25(1)$$

Weight Rate
GPM Water Equivalent

$$= \frac{(lbm/min)(0.318)}{\sqrt{(\rho_f - \rho)\rho}} \qquad 25(2)$$

Base or Contract Volume Rate
GPM Water Equivalent

$$= \frac{(GPM_b)(\rho_b)(2.65)}{\sqrt{(\rho_f - \rho)\rho}} \qquad 25(3)$$

GASES OR VAPORS

Standard Volume Rate
SCFM Air Equivalent

$$= \frac{(SCFM)(\rho_g\ std)(10.34)}{\sqrt{\rho_f(\rho_g\ act)}} \qquad 25(4)$$

Weight Rate
SCFM Air Equivalent

$$= \frac{(lbm/min)(10.34)}{\sqrt{\rho_f(\rho_g\ act)}} \qquad 25(5)$$

Operating or Actual Volume Rate
SCFM Air Equivalent

$$= \frac{(ACFM)(\rho_g\ act)(10.34)}{\sqrt{\rho_f(\rho_g\ act)}} \qquad 25(6)$$

GPM = maximum flow of liquid at metering condition in units of gallons per minute

GPM_b = maximum flow of liquid at base or contract condition in units of gallons per minute

lbm/min = maximum flow of fluid at metering condition in units of pounds per minute

SCFM = maximum flow of gas referred to a base or standard condition in units of cubic feet per minute

ACFM = maximum flow of gas at operating conditions in units of cubic feet per minute

ρ = density of flowing liquid at metering conditions in units of grams per cubic centimeter

ρ_b = density of flowing liquid at base or contract conditions in units of grams per cubic centimeter

ρ_f = density of float in units of grams per cubic centimeter

ρ_g std = density of gas at 14.7 PSIA and 70°F or 14.4 PSIA and 60°F in units of pounds per cubic foot

ρ_g act = density of gas at metering conditions in units of pounds per cubic foot

To facilitate these computations, manufacturers offer slide rules or nomographs specifically designed for rotameter sizing.

Rotameter Characteristics

A very wide range of liquids can be handled by the rotameter. A wide choice of tube, float, end fitting, and packing or O-ring materials are available for the particular service being considered. Even liquid metals like mercury and liquid lead can be metered. Since these metals are more dense than the stainless steel float, they are metered by an inverted rotameter. In this case, the flow is from top to bottom. When the meter is full of the liquid metal but there is no flow, the stainless steel float is buoyed up by the heavier liquid and rests at the inlet which is at the top. When there is flow, the flow forces the float down against the net buoyant force and the float takes a position related to the flow rate.

The rotameter is an inexpensive flowmeter for gas flow measurement. The pressure drop across the meter is essentially constant over the full 10:1 operating range. Pressure drop is low, generally less than 1 PSI (6.89 kPa). Special designs are available for even lower pressure drop.

The position of the float in the metering tube varies in a linear relationship with flow rate. This is true over ranges up to 10:1. Percent of maximum and direct reading scales are used. Rotameters can directly measure flows as high as 4000 GPM (920 l/hr). Even higher flow rates can be economically handled using the bypass-type rotameter. The capacity of the rotameter can be changed by changing the float. Various float configurations are available for higher capacities and generally permit a 2:1 change in capacity (Figure 25c). By using the same housing but changing both the metering tube and the float a gross change in capacity is possible. These changes can be necessitated by both a change in flow rate and a change in fluid density.

The rotameter tends to be self-cleaning. The velocity of the flow past the float and the freedom of the float to move vertically enable the meter to clean itself of some buildup of foreign material. Liquids with fibrous materials are one of the exceptions and should not be metered with rotameters. Generally, the size of particle, type of particle whether fibrous or particulate, and the abrasiveness of the particle determine the suitability of the rotameter for a given service. Also, the percent of solids by weight or by volume and the density of the solids influence the selection of the rotameter for this service.

Rotameters are relatively insensitive to viscosity variations. In the very small rotameters with ball floats, this is not the case and the meters do respond to Reynolds number changes, which makes them sensitive to changes in both viscosity and density. However, the larger size rotameters are less sensitive (Figure 25d). The viscosity immunity threshold can be as high as 100 centipoises (1 Pa·s). Meters can be operated above the viscosity limit; however, for these conditions the meter is calibrated for discrete viscosity conditions that are to be encountered, and correction curves are furnished to adjust the indicated flow to the actual flow for the given viscosity.

The rotameter can also be used to approximate mass flow rate since the float responds to changes in fluid density. For a fixed volumetric flow rate, the float position in the metering tube will change with changing fluid density. The effect of fluid density changes on float position is a function of the relative densities of the float and the fluid. The closer the float density approaches the fluid density, the greater the effect for a given fluid density change. It has been derived that if the float density is twice the fluid density then the compensation for fluid density change is exact and the rotameter is a mass flowmeter. However, fluid density normally varies, and since the float density is not adjustable to follow the fluid density changes, a compromise is made. The mean fluid density is used to establish the float density. A 10% fluid density change from the reference causes only a 0.5% inaccuracy in mass flow measurement. The mass rotameter can be used only for low-viscosity fluids such as

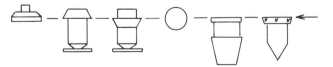

FIG. 25c
Variation in the shape of rotameter floats. The float on the right is provided with slots, which caused the early floats to rotate (for stabilizing and centering purposes); hence the name "rotameter."

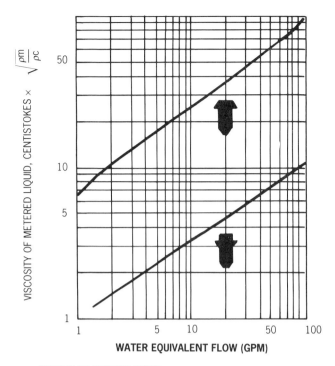

ρm = DENSITY OF METERED LIQUID
ρc = DENSITY OF CALIBRATING LIQUID (WATER)

FIG. 25d
Viscosity limits of rotameters depend on float shape. (Courtesy of Brooks Instrument Div. of Rosemount)

raw sugar juice, gasoline, jet fuels, and other light hydrocarbons.

Although the vast majority of rotameters operate at errors of 2 to 10% of full scale, some are available with percent of rate performance. Logarithmic scale meters are designed to give the same percent of rate accuracy at all scale positions over the 10:1 range of the meter. Accuracy statements of 0.5% of rate and 1% of rate are available. The high-accuracy-type rotameter finds greatest application in laboratory testing, development, and production, where best accuracy is mandatory.

The meter is not affected by upstream piping effects. The meter can be installed with practically any configuration of piping prior to the meter entrance.

The rotameter is a highly developed flowmeter. The meters are available with an extremely broad selection of alarms, indicators, transmitters (Figure 25e), totalizers, controllers, and recorders. A choice of totalizers, controllers, recorders, indicators, and alarms are available locally at the flowmeter. Practically any combination of system requirements can be handled by the accessories and instruments associated with rotameters.

Rotameter Types

Figure 25f shows a cross-sectional view of a representative general purpose rotameter. The meter is almost always used for flow indication only. A wide choice of materials is available for the float, packing, O-rings, and end fittings to

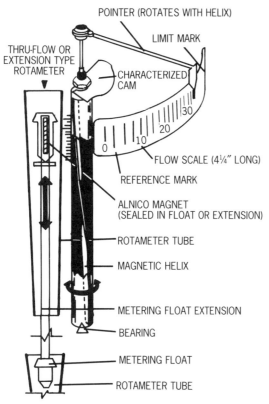

FIG. 25e
Magnetic coupling operates external indicator or transmitter across stainless steel rotameter tube.

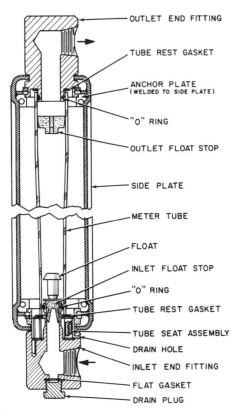

FIG. 25f
Glass tube rotameter.

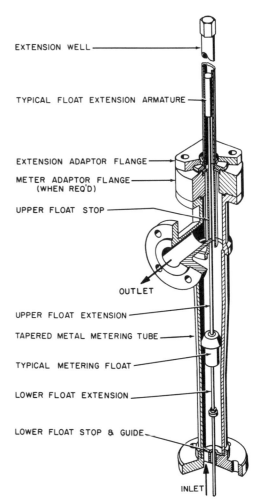

FIG. 25g
Metallic tube rotameter.

handle the widest selection of fluids. The only fluids which cannot be handled are those that attack the glass metering tube. The meters also are limited to the pressure and temperature extremes of the glass metering tube, and by safety considerations. The accuracy of these rotameters is usually ±1 to ±2% of full scale.

Metal tube meters are used when the general-purpose meters cannot be applied. They can be used for hot (above 100°F, or 38°C) and strong alkalies (above 20% concentration), fluorine, hydrofluoric acid, hot water (above 200°F, or 93°C), steam, slurries, or molten metals where glass cannot be used. This classification of meters is used where the operating temperature and pressure exceed the ratings of the glass tube or generally where transmission of electronic or pneumatic signals is needed. A typical metal tube meter is shown in Figure 25g.

Bypass and Pitot Rotameters

The cost of a rotameter installation can be reduced if, instead of a full pipe-size rotameter, an orifice or pitot tube is used in the main pipeline to develop a pressure drop, which in turn causes a related small flow that can be directed through an

Variable-Area, Gap, and Vane Flowmeters   **167**

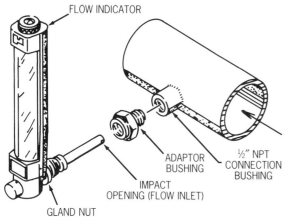

FIG. 25h
Pitot rotameter with bypass flow entering through impact opening (facing flow) and leaving through static port on opposite side (not shown). (Courtesy of Fischer & Porter Co.)

inexpensive bypass rotameter. Such units are illustrated in Figures 25h and 25i. The pitot-type can be used in pipe sizes 1.5 in. (38 mm) and larger, while the orifice bypass assemblies are available from ⅜ to 20 in. (1 to 51 cm) pipe diameter. In some designs the bypass rotameter is provided with a range orifice which is sized to lift the rotameter float to the maximum position when the flow in the main line is the maximum. The flow measurement is linear over a 10:1 range and is accurate to about 2% of full scale with the orifice and to 5 to 10% of full scale with the pitot design. These units are usually designed for clean process streams, such as water and air, and are provided with easily accessible filters for periodic cleaning. The bypass rotameters are also available with isolation valves in order to allow for their removal and maintenance while the process is in operation.

TAPERED PLUG AND PISTON METERS

The tapered-plug variable-area flowmeters are made with metallic meter bodies and are used on higher pressure applications, where errors of 5 to 10% full scale can be tolerated. They can be gravity-operated (Figure 25j) or spring-loaded (Figure 25k) and can handle pressures exceeding 1000 PSIG (70 bars). Their sizes range from ¼ to 4 in. (6 to 100 mm) and their body materials include brass, aluminum, steel, stainless steel, and PVC. The gravity-operated unit must be installed vertically, while the spring-loaded one can also be horizontal. One common application is to detect the flow rate of high-pressure oil.

There is a type of variable-area flowmeter that operates a piston in a perforated cylinder (see Figure 25l). This instrument is less expensive than a regular rotameter. It has been designed for clean liquid flows at rates up to 120 GPM (450 lpm) or gas flows up to 700 SCFM (20 SCMM) with pressures up to 100 PSIG (7 bars) and temperatures up to 400°F (205°C).

GATES AND VANES

A family of variable-area flowmeters operate by the flowing stream lifting hinged gates or forcing spring-loaded vanes to open. The variable gate is a mix of a variable-area and a flume-type flowmeter. It is used to measure wastewater or other liquids in open channels or in partially filled pipes. The meter can be inserted in 6 or 8 in. (150 or 200 mm) diameter pipes. A stainless steel ring holds it in place and an inflatable bladder seals the insert, so that all the flow will pass through the gate opening. The pivoted gate opening is pneumatically

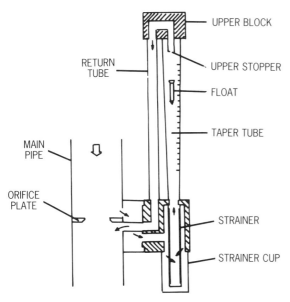

FIG. 25i
Bypass rotameter. (Courtesy of Aquamatic)

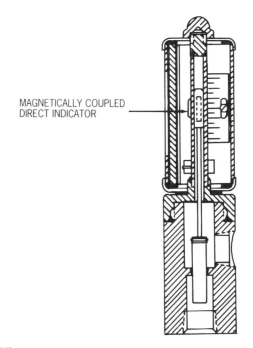

FIG. 25j
Tapered-plug variable-area flowmeter. (Courtesy of Brooks Instrument Div. of Rosemount)

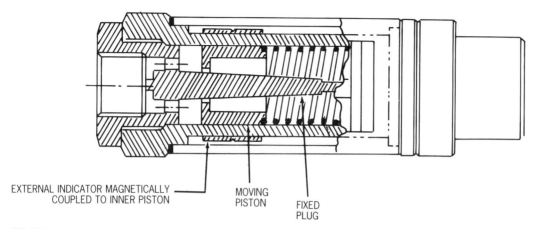

FIG. 25k
Tapered plug and spring-loaded piston. (Courtesy of Headland Div. of Racine Federated Inc.).

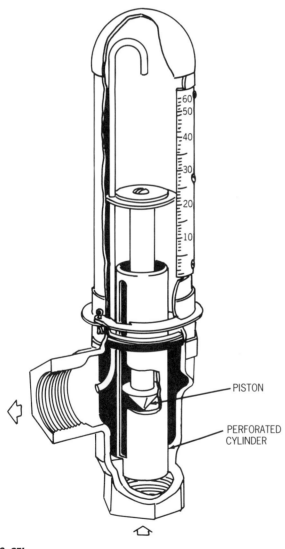

FIG. 25l
Piston in perforated cylinder variable-area flowmeter.

controlled and is measured together with the upstream level to arrive at the actual flow. The same 8 in. (200 mm) insert can measure the flow from 0.25 to 500 GPM (1 to 1900 lpm). Reasonable accuracy is obtained over this nearly 2000:1 flow range. Up to 10 GPM (38 lpm) the maximum error is 0.75 GPM (2.8 lpm), while at higher flows it is claimed to be 7.5% of actual reading. The meter is provided with data storage, printer, and local or remote monitoring capability.

One of the vane-type variable-area flowmeters resembles a butterfly valve (Figure 25m). The changing flow through the orifice area forces the spring-loaded vane to rotate. A shaft attached to the vane operates a pointer giving local flow rate indication. The measurement error is 2 to 5% of full scale. The meter can be used on oil, water, air, and other services and is available in 0.25 to 4 in. (6 to 100 mm) sizes and in most standard materials. This variable-area flowmeter is frequently used as an indicating flow switch for safety interlock purposes.

Another vane-type variable-area flowmeter is illustrated in Figure 25n. This unit can also measure the flow rates of liquids, gases, or steam and is available in sizes from ½ to 12 in. (12 to 300 mm). The flow indicator is magnetically

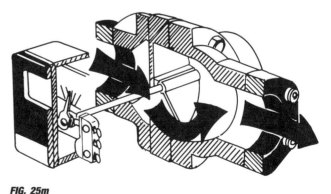

FIG. 25m
Rotary vane-type variable-area meter. (Courtesy of Universal Flow Monitors Inc.)

FIG. 25n
Magnetically coupled vane. (Courtesy of ERDCO Engineering Corp.)

coupled to the vane, and the indication is claimed to be accurate within 2% of full scale over a 10:1 range. The flow direction through the meter can be left-to-right, right-to-left, or vertical. Operating pressures are limited to 200 PSIG (1.4 MPa) and operating temperatures to 250°F (120°C) in standard and 400°F (205°C) in units with Viton O-rings. The housing can be aluminum, brass, copper/nickel, or stainless steel.

Bibliography

Blasso, L., "Flow Measurement Under Any Conditions," *Instruments and Control Systems,* February 1975.

Cross, D.E., "Rotameter Calibration Nomograph for Gases," *Instrumentation Technology,* April 1969, pp. 53–56.

Des Marais, P.O., "Variable-Area Meter for Viscous Service," *Instruments and Control Systems,* August 1961.

Gilmont, R., *Instruments and Control Systems,* November 1961.

Hall, J., "Solving Tough Flow Monitoring Problems," *Instruments and Control Systems,* February 1980.

Instrument Society of America, Recommended Practices RP16.1, RP16.2, RP16.3, RP16.4, RP16.5, and RP16.6. (These documents deal with the terminology, dimensions, installation, operation, maintenance, and calibration of rotameters.)

Polentz, L.M., "Theory and Operation of Rotameters," *Instruments and Control Systems,* June 1961.

"Rotameters/Variable-Area Flowmeters," *Measurements and Control,* September 1991.

Sanford, J., "What Should You Know About Flow Monitoring Devices," *Instruments and Control Systems,* September 1976.

26 V-Cone Flowmeter

B. G. LIPTÁK (1993)

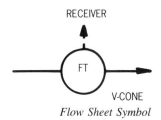

Flow Sheet Symbol

Applications:	Liquids, gases, and steam
Sizes:	From 0.5 to 72 in. (12 mm to 1.8 m)
Materials of Construction:	All 316 stainless steel or PVC construction with cone made out of stainless steel, PVC, or Teflon-coated aluminum
Design Pressure:	150 to 600 PSIG (10.3 to 41.4 bars) with flanged connections; higher with threaded connections
Design Temperature:	From cryogenic to 700°F (371°C)
Pressure Differential:	The low-range d/p cell can have a 0 to 2 in. H_2O (51 mm) range, while the high range can have 0 to 30 in. H_2O (762 mm). If even higher rangeability is needed, a third 0 to 250 in H_2O (6.35 m) d/p cell can be added
Reynolds Numbers:	Square root relationship is maintained down to Re = 8000
Inaccuracy:	Function of calibration and of accuracy of d/p cell used. With two transmitters, a 0.25% error of actual span used can be expected
Rangeability:	Over 10:1 if two transmitters are used, one for high the other for low pressure drop
Straight Pipe-Run Requirements:	Supplier recommends 2 diameters upstream and 5 diameters downstream
Beta Ratios:	From 0.35 to 0.85
Cost:	A ½ in. (12 mm) flow element costs $500. A high-quality d/p transmitter costs around $1800
Partial List of Suppliers:	Ketema Inc. McCrometer Div.

In a Venturi-Cone, or V-Cone, meter, a cone is positioned in the center of a metering tube (Figure 26a). This cone reduces the cross-sectional area available for the process flow and, in a way similar to an orifice restriction, generates a low-pressure region downstream of the flow element. The square root of the difference between the low pressure downstream and the upstream pressure is related to the flow through the meter.

The flow or pressure drop through the V-Cone meter can be calculated on liquid service using the following equations:

$$Q = 29.808 \times \frac{\beta^2 \times D^2}{\sqrt{1 - \beta^4}} \times \sqrt{\Delta P} \qquad 26(1)$$

$$\Delta P = \frac{Q^2 \times (1 - \beta^4)}{888.517 \times \beta^4 \times D^4} \qquad 26(2)$$

where:

Q = water flow in GPM
D = inside diameter of the process pipe in inches
d = outside diameter of the cone in inches
ΔP = pressure drop at full flow in PSID
β = beta ratio defined as: $\sqrt{D^2 - d^2}/D$

The main difference between an orifice plate and a V-Cone element is that at lower Reynolds numbers—where the velocity profile is no longer flat (as in the highly turbulent region) but starts to take on the shape of an elongated parabola, with the maximum velocity in the center of the pipe—the cone element tends to flatten the velocity profile. This is caused by the cone, which interacts with most of the flowing stream and tends to slow the flow velocity in the center while increasing it near the wall. This flow conditioning effect results in a velocity profile which is more uniform across the pipe and therefore closer to the fully developed turbulent behavior than it would be otherwise.

According to the supplier, this flattening of the velocity profile results in a true square-root relationship down to a Reynolds number of 8000. Below that, the transitional and later laminar flow behavior does develop, gradually changing the square root relationship into a linear one. As with all other d/p flow elements, the V-Cone is also usable in the transitional or laminar regions, but the interpretation of the pressure drop developed becomes more complex than just taking the square root. It requires the use of an accurately developed calibration curve, which can be read by the operator or can be placed into the computer's memory.

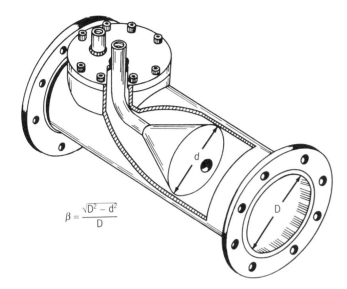

$\beta = \dfrac{\sqrt{D^2 - d^2}}{D}$

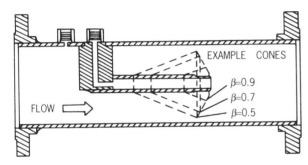

FIG. 26a
The V-Cone flowmeter requires less upstream straight pipe and maintains the square root relationship between flow and pressure drop at lower Reynolds numbers than does an orifice plate. (Courtesy of Ketema McCrometer Div.)

The manufacturer of the V-Cone flowmeter claims very high (30:1) rangeabilities and similarly high accuracies (0.5% of actual flow), presumably over such ranges. These claims are excessive. On the other hand, one should agree that 0.1% of actual span d/p cells are available, and if one uses two of them (a high and a low span), one can obtain a combined range of a 100:1 in terms of pressure drop, which corresponds to 10:1 in terms of flow, and the accuracy over that range can be 1% of actual flow. Two points should be made in this connection. One is that such performance assumes that the flow element is accurately calibrated over the complete flow range. The other is that this performance can be obtained from all well calibrated d/p flow elements, not just from the V-Cone.

In addition to maintaining turbulent conditions at lower Reynolds numbers, the V-Cone has the added advantage over the sharp-edged orifice of requiring less maintenance because the flow is directly away from the cone edge, and therefore the edge is not likely to wear. Because of the cone geometry, it also provides a sweeping action which eliminates stagnant areas and prevents gas accumulation or solids entrapment that can occur in front of sharp-edged orifices. The straight-pipe run requirements suggested by the manufacturer are substantially below that which are required by orifices: 2 diameters upstream and 5 downstream. This is because the cone reshapes the incoming nonuniform velocity profiles and thereby reduces the effect of upstream disturbances.

The V-Cone flowmeter should be installed horizontally, so that the two pressure taps are at the same elevation. This guarantees that the d/p cell will see zero pressure differential when there is no flow. If the unit is installed at a slope or vertically it is necessary to zero out the hydrostatic head difference between taps. The differential pressure generated by the flow element can be detected by any transmitting or indicating device. The manufacturer can also supply such devices, including "smart registers" that are capable of totalization and digital or analog retransmission.

Bibliography

De Boom, R.J., "Flow Meter Evaluation," 1991 ISA Conference, Paper #91-0509.
"Flow Measurement Solved With Venturi-Cone Meter," *InTech,* February 1989.
Husain, Z.D., "Flowmeter Calibration and Performance Evaluation," 1991 ISA Conference, Paper #91-0508.
Rusnak, J., "The Fundamentals of Flowmeter Selection," *InTech,* April 1989.

27 Venturi Tubes, Flow Tubes, and Flow Nozzles

W. H. HOWE (1969)   **J. B. ARANT** (1982)   **L. D. DINAPOLI** (1993)

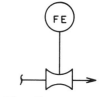

Flow Sheet Symbol

Design Types:	A. Venturi tubes B. Flow tubes C. Flow nozzles
Design Pressure:	Usually limited only by readout device or by pipe pressure ratings
Design Temperature:	Limited only by readout device, if operation is at very low or high temperature
Sizes:	A. 1 in. (25 mm) up to 120 in. (3000 mm) B. 4 in. (100 mm) up to 48 in. (1200 mm) C. 1 in. (25 mm) up to 60 in. (1500 mm)
Fluids:	Liquids, gases, and steam
Flow Range:	Limited only by minimum and maximum beta (β) ratio and available pipe size range
Inaccuracy:	Values given are for flow elements only; d/p cell and readout errors are additional A. ±0.75% of rate uncalibrated, to ±0.25% of rate calibrated in a flow laboratory B. May range from ±0.5 to ±3% of rate depending upon the particular design and variations in fluid operating conditions C. ±1% of rate uncalibrated to ±0.25% calibrated
Materials of Construction:	Virtually unlimited. Cast venturi tubes are usually cast iron, but fabricated venturi tubes can be made from carbon steel, stainless steel, most available alloys, and fiberglass plastic composites. Flow nozzles are commonly made from alloy steel and stainless steel
Pressure Recovery:	90% of the pressure loss is recovered by a low loss venturi when the beta (β) ratio is 0.3, while an orifice plate recovers only 12%. (The corresponding energy savings in a 24 in. (600 mm) waterline is about 20 HP.)
Reynolds Numbers:	Venturi and flow tube discharge coefficients are constant at Re > 100,000. Flow nozzles are used at high pipeline velocities (100 ft/s or 30.5 m/s), usually corresponding to Re > 5 million. Critical-flow venturi nozzles operate under choked conditions at sonic velocity
Costs:	Flow nozzles are less expensive than venturi or flow tubes, but cost more than orifices. ASME gas flow nozzles in aluminum for 3 to 8 in. (75 to 200 mm) lines cost from $200 to $750. Epoxy-fiberglass nozzles for 12 to 32 in. (300 to 812 mm) lines cost from $750 to $2500. The relative costs of Herschel venturis and flow tubes in different sizes and materials are given below:

	6" Stainless Steel	8" Cast Iron	12" Steel
Herschel Venturi	$8000	$5500	$6000
Flow Tube	$3600	$2100	$2900

Partial List of Suppliers:	ABB Kent Taylor (B); Badger Meter Inc. (A,B); Bethlehem Corp. (B); BIF Products of Leeds & Northrup (A,B,C); Daniel Flow Products Inc. (A,C); Delta-T Co. (C); Digital Valve Co. (critical-flow venturi nozzles); Fielding Crossman Div. of Lisle Metrix Ltd. (A,C); Fischer & Porter Co. (B); Flow Systems Inc. (B); Fluidic Techniques Inc. (A); Fox Valve Development Corp. (A); F.B. Leopold Co. (A,B); Permutit Co. Inc. (A,C); Perry Equipment Corp. (B); Henry Pratt Co. (A,B); Preso Industries (A,B); Primary Flow Signal Inc. (A,C); STI Manufacturing Inc.; Tri-Flow Inc. (A); Vickery-Simms Div. of FTI Industries (A); West Coast Research Corp.

Venturi tubes, flow nozzles, and flow tubes, like all differential pressure producers, are based upon Bernoulli's Theorem. General performance and calculations are similar to those for orifice plates. In these devices, however, there is continuous contact between the fluid flow and the surface of the primary device, in contrast to the pure line contact between the orifice plate edge and main flow. Surface finish of the devices can have some effect on the meter coefficient, although the venturi tube has a relatively constant coefficient, seldom varying more than a fraction of 1%. Today's modern precision manufacturing techniques allow much greater accuracy of the coefficient for venturi tubes and flow nozzles computed from dimensions, and the coefficients are only moderately less reliable than those for orifice plates. The C (meter coefficient) values for venturi tubes and flow nozzles have been well established with years of test data and are tabulated in reference sources such as the handbook called *Fluid Meters—Their Theory and Application*, published by the ASME (Reference 1). In general, this is not true of the proprietary flow tubes, and flow calibration is required to establish the actual meter coefficient. Meter coefficients for venturi tubes and flow nozzles are approximately 0.98 to 0.99 and for orifice plates average about 0.62. Therefore, almost 60% (98/62) more flow can be obtained through these elements for the same differential pressure.

THE CLASSIC VENTURI

The venturi tube as designed by Clemens Herschel in 1887, and described in the ASME handbook, *Fluid Meters*,[1] is shown in Figure 27a. It consists of a cylindrical inlet section equal to the pipe diameter; a converging conical section in which the cross-sectional area decreases causing the velocity to increase with a corresponding increase in the velocity head and a decrease in the pressure head; a cylindrical throat section where the velocity is constant so the decreased pressure head can be measured; and a diverging recovery cone where the velocity decreases and almost all of the original pressure head is recovered. The unrecovered pressure head is commonly called head loss.

The classic venturi is always manufactured with a cast iron body and a bronze or stainless steel throat section. At the midpoint of the throat, six to eight pressure taps connect the throat to an annular chamber so the throat pressure is averaged. The cross-sectional area of the chamber is 1.5 times the cross-sectional area of the taps. Since there is no movement of fluid in the annular chamber, the pressure sensed is strictly static pressure. Usually four taps from the external surface of the venturi into the annular chamber are made. These are offset from the internal pressure taps. It is through these taps that throat pressure is measured. This flow meter is limited to use on clean, noncorrosive liquids and gases, because it is impossible to clean out or flush out the pressure taps if they clog up with dirt or debris. The flow coefficient for the classic venturi is .984, with an uncertainty tolerance of ±0.75%.

SHORT-FORM VENTURIS

In the 1950s, in an effort to reduce costs and laying length, manufacturers developed a second generation, or short-form, venturi, shown in Figure 27b. There were two major differences in this design. The internal annular chamber was replaced by a single pressure tap or in some cases an external pressure averaging chamber, and the recovery cone angle was increased from 7 degrees to 21 degrees. The short-form venturi can be manufactured from cast iron or welded from a variety of materials compatible with the application. The flow coefficient for the short-form venturi is .985, with an uncertainty tolerance of ±1.5%.

The pressure taps are located one-quarter to one-half pipe diameter upstream of the inlet cone and at the middle of the throat section. A piezometer ring is sometimes used for differential pressure measurement. This consists of several holes in the plane of the tap locations. Each set of holes is connected together in an annulus ring to give an average pressure. Venturis with piezometer connections are unsuitable for use with purge systems used for slurries and dirty fluids since the purging fluid tends to short circuit to the nearest tap holes. Piezometer connections are normally used only on very large tubes or where the most accurate average pressure is desired to compensate for variations in the hydraulic profile of the flowing fluid. Therefore, when it is necessary to meter dirty fluids and use piezometer taps, sealed sensors which mount flush with the pipe and throat inside wall should be used. These sensors function as independent measuring devices at each tap connection, yet function together to read differential pressure only, while automatically compensating for static pressure changes within the pipe. Single pressure tap venturis can be purged in the normal manner when used with dirty fluids. Because the venturi tube has no sudden changes in contour, no sharp

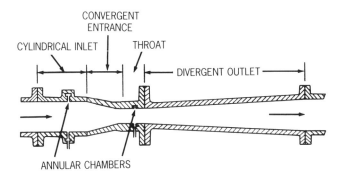

FIG. 27a
Classic Herschel venturi with annular pressure chambers.[1]

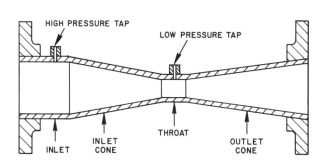

FIG. 27b
Short-form venturi tube.

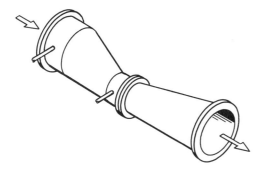

FIG. 27c
Eccentric venturi tube.

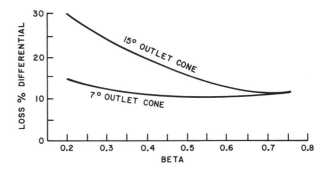

FIG. 27e
Venturi pressure loss.

corners, and no projections or stagnant areas, it is often used to measure slurries and dirty fluids which tend to build up on or clog other primary devices.

Venturis are built in several forms. These include the standard long-form or classic venturi (Figure 27a); a modified short form where the outlet cone is shortened (Figure 27b) an eccentric form (Figure 27c) to handle mixed phases or to minimize buildup of heavy materials; and a rectangular form (Figure 27d) used in ductwork. If a rectangular venturi is substantially square, it is customary to converge-diverge all four sides with angles the same as for the circular form. Where duct width is different from height, the short sides are kept parallel with the long sides converging-diverging. A converging angle of 21 degrees and a diverging angle of 15 degrees gives satisfactory operation. Throat length should be equal to minimum throat height or width, whichever is smaller. Tap locations are the same as for the circular form.

The angle of convergence, which may range from 19 to 23 degrees, is the classical value established by Herschel in 1887. This angle is not particularly critical and 21 ±1 degrees is commonly used. The recovery cone provides pressure recovery with its smooth flow transition. The classic long cone form is 7½ ±½ degrees on the divergence, but up to 15 degrees is allowed, and the sharper angle allows the short-form version to be fabricated. The 15-degree outlet cone sacrifices a modest amount of pressure recovery (Figure 27e). The venturi pressure loss of 10 to 25% is the lowest of the standard primary head measurement elements. The long-cone form develops up to 89% pressure recovery at 0.75β ratio, decreasing to 86% at 0.25β ratio. The short-cone form develops up to 85% recovery at 0.75β, decreasing to 75% at 0.25β ratio. As an example of the power savings to be obtained in an energy-short era, an added pressure recovery of 50 in. H_2O (1270 mm H_2O) differential pressure can represent a 10-horsepower savings in a 24 in. (610 mm) waterline flowing at a velocity of 6 ft/s (1.829 m/s). For a comparison of various head-meter elements from the pressure recovery point of view, see Figure 27f.

Installation

A venturi tube may be installed in any position to suit the requirements of the application and piping. The only limitation is that with liquids the venturi is always full. In most

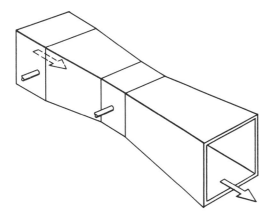

FIG. 27d
Rectangular venturi tube.

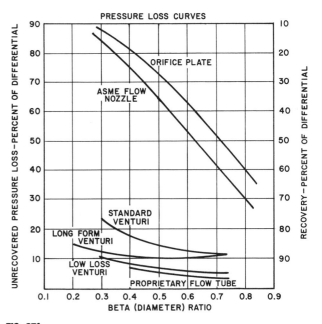

FIG. 27f
Pressure loss curves.

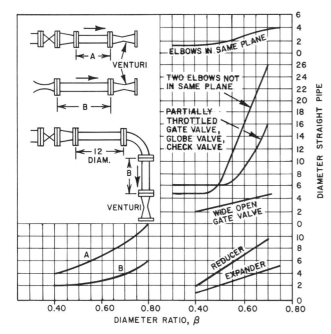

FIG. 27g
Venturi piping requirement.

cases, the valved pressure taps will follow the same installation guidelines as for orifice plates.

Upstream piping should be as long as needed to provide a proper velocity profile (Figure 27g). However, in most installations shorter upstream piping is required than for orifices, nozzles, or pitot tubes because the venturi hydraulic shape itself provides some flow conditioning. Often, the combined length of a venturi and its upstream piping is less than the overall amount of piping required for an orifice or nozzle. Figure 27h shows typical upstream pipe diameters required for various elements at 0.7β ratio and one elbow

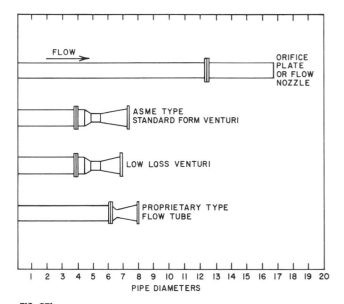

FIG. 27h
Typical installation piping comparison.

upstream. Straightening vanes can be used upstream to reduce the inlet pipe length.

In *Fluid Meters*,[1] the ASME recommends the use of tubular straightening vanes (19 tubes and 2 diameters long) upstream of the venturi to reduce the inlet pipe length. The vane installation should have a minimum of 2 diameters upstream and 2 diameters downstream before entering the venturi.

There is no limitation on piping configuration downstream of the venturi except that a valve should be no closer than 2 diameters. Valves on other devices that protrude into the flow stream should not be mounted upstream of the venturi, if possible.

Flow Calculations

The American Society of Mechanical Engineers Fluid Research Committee has adopted a general coefficient of discharge of 0.984 for the classic rough-cast entrance cone venturi tube from 4 in. (100 mm) through 32 in. (813 mm) and for β ratios between 0.3 and 0.75. For tubes with machined entrance cones, the general coefficient is 0.995. Reynolds number must be 200,000 or greater. Approximate flow rates can be calculated from a working equation:

$$W = 353 d^2 \sqrt{\frac{h\rho}{1-\beta^4}} \qquad 27(1)$$

and, for approximate venturi tube design:

$$\beta = \frac{1}{\sqrt[4]{1 + \dfrac{125{,}000\, h\rho D^4}{W^2}}} \qquad 27(2)$$

For Reynolds numbers between 50,000 and 200,000 substitute 344 instead of 353. Below 50,000, reliable data is not available. It should be noted that, in contrast to an orifice, a decrease in Reynolds number results in a decrease of flow corresponding to a given differential pressure. Other correction factors such as temperature coefficient of expansion, gas expansion factor, and so on are similar to those for orifices and flow nozzles.

FLOW TUBES

There are several proprietary primary-head-type devices which have a higher ratio of pressure-developed-to-pressure-lost than a venturi tube (Figure 27i). They are all considerably more compact than the classical venturi tube with its long recovery cone, although the short-form venturi can come close to some types of these tubes.

These designs are available in cast iron, can be welded from various materials, or in some cases can have insert-type units in fiberglass-reinforced plastic or metal. The flow coefficient ranges from .9797 for an all static tap "near venturi" design to .75 for an all corner tap "flow tube" design. All of these proprietary units are available in the United States except the Dall tube, which was developed in

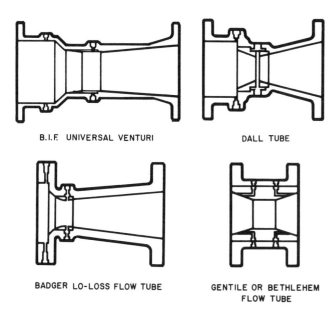

FIG. 27i
Proprietary flow tubes.

England. All of these tubes vary in contour used, tap locations, and differential pressure and pressure loss for a given flow. All have a laying length less than 4 diameters long. The shortest are the corner tap designs, with lengths equaling 2 to 2.5 diameters.

A flow tube is broadly defined by the ASME as any differential-pressure-producing primary whose design differs from the classic venturi. Flow tubes fall into three main classes, depending upon the hydraulic position of the inlet and throat pressure tap. Type 1 has static pressure taps at both the inlet and outlet, Type 2 has a corner tap in the inlet and a static tap in the throat, and Type 3 has a corner tap at both the inlet and outlet.

The classic venturi had static pressure taps which provided a section where the velocity is not changing direction and is parallel to the pipe wall. A corner tap senses pressure in a section where the velocity is changing direction and is not parallel to the pipe wall. Figure 27i shows examples of several flow tubes.

While Type 3 flow tubes can be useful in larger sizes because of their shorter lay length, they may also require longer upstream pipe runs for proper performance. They can be subject to coefficient change due to variations in Reynolds number, line size, and beta ratio; the manufacturers can provide data on these effects.

The B.I.F. Universal Venturi is the product (Type 1) that most closely approaches the Herschel design classic venturi. The inlet cone has two vena-contracta angles which condition the fluid as it enters the throat that is claimed to reduce the sensitivity to upstream piping configuration and give higher accuracy. Also claimed are a stable coefficient (0.9797) unaffected by internal surface roughness, lower Reynolds number application (90,000), low head loss (4 to 18%), and extensive documentation including expansion factors.

While flow tubes can be useful in larger sizes because of their shorter lay length, they may also require longer upstream pipe runs than the venturi for proper performance and thus lose any real advantage. They can be subject to coefficient change with viscosity and Reynolds number; the manufacturers can provide data on these effects. None has the smooth contour and resistance to clogging of the venturi meter; however, some are claimed to operate satisfactorily on wastewater and sewage flow measurement.

In general, these devices are available only in 4 in. (100 mm) and larger sizes up to 48 in. (1219 mm). There is little justification for their use in small flow, small pipe applications. In the larger sizes, their installed cost may be less than that of the venturi tube. Accuracy depends basically on the manufacturer's calibration data. Derivation of the flow coefficient by extrapolation from theory and tests on smaller sizes is much less direct than in the simple structure of the venturi tube; actual flow calibration, particularly in sizes above 24 in. (610 mm), can be difficult and expensive. While these devices generally have a better pressure recovery than the venturi (expressed as a percentage of the differential), most flow tubes have a lower coefficient of discharge (less efficient). As a result, there is often very little difference in the actual head loss.

In selecting a primary flow element, the possible advantages of slightly lower pressure loss and shorter laying length of the flow tubes should be carefully weighed against the metering accuracy and well-established flow data available on the Herschel-form venturi. The ASME recommends that if a proprietary flow tube is used, it should be calibrated with the piping section in which it is to be used and over the full range of flows to which it will be subjected. The only possible exception to this is the Universal Venturi, and it must be carefully evaluated. The background of extensive tests under a wide range of conditions which supports orifice meters does not exist for these proprietary devices.

FLOW NOZZLES

There are two types of flow nozzles. The 1932 ISA nozzle is a European design that has not seen use in the United States. A special variation known as a venturi nozzle is a hybrid combination of a 1932 ISA nozzle inlet profile combined with the divergent cone of a venturi tube. The common nozzle used in the United States is the so-called long radius or ASME flow nozzle. This nozzle comes in two versions known as low beta ratio and high beta ratio designs. This flow nozzle, shown in Figure 27j, is a metering primary whose shape consists of a quarter ellipse convergence section and a cylindrical throat section. In the United States the nozzle generally used is the long radius ASME flow nozzle. The ASME Fluid Meters Research Committee has investigated various configurations and has developed the geome-

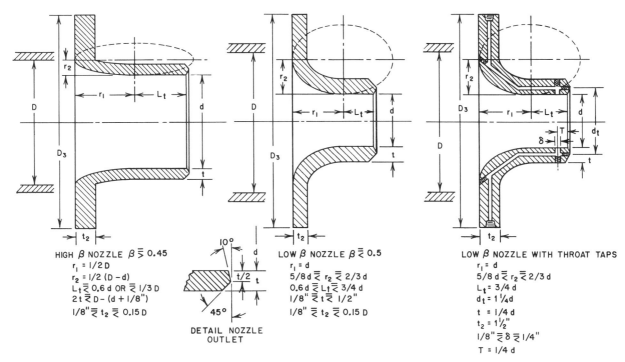

FIG. 27j
ASME nozzle construction.

try for these nozzles based on the required beta ratio for the application. High beta nozzles are recommended for diameter ratios between 0.45 and 0.80. Low beta nozzles are recommended for diameter ratios between 0.20 and 0.50. For beta values between 0.25 and 0.5, either design may be used.

The differences between the two nozzles is basically a flattening of the ellipse in the high beta ratio version. The power test code, PTC-6, requires that the low beta ratio version be used in their test section for turbine acceptance.

Both types of nozzles may be either welded in the pipe line or provided with a holding ring for mounting between flanges. The latter design, shown in Figure 27k, is preferred when frequent inspection of the nozzle is required.

Nozzles may be manufactured from any material that can be machined; typically, they are fabricated from aluminum, fiberglass, stainless steel, or chrome-moly steel. Modern manufacturing methods and fluid contact surface finishes in the order of 6 to 10 microinches result in more predictable nozzle coefficients and highly repeatable data. The standard surface finish is 16 RMS. Flow nozzle inaccuracy of $\pm 1\%$ is standard with $\pm .25\%$ flow calibrated. The standard coefficient, as published in the ASME's *Fluid meters* (Ref. 1), is .9962 with correction factors for beta ratio and throat Reynolds number. ASME gives an uncertainty of $\pm -2\%$ for nozzles having a beta ratio between .2 and .8 and throat Reynolds numbers between 1×10^5 and 2.5×10^6.

Most ASME nozzles are calibrated as meter sections 20 pipe diameters long. As with venturi tubes, the uncertainty of calibrated units depends upon the uncertainty of the hydraulic laboratory. Generally one could expect $\pm -.25\%$ uncertainty on the calibration.

The outlet or discharge side of the nozzle is normally beveled and is one of the more critical points of manufacture. Where the 10-degree back angle meets the throat bore, the edge must be sharp. Particular care must be taken to avoid taper and out-of-roundness of the throat.

Flow nozzles are made in various configurations. The most common is the flange-type (Figure 27k) but others are the holding-ring type, the weld-in type, and the throat-tap type. Differential pressure measurement taps are commonly located one pipe diameter upstream and one-half pipe diameter downstream from the inlet face (United States practice), except for the throat-tap type, which has a special down-

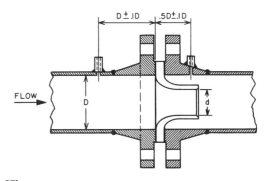

FIG. 27k
Typical nozzle installation.

stream construction. The PTC-6 nozzle uses throat taps to sense the low pressure and standard pipe wall taps for the high pressure.

Application Considerations

Tap installation precautions are the same as for orifice plates. The preferred installation position for flow nozzles is horizontal, but they can be installed in any position. However, a vertical downflow position is preferred for wet steam or gases and liquids with suspended solids. In general, upstream and downstream piping requirements are similar to those required for orifices. Because of width, nozzles installed between flanges are difficult to remove. Common practice is to provide a flange in the downstream piping to allow the nozzle to be removed as part of a spool section for inspection at regular intervals. Sometimes inspection openings are placed just upstream of the nozzle so that frequent inspections can be made without removing the nozzle from service.

Flow nozzles are particularly suited for measurement of steam flow and other high-velocity fluids, fluids with some solids, wet gases, and similar materials. Since the exact contour is not critical, the flow nozzle can be expected to retain good calibration for a long time under erosion or other hostile conditions. Because of its streamlined contour, it tends to sweep solids or moisture through the throat and is far superior to orifice plates in these services. Because tap geometry and contour is critical in order to maintain calibration, it is not recommended to use flow nozzles on slurries or dirty fluids.

A flow nozzle will pass about 60% more flow than an orifice plate of the same diameter and differential pressure. It also has the advantage of operating acceptably over a wide beta ratio range of 0.2 to 0.8. For the same flow and differential pressure, the flow nozzle has a similar but slightly lower pressure loss than an orifice plate. This becomes apparent when it is recognized that the area of the throat and the velocity in the throat of a flow nozzle must be approximately the same as the area of flow and velocity at the vena contracta following an orifice in order to develop the same differential pressure from the same flow. The slightly lower pressure loss of the flow nozzle is due to its streamlined entrance.

On the other hand, because the ASME flow nozzle does not utilize a recovery cone, the permanent head loss can still be as much as 40% of the differential pressure (Figure 27f). In an effort to reduce these losses, particularly in applications for PTC-6 testing of turbines, a recovery cone may be added. In this design the permanent head losses can be substantially reduced. The actual amount of reduction should be determined through testing.

While nozzles should be used at Reynolds numbers of 50,000 or above, data is available for Re down to 6,000, so it is possible to use nozzles with more viscous fluids. Yet, work published by the ASME Fluid Research Committee suggests that the most stable flow coefficients are seen at

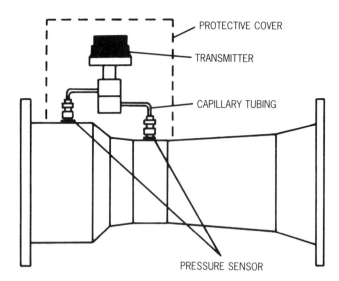

FIG. 27l
A variable capacitance flow transmitter can be mounted integrally to the flow tube and provided with chemical seals for protection against plugging or corrosion. (Courtesy of BIF Products of Leeds & Northrup)

throat Reynolds number of 1×10^6. For throat Reynolds numbers below 1×10^5, the shift in flow coefficient can be as high as 6%.

Flow nozzles have very high coefficients of discharge, typically 0.99 or greater. Using a typical value of 0.993, approximate flow rates can be calculated from a working equation:

$$W = 358d^2 \sqrt{\frac{h\rho}{1-\beta^4}} \qquad 27(3)$$

and, for approximate flow nozzle design:

$$\beta = \frac{1}{\sqrt[4]{1 + \frac{128{,}000h\rho D^4}{W^2}}} \qquad 27(4)$$

Critical-Velocity Venturi Nozzles

One of the most accurate ways to measure gas flow is to cause "choked" flow (sonic velocity flow) through a venturi nozzle. These units are also used as secondary flow standards in calibrating other flowmeters. The nozzle can be the ASME long-radius, elliptical inlet, wall-tap nozzle; the ISA 1932 nozzle; or the ASME throat-tap nozzle used in steam turbine testing.

One of the highest rangeability and most accurate gas flowmeters has been devised by combining the sonic venturi nozzles with the digital control valve (Figure 1b).

ACCURACY

Operation and calibration of venturi tubes over a period of many years has resulted in extensive documentation. As a result, most manufacturers will guarantee a standard design inaccuracy of ±0.75% of actual flow. This can be reduced to ±0.25% by calibration at a recognized hydraulics laboratory. Modern manufacturing techniques have led to predictable discharge coefficients and a repeatability of ±0.2% for venturi tubes of the same size and design.

For very small (under 4 in., or 100 mm) and very large (over 32 in., or 813 mm) venturi tubes and for very high (over 2,000,000) or very low (under 150,000) Reynolds numbers, flow calculations for venturi tubes have about a 50% greater uncertainty than a corresponding sharp-edged orifice plate. However, fluid flow calibration, particularly when made under conditions closely approximating service values, can provide a coefficient with practically the same accuracy as that of the calibration facilities.

The error contribution of the d/p generating flow sensor is defined as the uncertainty tolerance of the flow coefficient. The inaccuracy values can range from as low as 0.25% of rate for calibrated units to 1.5% of rate for uncalibrated welded units and can be expected to hold true only for a limited range of Reynolds numbers (see Figure 1i) and beta ratios.

The overall performance of the total flow measurement system therefore will be the sum of the transmitter and sensor errors. This sum will hold true only over the flow range between the maximum flow and the flow rate corresponding to the minimum Reynolds number for which the sensor error is still guaranteed. This minimum Reynolds number for venturis and flow tubes is around 100,000, and for flow nozzles it is over a million. Consequently, the rangeability of these devices, if defined in terms of actual flow error, can be rather low.

DIFFERENTIAL PRESSURE MEASUREMENT

The differential pressure generated by these primary devices (venturi, flow tubes, and flow nozzles) can be measured by manometers, gauges, or electronic pressure transmitters. Accuracy of analog electronic transmitters varies from ±0.1 to ±0.5% of calibrated span depending on the manufacturer. When square root circuitry is added there is usually a ±.05% increase. Microprocessor-based, or "smart," transmitters have an inherent accuracy of ±0.1% of calibrated span or less, whether or not the square root function is used.

Overall accuracy of the entire flowmeter system is a function of the transmitter and other instruments in the loop. The most common method for determining accuracy is to root mean square (RMS) the errors to calculate the total error. Let's look at an example with only the primary device and an electronic transmitter. Recognize that transmitter accuracy is specified as a percent of calibrated span. Thus, at 25% of span the error will be four times the error at full scale. The venturi, flow tube, and flow nozzle accuracy is specified as a percent of rate. Thus, the sensor accuracy is the same throughout its usable range.

Assume the venturi and the electronic transmitter are both set up such that there is a 100 in. H_2O differential pressure at full scale flow. Assume the venturi has an accuracy of 0.75% of rate and the transmitter an accuracy of 0.25% of calibrated span. At full scale flow, the total RMS uncertainty will be $\sqrt{(.75)^2 + (.25)^2} = 0.79\%$. At 50% flow the total RMS uncertainty will be $\sqrt{(.75)^2 + (.50)^2} = 0.90\%$. At 25% of flow the total RMS uncertainty will be $\sqrt{(.75)^2 + (.50)^2} = 1.25\%$. At 10% flow the total RMS uncertainty will be $\sqrt{(.75)^2 + (2.5)^2} = 2.61\%$. Therefore, the transmitter can contribute significantly to the total error of the system when used over a wide range, even though the primary device maintains its accuracy over that wide range.

Some manufacturers of smart transmitters have routines that reduce the full scale value of the transmitter as the differential pressure signal from the primary decreases in order to increase the total accuracy. This requires that the transmitter communicate digitally to the receiver so the reduced full scale and the measured differential pressure can be transmitted to the receiver. The total error in a flow measurement is the sum of two errors: that of the sensor and that of the transmitting or readout device. The error contribution of the best d/p transmitters is about 0.1% of span. In order to cover a flow range of 10:1, a d/p range of 100:1 needs to be covered, which requires either an extremely wide-range d/p cell or, more likely, two d/p transmitters (a high span and a low span one). If such a dual transmitter configuration is used and if the transmitters are switched as needed, the actual error contribution of the transmitter can be limited to 1% of actual flow.

CONCLUSIONS

The main limitation of venturi tubes is cost, both for the tube itself and often for the piping layout required for the length necessary in the larger sizes. However, the energy cost savings attributable to its higher pressure recovery and reduced pressure loss usually justify the use of venturi tubes in larger pipes.

Another limitation is the relatively high minimum Reynolds number required to maintain accuracy. For venturis and flow tubes this minimum is around 100,000, while for flow nozzles it is over 1 million. Naturally, correction data is available for Reynolds numbers below these limits, but measurement performance will suffer.

Cavitation can also be a problem. At the high flow velocities (corresponding to the required high Reynolds numbers) at the vena-contracta, the static pressure will be low, and when it drops below the vapor pressure of the flowing fluid, cavitation occurs. This, if present, will destroy the throat section of the tube, as no material can stand up to cavitation. The possible ways to eliminate cavitation include: relocating the meter to a point in the process where the pressure is higher and the temperature is lower; reducing the pressure drop across the sensor; or replacing the sensor with one that has less pressure recovery.

Due to their construction, venturis, flow tubes, and flow nozzles are relatively difficult to inspect. This problem can

TABLE 27m
Venturi, Flow Tube, and Flow Nozzle Inaccuracies (Errors) in Percent of Actual Flow for Various Ranges of Beta Ratios and Reynolds Numbers

Flow Sensor		Line Size In inches (1 in. = 25.4 mm)	Beta Ratio	Pipe Reynolds Number Range for Stated Accuracy	Inaccuracy in % of Actual Flow
Herschel Standard	Cast[1]	4–32	.30–.75	2×10^5 to 1×10^6	±0.75%
	Welded	8–48	.40–.70	2×10^5 to 2×10^6	±1.5%
Proprietary True Venturi	Cast[2]	2–96	.30–.75	8×10^4 to 8×10^6	±0.5%
	Welded	1–120	.25–.80	8×10^4 to 8×10^6	±1.0%
Proprietary Flow Tube	Cast[3]	3–48	.35–.85	8×10^4 to 1×10^6	±1.0%
ASME Flow Nozzles[4]		1–48	.20–.80	7×10^6 to 4×10^7	±1.0%

[1] No longer manufactured because of long laying length and high cost
[2] Badger Meter Inc.; BIF Products of Leeds & Northrup; Fluidic Techniques, Inc.; F.B. Leopold Co.; Permutit Co., Inc.; Henry Pratt Co.; Primary Flow Signal, Inc.; Tri-Flow Inc
[3] Badger Meter Inc.; Bethlehem Corp.; BIF Products of Leeds & Northrup; Fischer & Porter Co.; F.B. Leopold Co.; Henry Pratt Co.; Preso Industries
[4] BIF Products of Leeds & Northrup; Daniel Flow Products, Inc.; Permutit Co., Inc.; Primary Flow Signal, Inc

be solved by providing an inspection port on the outlet cone near the throat section. This can be an important factor when metering dirty (erosive) gases, slurries, or corrosive fluids. On dirty services where the pressure ports are likely to plug, the pressure taps on the flow tube can be filled with chemical seals having stainless steel diaphragms that are installed flush with the tube interior (Figure 27l).

The main advantages of these sensors include their relatively high accuracy, good rangeability (on high Reynolds number applications), and energy-conserving high pressure recovery. For these reasons, in higher velocity flows and in larger pipelines (and ducts), the venturis are still favored by many users in spite of their high costs. Their hydraulic shape also contributes to greater dimensional reliability and therefore to better flow-coefficient stability than that of the orifice-type sensors, which depend on the sharp edge of the orifice for their flow coefficient.

The accuracy of a flow sensor is defined as the uncertainty tolerance of the flow coefficient. Accuracy can be improved by calibration. Table 27m gives some accuracy data in percentage of actual flow, as reported by various manufacturers. These values are likely to hold true only for the stated ranges of beta ratios and Reynolds numbers, and they do not include the added error of the readout device or d/p transmitter.

References

1. Bean, H.S., ed., *Fluid Meters—Their Theory and Application, ASME Research Committee on Fluid Meters,* 6th edition, New York: ASME, 1971.

Bibliography

ANSI/ASME MFC, "Differential Producers Used for the Measurement of Fluid Flow in Pipes," Draft 8, New York, December 1983.
Bean, H.S., ed., "Fluid Meters—Their Theory and Application," ASME Research Committee on Fluid Meters, 6th edition, New York: ASME, 1971.
Halmi, D., "Metering Performance Investigation and Substantiation of the Universal Venturi Tube," *Journal of Fluids Engineering,* February 1974.
Herschel, C., "The Venturi Water Meter," Trans. of American Society of Civil Engineers, Vol. 17, 1987, p. 228.
ISO/TC 30 231 E. Draft ISO Recommendation No. 1–157, November 1966
ISO Standard R781, "Measurement of Fluid Flow by Means of Venturi," ISO, Geneva, 1968.
Jones, J.T., "Field Experience with Sonic Nozzle," AGA Operating Section Proceedings, AGA, Arlington, Va, 1976, pp. 315–319.
Kochen, G., Smith, D.J.M., and Umbach, H., "Installation Effects on Venturi Tube Flowmeters," *InTech,* October 1989.
Kopp, J.G., "How to Get Accuracy You Expect from Flowmeters," *Instruments and Control Systems,* September 1989.
Lipták, B.G., ed., *Instrument Engineer' Handbook,* Vol. I, Radnor, Pennsylvania: Chilton, 1969.
Rudback, S., "Optimization of Orifice Plates, Venturis, and Nozzles," *Measurements and Control,* June 1991.
Shell Flow Meter Engineering Handbook, Royal Dutch/Shell Group. Delft, The Netherlands: Waltman Publishing Co., 1968 (now out of print).
Spink, L.K., *Principles and Practice of Flow Meter Engineering,* 9th edition, The Foxboro Co., 1967.

28 Vortex and Fluidic Flowmeters

J. G. KOPP (1969) **D. J. LOMAS** (1982) **B. G. LIPTÁK** (1993)

Flow Sheet Symbol

Types:	A. Vortex
	B. Fluidic shedding coanda effect
	C. Oscillating vane in orifice bypass
Services:	A. Gas, steam, clean liquids
	B and C. Clean liquids
Size Ranges Available:	A. 0.5 to 12 in. (13 to 300 mm), also probes
	B. 1 to 4 in. (25 to 100 mm)
	C. 1 to 4 in. (25 to 100 mm)
Detectable Flows:	A. Water—2 to 10,000 GPM (8 lpm to 40 m^3/hr)
	Air—3 to 12,000 SCFM (0.3 to 1100 SCMM)
	Steam (D&S at 150 PSIG [10.4 bars])—25 to 250,000 lbm/hr (11 to 113,600 kg/hr)
	B. Water—1 to 1000 GPM (4 to 4000 lpm)
	C. Water—5 to 800 GPM (20 to 3024 lpm)
Flow Velocity Range:	A. Liquids—1 to 33 ft/s (0.3 to 10 m/s)
	Gas and steam—20 to 262 ft/s (6 to 80 m/s)
Minimum Reynolds Numbers:	A. Under Re of 8000 to 10,000, meters do not function at all; for best performance Re should exceed 20,000 in sizes under 4 in. (100 mm) and exceed 40,000 in sizes above 4 in.
	B. Re = 3000
Output Signals:	A, B, and C. Linear pulses or analog
Design Pressure:	A. 2000 PSIG (138 bars)
	B. 600 PSIG (41 bars) below 2 in. (50 mm); 150 PSIG (10.3 bars) above 2 in.
	C. 300 PSIG (30.6 bars)
Design Temperature:	A. −330 to 750°F (−201 to 400°C)
	B. 0 to 250°F (−18 to 120°C)
	C. −14 to 212°F (−25 to 100°C)
Materials of Construction:	A. Mostly stainless steel; some in plastic
	B. 316 stainless steel with Viton A O-rings
	C. Wetted body is Kynar, sensor is Hastelloy C
Rangeability:	A. Reynolds number at maximum flow divided by minimum Re of 20,000 or more
	B. Reynolds number at maximum flow divided by minimum Re of 3000
	C. 10:1 for Reynolds number at maximum flow divided by minimum Reynolds numbers of 14,000 for 1 in. 28,000 for 2 in., 33,000 for 3 in., and 56,000 for 4 in.
Inaccuracy:	A. 0.5 to 1% of rate for liquids, 1 to 1.5% of rate for gases and steam with pulse outputs; for analog outputs add 0.1% of full scale
	B. 1 to 2% of actual flow
	C. 0.5% of full scale over 10:1 range
Cost:	A. Plastic and probe units cost about $1500; stainless steel units in small sizes cost about $2500; insertion-types cost about $3000
	C. The sensor with unscaled pulse output only in 1, 2, 3, and 4 in. sizes respectively costs $535, $625, $875, and $1295. The additional cost of a scaler is $250, and of a 4–20 mA transmitter, $350.
Partial List of Suppliers:	ABB Kent (A); Alphasonics Inc. (A); Badger Meter Inc. (C—proximity switch sensor); Brooks Div. of Rosemount (A—ultrasonic); EMC Co. (A—dual piezoelectric sensor); Endress + Hauser Instruments (A—capacitance sensor); Fischer & Porter Co. (A—internal strain gauge sensor); Fisher Controls (A—dual piezoelectric sensor); Flowtec AG of Switzerland (A); Foxboro Co. (A—piezoelectric sensor); Johnson Yokogawa Corp. (A—dual piezoelectric sensor); J-Tec Associates Inc. (A—retractable design available, ultrasonic sensor); MCO/Eastech (A—including insertion-type, mechanical, thermal, or piezoelectric sensors); Moore Products Co. (B); Nice Instrumentation Inc. (A—dual piezoelectric sensor); Oilgear/Ball Products (A—vortex velocity);

Sarasota Automation Inc. (A); Schlumberger Industries Inc. (A—dual piezometric sensor); Turbo Instruments Inc. (A); Universal Flow Monitors Inc. (A—plastic body, piezoelectric sensor); Universal Vortex (A—piezoelectric sensor)

This section is devoted mainly to the vortex shedding flowmeter and its variations, including the earlier designs of vortex precession (swirl) meters and the recent combination designs of vortex bypass elements around orifices. Included in this category of devices is also an oscillating fluidic flowmeter utilizing the Coanda effect.

THE VORTEX SHEDDING PHENOMENON

It was Tódor von Kármán who discovered that when an obstruction (a non-streamlined object) is placed in the path of a flowing stream, the fluid is unable to remain attached to the object on its downstream sides and will alternately separate (shed) from one side and then the other. The slow-moving fluid in the boundary layer on the bluff body becomes detached on the downstream side and rolls into eddies and vortices (Figure 28a). Von Kármán also noticed that the distance between the shed vortices *is constant,* regardless of flow velocity. Stated in terms of a flag fluttering in the wind, what von Kármán discovered was that the intervals between vortices (l) (or the wavelength of fluttering) *is constant* and is only a function of the diameter of the flag pole (d). Therefore, the faster the wind, the faster the vortices are formed and the faster the flag flutters as a consequence—*but without changing its wavelength.*

Later, Strouhal determined that as long as the Reynolds number of the flowing stream is between 20,000 and 7 million, the ratio between the shedder width (d) and the vortex interval (l) is 0.17. This number is called the Stouhal number. Therefore, if one knows the vortex shedder width (d) and has a detector that is sensitive enough to count the vortices and determine the vortex frequency (f), one can measure the flowing velocity of any substance as:

$$\text{flow velocity} = (f \times d)/(0.17) = kfd \qquad \textbf{28(1)}$$

In building a flowmeter based on Kármán's principle, the manufacturer usually selects an obstruction width (d) that is one-quarter of the pipe diameter (ID). As long as the obstruction is not eroded or coated, as long as the pipe Reynolds number is high enough to produce vortices, and as long as the detector is sensitive enough to detect these vortices (for gases such as hydrogen the forces produced by the vortices are very small), what results is a flowmeter that is sensitive to flow velocity and is insensitive to the nature of the flowing media (liquid, gas, steam), the density, the viscosity, the temperature, the pressure, and any other property.

THE DETECTOR

As a vortex is shed from one side of the bluff body, the fluid velocity on that side increases and the pressure decreases. On the opposite side the velocity decreases and the pressure increases, thus causing a net pressure change across the bluff body. The entire effect is then reversed as the next vortex is shed from the opposite side. Consequently, the velocity and pressure distribution adjacent to the bluff body change at the same frequency as the vortex shedding frequency.

Various detectors can be used to measure one of the following:

1. The oscillating flow across the face of the bluff body
2. The oscillating pressure difference across the sides of the bluff body
3. A flow through a passage drilled through the bluff body
4. The oscillating flow or pressure at the rear of the bluff body
5. The presence of free vortices in the downstream to the bluff body

A flow-sensitive detector can be either a heated thermistor element or a spherical magnetic shuttle (with the movement of the shuttle measured inductively). Detectors that are sensitive to pressure use either metal diaphragms or vanes. Pressure exerted on diaphragms can be converted into a variable capacitance, or a variable strain on a piezo-resistive, piezoelectric, or inductive sensor. Pressure exerted on vanes can similarly be converted into an electrical signal through any of the aforementioned sensors. Alternatively, the velocity components in the free vortices downstream of the bluff body can be used to modulate an ultrasonic beam diametrically traversing the meter housing. Depending on the characteristics of the sensing system, the flowmeter will be suitable for liquid or gas or both.

The earliest detector designs were highly sensitive to plugging and required frequent maintenance (Figure 28b). These devices were later replaced by units that could not plug and were solid state in design (Figure 28c). The majority of these designs are still marketed and are well received

FIG. 28a
The distance between the Kármán vortices (l) is only a function of the width of the obstruction (d), and therefore the number of vortices per unit of time gives flow velocity (V).

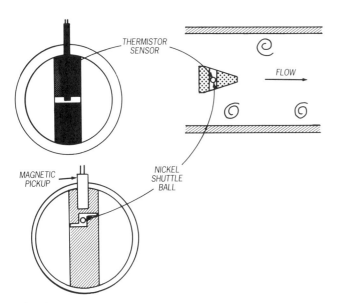

FIG. 28b
Shuttle-ball and shuttle-flow-type early vortex flowmeter detectors.

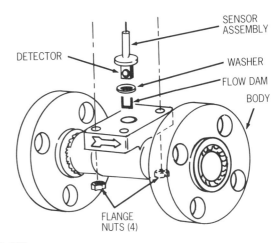

FIG. 28d
Piezoelectric capsule detector element is removable from flow element. (Courtesy of The Foxboro Co.)

by users who are not concerned about quick and convenient access to and replacement of the detector or about the reliability and sensitivity of heat transfer or ultrasonic detectors. Yet the trend seems to be toward the type of detectors that are modular, inexpensive, and interchangeable, so that they can be quickly replaced when necessary. Several vortex flowmeter detectors available on today's market can be replaced easily (Figure 28d). In this design the detector is a liquid-filled, double-faced diaphragm capsule with a piezoelectric crystal in the center which detects the vortex-produced pressure changes as they are transmitted through the filling liquid.

Other design modifications aim at compensating for background noise by using two detectors, of which one is exposed to vortex forces and the other is not, and using their difference as the measurement signal (Figure 28e). Yet other design modifications aim at amplifying the signal generated by low-energy vortices, such as by low-density gases. One approach is to use two detector elements (capacitance or piezoelectric) and measure the difference between their signals. This tends to amplify the detector output, because as the vortices emerge on alternate sides of the flow element, the two detectors sense the forces acting on the two different sides of the element. Still another method of amplifying the vortex forces is by physically separating the vortex shedding element and the vortex force detector (Figure 28f). If the vortex forces are amplified, the force detectors can be made less sensitive and therefore more rugged and reliable.

The types of detectors that are in use as of this writing and the approximate number of manufacturers marketing them are listed below: mechanical (2), thermal (2), ultrasonic (2), strain gauge (1), capacitance (1), and piezoelectric (7). While based on these numbers it would seem that the piezoelectric designs (particularly their dual or differential versions) dominate the market, other designs do claim superior performance under certain operating conditions. The manufacturers of the capacitance design, for example, claim superior immunity to pipe vibration effects.

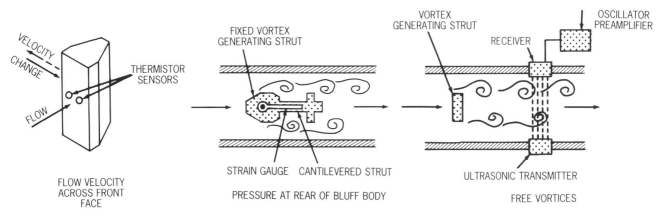

FIG. 28c
Solid-state vortex flowmeter designs with limited accessibility to their sensors.

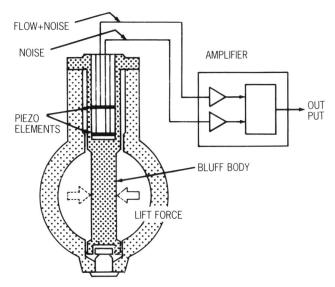

FIG. 28e
Dual detector serves noise compensation. (Courtesy of Johnson Yokogawa)

The fundamental meter output is a frequency signal in all cases, which can be fed directly into digital electronic units for totalization and/or preset batching, into computers, or into data loggers. The frequency signal also can be converted into a conventional 4–20mA DC analog signal for flow rate indication, recording, or control purposes. Most meters are available in either a standard form or in a design to satisfy Division 1 explosion-proof-area requirements.

Features

The vortex shedding meter provides a linear digital (or analog) output signal *without* the use of separate transmitters or converters, simplifying equipment installation. Meter accuracy is good over a potentially wide flow range, although this range is dependent upon operating conditions. The shedding frequency is a function of the dimensions of the bluff body and, being a natural phenomenon, ensures good long-term stability of calibration and repeatability of better than $\pm 0.15\%$ of rate. There is no drift because this is a frequency system.

The meter does not have any moving or wearing components, providing improved reliability and reduced maintenance. Maintenance is further reduced by the fact that there are no valves or manifolds to cause leakage problems. The absence of manifolds and valves results in a particularly safe installation, an important consideration when the process fluid is hazardous or toxic.

If the sensor utilized is sufficiently sensitive, the same vortex sheddding meter can be used on both gas and liquid. In addition, the calibration of the meter is virtually independent of the operating conditions (viscosity, density, pressure, temperature, and so on) whether the meter is being used on gas or liquid (see Figure 28g).

The vortex shedding meter also offers a low installed

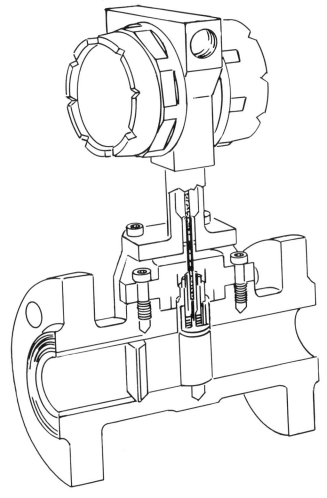

FIG. 28f
Separating the rugged obstruction and the detector allows the detector to be much more sensitive to the pressure waves. The increases in the forces detected allows for the use of more rugged (less sensitive and therefore less fragile) sensors. (Courtesy of EMC Co.)

cost, particularly in pipe sizes below 6 in. (152 mm) diameter, which compares competitively with the installed cost of an orifice plate and differential pressure transmitter.

The limitations include meter size range. Meters below 0.5 in. (12 mm) diameter are not practical, and meters above 12 in. (30.0 mm) have limited application due to their high cost compared to an orifice system and their limited output pulse resolution. The number of pulses generated per unit volume decreases on a cube law with increasing pipe diameter. Consequently, a 24 in. (610 mm) diameter vortex shedding meter with a typical blockage ratio of 0.3 would only have a full scale frequency output of approximately 5 Hz at 10 ft/s (3 m/s) fluid velocity.

Selection and Sizing

As the first step in the selection process, the operating conditions (process fluid temperature, ambient temperature,

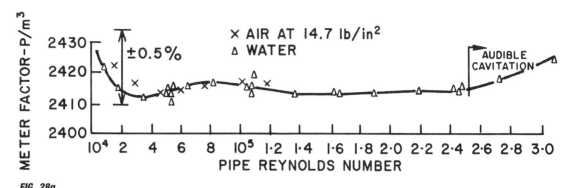

FIG. 28g
Typical calibration curves for a 3 in. (76 mm) vortex meter showing the close correlation between water and atmospheric air calibrations.

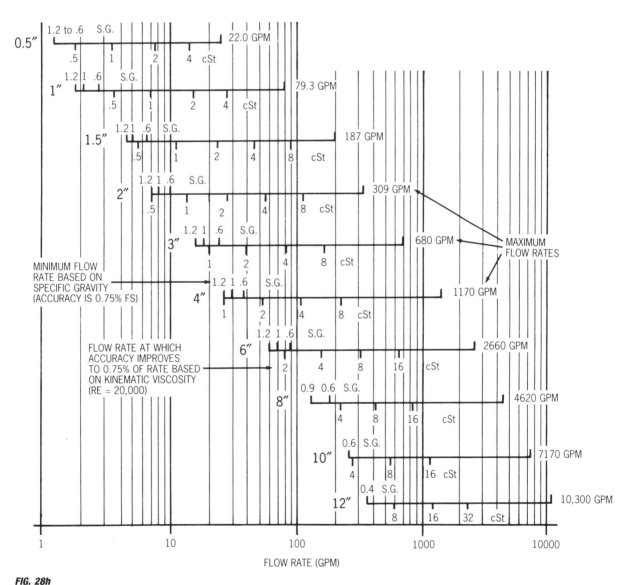

FIG. 28h
Sizing chart for liquid flow measurement. Note that minimum flows are limited by both specific gravity (water SG = 1) and viscosity limitations. (To convert to metric units use: 1 in. = 25.4 mm, 1 GPM = 3.78 lpm). (Courtesy of Endress + Hauser Instruments)

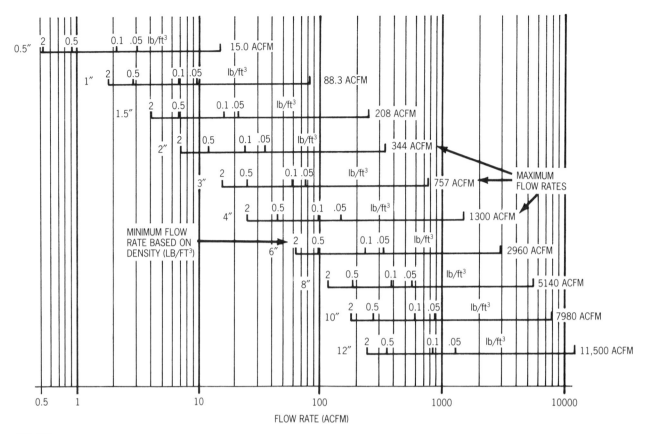

FIG. 28i
Sizing chart for gas and vapor flow detection: For extremely dense gases, the maximum flow may be less than shown. Gases with extremely low densities (eg. hydrogen, helium) may not be measurable. Note that minimum flows are a function of flowing density. To convert to metric units use: 1 in. = 25.4 mm, 1 ACFM = 0.02832 ACMM, and 1 lb/ft³ = 16 kg/m³. (Courtesy of Endress + Hauser Instruments)

line pressure, and so on) should be compared with the meter specification. The meter wetted materials (including bonding agents) and sensors should then be checked for compatability with the process fluid both with regard to chemical attack and safety. On oxygen, for example, nonferrous materials should be used due to the reactive nature of oxygen. Applications where there are large concentrations of solids, two-phase flow, or pulsating flow should be avoided or approached with extreme caution. The meter minimum and maximum flow rates for the given application should then be established. (See Figures 28h and 28i, and Table 28j.)

A typical performance curve for a vortex shedding flowmeter is shown in Figure 28g. The meter minimum flow rate is established by a Reynolds number of 10,000 to 10,500, the fluid density, and a minimum acceptable shedding frequency for the electronics. The maximum flow rate is governed by the meter pressure loss (typically two velocity heads), the onset of cavitation with liquids, and sonic velocity flow (choking) with gases. Consequently, the flow range for any application depends totally upon the operating fluid viscosity, density, and vapor pressure, and the application's maximum flow rate and line pressure. On low-viscosity products such as water, gasoline, and liquid ammonia, and with an application maximum velocity of 15 ft/s (4.6 m/s), vortex shedding meters can have a rangeability of about 20:1 with a pressure loss of approximately 4 PSIG (27.4 kPa).

The meter's good (''of rate'') accuracy and digital linear output signal make its application over wide flow ranges a practical proposition. The rangeability declines proportionally with increases in viscosity, decreases in density, or reductions in the maximum flow velocity of the process. Vortex shedding meters are therefore *unsuitable* for use on high-viscosity liquids.

On liquid applications, it is necessary to verify that sufficient line pressure exists to prevent cavitation in the vortex meter. The maximum pressure drop in a vortex shedding meter is in the region of the bluff body and there is a considerable pressure recovery by the meter outlet. Upstream line pressure requirements vary from one meter design to another, but a typical minimum acceptable upstream pressure requirement (to protect against cavitation) is given by the expression: Upstream pressure ≥ 1.3 (vapor pressure plus 2.5 times net pressure loss across the meter). Cavitation conditions must be avoided at all costs, as no material can stand up to the damage caused by cavitation. One might

TABLE 28j
*Sizing for Steam Flow in Lbm/Hr Units**[*][†]

Meter Size (in.)		\multicolumn{18}{c}{Steam Pressure (PSIG)}
		10	20	30	40	50	60	80	100	150	200	250	300	350	400	500	600	700	800	900
0.5	min	10	12	13	15	16	17	19	21	25	28	31	34	36	39	40	46	51	57	63
	max	55	75	95	115	134	154	193	231	326	421	516	610	707	803	997	1197	1401	1611	1826
1	min	30	36	40	44	48	51	57	63	75	85	94	102	110	117	130	143	154	166	176
	max	322	442	560	677	792	907	1140	1360	1920	2490	3040	3600	4170	4740	5880	6440	6970	7470	7950
1.5	min	72	84	95	104	113	121	135	148	176	200	221	241	259	276	308	337	365	391	417
	max	761	1040	1320	1600	1870	2150	2690	3220	4550	5880	7190	8510	9850	11,200	13,900	15,200	16,500	17,700	18,800
2	min	119	139	156	172	186	199	223	244	290	330	365	397	427	455	507	556	601	645	686
	max	1250	1720	2180	2640	3090	3530	4420	5310	7490	9680	11,900	14,000	16,200	18,500	22,900	25,100	27,100	29,100	31,000
3	min	261	306	344	379	410	439	491	537	639	726	803	873	940	1000	1120	1220	1320	1420	1510
	max	2760	3790	4800	5800	6800	7780	9740	11,700	16,500	21,300	26,100	30,900	35,800	40,600	50,400	55,200	59,800	64,100	68,200
4	min	450	528	594	653	707	756	846	927	1100	1250	1390	1510	1620	1730	1930	2110	2280	2450	2610
	max	4760	6530	8280	10,000	11,700	13,400	16,800	20,200	28,500	38,800	45,000	53,200	61,700	70,100	86,900	95,200	103,000	110,000	118,000
6	min	1020	1200	1350	1480	1600	1720	1920	2100	2500	2840	3140	3420	3680	3920	4370	4790	5180	5550	5910
	max	10,800	14,800	18,800	22,700	26,600	30,500	38,100	45,700	64,600	83,400	102,000	121,000	140,000	159,000	197,000	216,000	234,000	251,000	267,000
8	min	1780	2080	2340	2570	2790	2980	3340	3650	4340	4930	5460	5940	6470	7120	8370	9600	10,800	12,000	13,200
	max	18,800	25,700	32,600	39,400	46,200	52,900	66,200	79,400	112,000	145,000	177,000	210,000	243,000	276,000	343,000	375,000	406,000	435,000	464,000
10	min	2750	3230	3630	3990	4320	4630	5180	5670	6740	7660	8470	9210	10,000	11,000	13,000	14,900	16,800	18,600	20,500
	max	29,100	39,900	50,600	61,200	71,700	82,100	103,000	123,000	174,000	225,000	275,000	326,000	377,000	429,000	532,000	582,000	630,000	676,000	720,000
12	min	3970	4660	5240	5760	6240	6570	7470	8180	9720	11,000	12,200	13,300	14,500	15,900	18,700	21,500	24,200	26,900	29,500
	max	42,000	57,600	73,000	88,300	103,000	118,000	148,000	178,000	251,000	324,000	397,000	470,000	544,000	618,000	767,000	840,000	909,000	975,000	1,040,000
Temp.$_{sat.}$	°F	239	259	274	287	298	307	323	338	366	388	406	422	436	448	470	489	506	520	534
Density$_{sat.}$	lb/ft³	0.061	0.083	0.106	0.128	0.150	0.171	0.214	0.257	0.363	0.469	0.574	0.679	0.787	0.894	1.11	1.33	1.56	1.79	2.03

*To convert to metric units use: 1 inch = 25.4 mm, 1 PSIG = 0.069 bars, and 1 lbm = 0.454 kg.
†Courtesy of Endress + Hauser Instruments.

approximate the minimum upstream pressure required to avoid cavitation (P_{min}) on the basis of the maximum velocity expected in the pipeline (V_{max}) as follows:

$$P_{min} = (1.3)P_v + (2.5)(V_{max})^2/g \qquad 28(2)$$

where

P_{min} = minimum required upstream pressure in feet of liquid head
P_v = vapor pressure of the flowing liquid at maximum operating temperature in feet of liquid head
V_{max} = maximum anticipated flowing velocity in ft/s
g = gravitational acceleration constant of 32.2 having the units ft/s^2

Vortex shedding flowmeters cannot survive cavitation, but they can survive episodes of "flashing"—when some of the incoming liquid stream is permanently vaporized in the flowmeter. If the liquid "gases," the vortex shedding flowmeter will not be mechanically damaged (although the meter output will be seriously in error).

Installation Requirements

Vortex shedding meters require a fully developed flow profile. The length of upstream pipework necessary to ensure satisfactory approach conditions depends on the specific design of meter, the type of upstream disturbance present, and the level of accuracy required. Typical upstream and downstream pipework requirements for a variety of disturbances are given in Figure 28k.

Where there is a severe upstream disturbance, the resulting long straight lengths of pipe can be reduced by fitting a radial vane or bundle-of-tubes flow-straightening element in the upstream pipework. Wherever possible, however, the meter should be installed upstream of any severe source of disturbance such as regulating control valves. The downstream straight pipe requirement is 5 times nominal meter diameter. The meter can be installed in any attitude (horizontal or vertical) but it is not suitable for reverse flowmetering.

Other instrument connections (pressure, temperature) should all be located downstream of the flowmeter and more than 5 diameters away from it. The flowmeter should be the same size as or smaller than the pipeline, but never larger. The unit can be insulated for cryogenic or high-temperature services and can be provided with extension bonnets. It should be installed in self-draining low points in the piping or in vertical upward flows to keep the meter flooded and avoid air bubbles or standing liquid pools. Block and bypass valves should be provided if the meter is to be maintained while the process is in operation. There should be no excessive pipe vibration in the area where the meter is installed, and gaskets should not protrude into the pipeline.

VORTEX PRECESSION (SWIRL) METERS

These units were the predecessors of the vortex shedding flowmeters and are described for reason of completeness.

Construction of a typical vortex precession (swirl) meter and the operating principles are illustrated in Figure 28l. The fixed swirl-inducing helical vanes at the entrance to the meter introduce a spinning or swirling motion to the fluid. After the exit of the swirl vanes the bore of the meter contracts progressively, causing the fluid to accelerate, but with the axis of rotation still on the centerline of the meter. The swirling fluid then enters an enlarged section in the meter housing which causes the axis of fluid rotation to change from a straight to a helical path. The resulting spiraling vortex is known as vortex precession. The frequency of precession is proportional to velocity and, hence, volumetric flow rate above a given Reynolds number.

The velocity of fluid in the vortex is higher than that of the surrounding fluid. Consequently, as each vortex passes the sensor, there is a change in the local fluid velocity. The

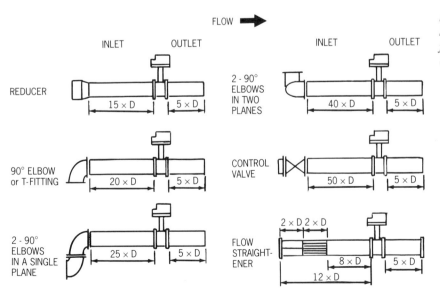

FIG. 28k
Straight pipe-run requirements as a function of upstream disturbance. (Courtesy of Endress + Hauser)

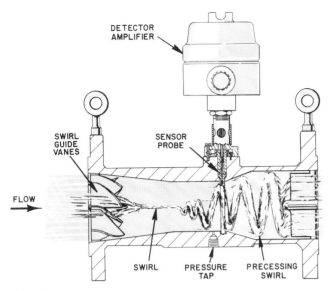

FIG. 28l
Construction of a typical vortex precession (swirl) meter.

frequency at which the velocity changes occur is proportional to volumetric flow rate and can be detected by piezoelectric or thermistor sensors. The thermistor, which changes its resistance with temperature, is heated by a constant electrical current. The amount of heat extracted from the thermistor by the passing fluid is dependent upon the fluid velocity. Consequently, each "high velocity" vortex that goes past the thermistor changes the resistance and, since a constant current is applied, the resistance change is detected as a voltage change. After amplification and filtering, a square wave voltage output signal is provided, linear with flow rate.

A flow-straightener is fitted at the meter outlet in order to isolate the meter from downstream piping effects which might otherwise impair the development of the precessing vortex.

The internal components of the swirl meter required a significant amount of complex machining; thus, it was more expensive than some other meter types.

FLUIDIC (COANDA EFFECT) METER

Fluid entering the meter is entrained into a turbulent jet from its surroundings, causing a reduction in pressure. The internal geometry of the meter body causes the jet to be deflected from its central position and initially attach itself to one of the side walls. The jet curvature is sustained by the pressure differential across the jet. If sufficient volume of fluid is then introduced into the control port on that side, it will cause the jet to switch to the opposite side wall. This is known as a Coanda Effect. The jet can be made to oscillate by one of two methods. The simplest method is a relaxation oscillator. In this system, the two ports are connected together. Fluid is sucked from the high-pressure side to the low-pressure side causing the jet to switch to the other wall. The jet thus continues to oscillate as the fluid is sucked alternately from one side to the other. The more commonly used system is the feedback oscillator (see Figure 28m). The deflected jet causes a low-pressure area at the control port. At the upstream feedback passage the pressure is higher due to a combination of the jet expanding and the stagnation pressure. Thus, a small portion of the main stream of fluid is diverted through the feedback passage to the control port. The feedback flow intersects the main flow and diverts it to the opposite side wall. The whole feedback operation is then repeated, resulting in a continuous self-induced oscillation of the flow between the side walls of the meter body. The frequency of oscillation is linearly related to the volumetric flow rate above a minimum Reynolds number. As the main flow oscillates between the side walls, the flow in the feedback passages oscillates between zero and a maximum value. This frequency is detected by means of a thermistor sensor, providing a frequency output signal.

Characteristics

The principal features include no moving components, fixed calibration based on the geometry of the housing, linear digital or analog output, and good rangeability. One advantage over vortex meters is that fluidic meters can operate down to a Reynolds number of 3000. The maximum flow range (dependent on size and viscosity) is 30:1. The operating pressure and maximum practical pipe diameter are largely dictated by the complex housing shape. In practice, a 4 in. (100 mm) diameter unit is the largest commercially available, and the operating pressure in this diameter is typically limited to 150 PSIG (1.03 MPa). Although theoretically suitable for gaseous applications, fluidic meters are almost exclusively used on liquid applications.

A special, separate converter is required for the meter, which, in some instances, can incorporate a pneumatic output.

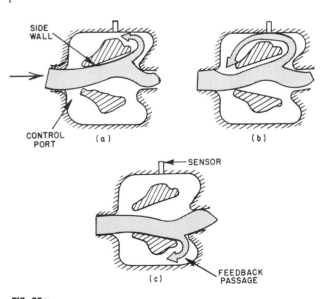

FIG. 28m
Diagram of the mode of operation of a feedback oscillator.

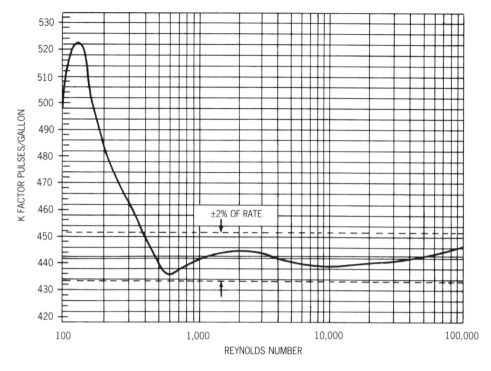

FIG. 28n
The meter factors of a 1 in. (25.4 mm) fluidic flowmeter stay accurate at lower values of Reynolds numbers than they do for vortex shedding flowmeters. (Courtesy of Moore Products Co.)

As shown in Figure 28n, the meter factor in pulses per volume of flow passed remains within 1%, and therefore the measurement error remains well within 2% of actual flow between the Reynolds numbers of 3000 and 100,000.

OSCILLATING VANE IN ORIFICE BYPASS

Some of the limitations of an orifice-type installation include the low rangeability (3:1) and the relatively high installed cost in smaller pipe sizes. The d/p cell manifold causes the high installation cost, and it is the square root characteristics which limit the flow rangeability. The oscillating vane bypass aims to overcome these limitations.

In this design (Figure 28o), the differential pressure developed by the orifice is used to generate a small bypass flow, which is measured by a vortex-like oscillating vane element. The frequency of vane oscillations is detected by a proximity switch and can be used as an unscaled digital flow rate indication; or the unscaled pulses can be scaled or converted to 4 to 20 mA DC. The flowmeter is accurate to 0.5% of full scale over a good rangeability, and on an installed basis it is less expensive than a combination of orifice and d/p cell. The actual rangeability is the ratio of Reynolds number at maximum flow divided by the minimum Re given in the feature summary at the front of this section. The oscillating vane bypass meter is also suited for corrosive services because its wetted parts, including the orifice, are made out of Kynar, while the sensor is Hastelloy C.

The oscillating vane meter is limited to use on clean and

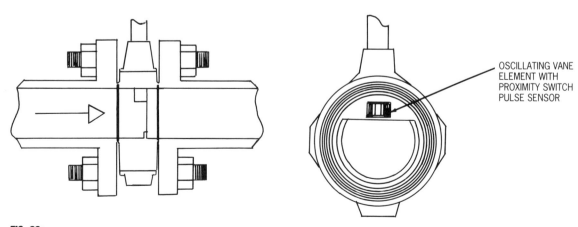

FIG. 28o
Oscillating vane flowmeter mounted in bypass around orifice. (Courtesy of Badger Meter Inc.)

cold liquids (-14 to 212°F, or -25 to 100°C), and it is likely to require maintenance due to the continuous motion of the oscillating vane and the associated bearing components (tungsten carbide or ceramic pivots riding on sapphire end stones). Installation considerations, including straight pipeline requirements, are similar to those of regular vortex shedding flowmeters.

CONCLUSIONS

The advantages of vortex shedding flowmeters include their suitability for liquid, gas, and steam service; independence from viscosity, density, pressure, and temperature effects; low installed cost in smaller sizes; good accuracy and linearity without requiring calibration; wide rangeability; low maintenance using simple, easily accessible and interchangeable spare parts; simple installation; and direct pulse output capability.

The disadvantages include the following: they are not suitable for services that are dirty, abrasive, viscous, or mixed-flow (gas with liquid droplets, liquid with vapor bubbles), or that have low Reynolds numbers (below 20,000); the available choices in materials of construction are limited; the pulse resolution (number of pulses per gallon or liter) drops off in larger sizes; the pressure drop is high (two velocity heads); and substantial straight runs are required both upstream and downstream.

Bibliography

Biles, R., "Vortex Flowmeter Performance," *Measurements and Control*, September 1991.

"Choices Abound in Flow Measurement," *Chemical Engineering*, April 1991.

Cousins, T., "The Performance and Design of Vortex Meters," Fluid Flow Conference, East Kilbride, 1975.

Gotthardt, W.C., "Is It Real Vortex Flow or Not?" *Measurements and Control*, June 1991.

Herzl, J., "New Sensing Techniques and Modular Construction as Applied to the Swirl Meter," ISA 28th Annual Conference, Pittsburgh.

Lomas, D.J., "Vortex Meters—A Practical Review," Measurement Technology for the 80's, ISA Symposium, Delaware, 1979.

Medlock, R.S., "Vortex Shedding Meters," Liquified Gas Symposium, London, 1978.

"New Design Flowmeters Boost Accuracy," *Power*, December 1976.

Nissen, C., "HPV Meter," *Measurements and Control*, February 1989.

O'Brien, C.J., "Fueling Flowmeter Accuracy, Reliability," *InTech*, April 1989.

Satori, T., "Vortex Flowmeter Application Report," 1984 ISA Conference, Houston.

Within, W.G., "Theory, Design and Application of Vortex Shedding Flowmeters," Measurement Technology for the 80's, ISA Symposium, Delaware, 1979.

29 Weirs and Flumes

W. H. HOWE (1969, 1982)   **B. G. LIPTÁK** (1993)

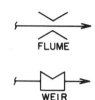

Flow Sheet Symbol

Types:	Open-channel flow can be measured by detecting level in front of primaries. Bubblers, capacitance, float, hydrostatic, and ultrasonic devices are used as level sensors. Open-channel flows can also be measured without primaries by calculating flow from depth and velocity using ultrasonic and magnetic sensors
Operating Conditions:	Atmospheric
Applications:	Waste or irrigation water flows in open channels
Flow Range:	From 1 GPM (3.78 l/m)—no upper limit
Rangeability:	Most devices provide 75:1; V-notch weirs can reach 500:1
Inaccuracy:	2 to 5%
Costs:	Primaries used in pipe inserts cost under $1000. A 6 in. (150 mm) Parshall flume costs about $1500, and a 48 in. (1.22 m) one costs about $5000. Primaries for irrigation applications are usually field-fabricated. Manual depth sensors can be obtained for $200; local bubbler or float indicators for $750 to $1500; programmable transmitting capacitance, ultrasonic, or bubbler units from $1800 to $3000. Open-channel flowmeters calculating flow based on depth and velocity range from $5000 to over $10,000
Partial List of Suppliers:	ABB Kent Taylor Inc. (primaries); American Sigma Inc. (bubbler); Badger Meter Inc. (Parshall or manhole flume, ultrasonic and open-channel computing); Bernhar Inc. (ultrasonic for partially filled pipes); Bestobell/Mobrey (ultrasonic); BIF Unit of Leeds & Northrup (primary and detector); Drexelbrook Engineering Co. (capacitance for flumes); Endress + Hauser Inc. (ultrasonic and capacitance); Fischer & Porter Co. (ultrasonic); Free Flow Inc. (primaries); Greyline Instruments Inc. (ultrasonic); Inventron Inc. (ultrasonic); ISCO Inc. (bubbler, hydrostatic, ultrasonic); Key-Ray/Sensall Inc. (ultrasonic); Leeds & Northrup BIF (flow nozzles); Leupold & Stevens Inc. (float); Manning Environmental Corp. (primaries); Marsh-Mcbirney Inc. (electromagnetic); Mead Instruments Corp. (velocity probe); Milltronics Inc. (ultrasonic); Minitek Technologies Inc. (open-channel magmeter, ultrasonic); Montedoro-Whitney Corp. (open-channel flow by ultrasonics); MSR Magmeter Mfg. Ltd. (robotic magmeter probe for open channel); N.B. Instruments Inc. (computer monitoring of sewers); Plasti-Fab Inc. (primaries); Princo Instruments Inc. (capacitance); J.L. Rochester Co. (manual depth sensor); Sparling Instruments Co. (primaries); TN Technologies Inc. (ultrasonic)

WEIRS

Weirs are apertures in the top of a dam across a channel through which flows the liquid to be measured (Figure 29a). The aperture may be rectangular (Figure 29b), trapezoidal (Figure 29c), or V-notch (Figure 29d). The special case of a trapezoidal weir with side slopes of 1:4 (Figure 29c) is known as a Cippoletti weir; this form leads to a simplified

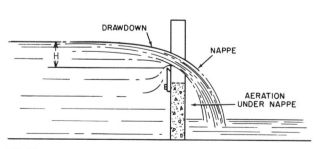

FIG. 29a
Flow over a weir.

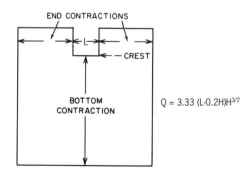

$Q = 3.33\,(L - 0.2H)H^{3/2}$

FIG. 29b
Rectangular weir.

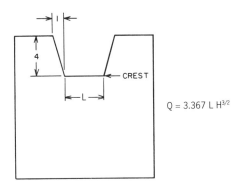

FIG. 29c
Cippoletti (trapezoidal) weir.

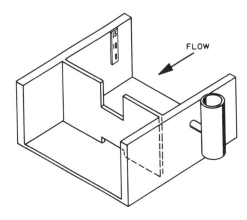

FIG. 29e
Weir box.

flow calculation. V-notch weirs generally have a notch angle from 30 degrees to 90 degrees depending on required flow capacity.

The head is measured as the difference in level of the pool at an adequate distance upstream from the weir, compared to the horizontal crest of a rectangular or trapezoidal weir, or the bottom point of the V of a V-notch weir. Heads less than 0.1 ft (30 mm) for minimum measured flow or more than 1.0 ft (300 mm) for maximum flow are generally to be avoided although 1.25 ft (380 mm) head can be tolerated under favorable conditions. These limits are easily met by practical design since a 30-degree V-notch will measure a minimum flow of 1 GPM (3.8 l/m) while the maximum value for a rectangular or trapezoidal weir is limited only by practical crest length.

V-notch weirs are used for smaller flows. A 30-degree V-notch weir has a practically constant coefficient from 3.0 to 300 GPM (11.4 to 1140 l/m) with flow proportional to the five-halves power of the head. Coefficient increases roughly 2% for flow down to 1 GPM (3.8 l/m) and changes relatively little for flow up to 500 GPM (1893 l/m). For notch angle up to 90 degrees, flow varies as the tangent of half the notch angle. Notch angle exceeding 90 degrees is not recommended.

Rectangular or Cippoletti weirs are used for larger flows. A rectangular weir with a crest 2 ft (0.6 m) long develops a head of about 0.2 ft (60 mm) for 250 GPM (946 l/m) and 1.0 ft (305 mm) for 2700 GPM (10,221 l/m). For this weir, flow is directly proportional to crest length and to the three-halves power of the head.

The weir plate may be located in a dam in a natural channel or in a weir box (Figure 29e). The stilling basin ahead of the weir should be large enough so that the upstream velocity does not exceed ⅓ ft/sec (.01 m/s). Width and depth immediately ahead of the weir should be sufficient so that the wall effect of the bottom and sides of the channel has negligible effect on the pattern of flow through the notch. It is important that the flow should break clear from the sharp edge of the notch with an air pocket maintained immediately beyond and below the weir plate. The channel downstream from the weir must be sufficiently wide and deep so that at maximum flow there is ample clearance between flow through the notch to downstream liquid level so that this air pocket is maintained (Figure 29a). The upstream edge of the weir should be sharp and straight. It is usual practice to bevel the downstream edge of the weir at 45 degrees to about a ¹⁄₃₂ in. (.8 mm) edge. For rectangular and Cippoletti weirs, the crest must be carefully leveled.

Accuracy of the relation between flow and head (level) to ±2% is attainable, based on the dimensions of the primary device. Reference 1 gives full data on installation and operation of weirs.

The following equations establish the relationships between flow and measured head, provided that the installation and operation of the weir are as recommended in this section and also in the cited references.

For a V-notch weir:

$$Q = 2.48 \tan \tfrac{1}{2} \theta \, H^{2.5} \qquad 29(1)$$

For a rectangular weir:

$$Q = 3.33 (L - 0.2H) \, H^{1.5} \qquad 29(2)$$

For a Cippoletti weir:

$$Q = 3.367 \, L H^{1.5} \qquad 29(3)$$

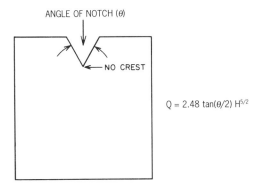

FIG. 29d
V-notch weir.

where

- Q = rate of flow in cubic feet per second
- θ = V-notch angle in degrees
- H = head* in feet of flowing liquid
- L = crest length in feet

For conditions other than exactly as recommended, see references for correction factors.

THE PARSHALL FLUME

Developed by R.L. Parshall at the Colorado Experiment Station of the Colorado Agricultural College in cooperation with the Division of Irrigation of the U.S. Department of Agriculture,[2] this device is a special type of venturi flume (Figure 29f). The loss of head is about one-quarter of that for a weir of equal capacity. The effect of velocity of approach is practically eliminated so that a large upstream stilling basin is not required. The relatively high velocities in the system tend to flush away deposits of silt and other solids which might accumulate and alter measurement. There are no sharp edges, no pockets, and few critical dimensions; also, the device can be locally fabricated from available materials. Calibration data based on physical dimensions are available from 3 in. (76 mm) throat width with minimum range of 0.03 second feet (13 GPM/49 l/m) up to 40 ft (12.2 m) throat width with maximum capacity of 2000 second feet (900,000 GPM/3,406,000 l/m). Flow is approximately proportional to the three-halves power of level with stated flow rangeability for a single unit of 35:1 or more, depending on size.

*Head is measured between level in stilling pond and crest of a rectangular or Cippoletti weir, or bottom of V of a V-notch weir.

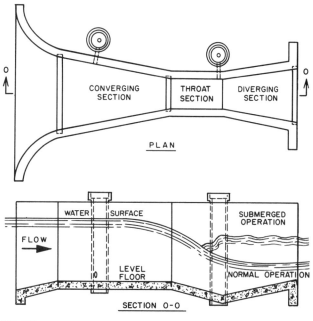

FIG. 29f
Parshall flume.

Extreme accuracy is not claimed for flow measurement using this device; however, measurement is very dependable with minimum maintenance and good repeatability. Accuracy is adequate for most applications to irrigation, waste, and sewage flows.

Downstream level has no effect on the measurement so long as the level near the downstream end of the throat does not exceed 70% of the level measured near the upstream end of the converging section (Figure 29f). (Both levels are referred to the floor section of the flume.) For flumes less than 1 ft (305 mm) wide, the ratio of levels is 60% maximum. This is the preferred and more usual mode of operation. It provides best accuracy. Only one measurement of level is required with flow computed directly from this upstream level measurement; direct, continuous readout of flow rate is readily provided.

Where operating conditions (available head, maximum flow rate, weir size, etc.) result in a throat level greater than 70% of upstream level, so-called submersion results. Measurement can be obtained with a downstream level as great as 95% of upstream level. However, this requires a correction factor based on both upstream level and downstream level in the flow computation, and accuracy suffers. Standard equipment for direct readout in flow is not available.

The simplified equations based on a single measurement at the upstream location are as follows:

For L = ¼ foot:
$$Q = 3.97 \, L \, H^{1.547} \quad \quad 29(4)$$

For L = ½ foot:
$$Q = 4.12 \, L \, H^{1.58} \quad \quad 29(5)$$

For L = ¾ foot:
$$Q = 4.10 \, L \, H^{1.53} \quad \quad 29(6)$$

For L = 1 foot to 8 feet:
$$Q = 4.0 \, L \, H (1.522 L)^{0.026} \quad \quad 29(7)$$

For L over 8 feet:
$$Q = (2.5 + 3.69 \, L) H^{1.6} \quad \quad 29(8)$$

where

- L = width of throat section in feet
- Q = volume flow rate in cubic feet per second
- H* = head in feet

Parshall flumes are available in plastic construction. One variation of the plastic units is the nested, dual-range configuration, in which two flumes are nested inside each other. This configuration is used in installations where the start-up conditions are substantially lower than the final operating flow rates (Figure 29g). With these units, initially the flow passes through the inner flume; then, when the flow exceeds its capacity, the inner flume is removed, while the outer

*H (head) is measured at a designated point in the upstream converging section, referred to the level floor of this section.

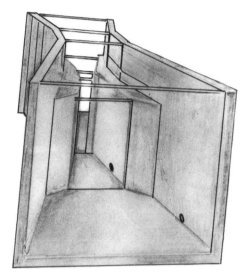

FIG. 29g
Dual-range Parshall flume. (Courtesy of Fischer & Porter Co.)

TABLE 29h
*Dimensions and Capacities of One-Piece Parshall Flumes**†*

Throat Width	Depth (inches)	Length	Weight (pounds)	Free Flow (GPM)	
				Minimum	Maximum
2"	12	2'6½"	35	9.0	210
3"	24	3'0"	40	13.5	494
6"	24	5'0"	100	22.4	1,750
9"	30	5'4"	130	40.4	3,950
12"	36	9'4-⅞"	280	157.0	7,225
18"	36	9'7-⅞"	305	228.9	11,040
24"	36	9'10-⅞"	330	296.2	14,855
3'0"	36	10'4-¾"	385	435.3	22,619
4'0"	36	10'10-⅝"	450	565.5	30,473
5'0"	36	11'4-¼"	515	996.3	38,417
6'0"	36	11'10-⅜"	575	1,180.3	46,450
7'0"	36	12'4-¼"	650	1,831.1	54,484
8'0"	36	12'10-⅛"	730	2,073.5	62,607

*Units in table can be converted using 1 inch = 25.4 mm, 1 pound = 0.45 kg, 1 in. H_2O = 249 Pa, and 1 GPM = 3.785 lpm.
†Courtesy of ABB Kent-Taylor.

flume remains in place permanently. Dimensions of fiberglass-reinforced resin Parshall flumes are given in Table 29h.

THE PALMER BOWLUS FLUME

This is actually a set of principles rather than a specific design. A constriction causes the liquid to flow at critical depth with parallel filaments. This approach is applicable to situations where space or other limitations preclude application of standard devices. Reference 3 provides data on this.

These flumes are available for installation in existing round pipe using the type of insert shown in Figure 29i.

THE KENNISON NOZZLE, PARABOLIC FLUME, AND LEOPOLD LAGCO FLUME

These are typical proprietary products designed primarily for end of pipe flow measurement of waste, sewage, and the like, where the liquid flow to be measured emerges from a cylindrical pipe or conduit which usually is not completely full of liquid. All are designed to flush solids through the device without accumulations and also with accessibility for inspection and cleaning if necessary.

These devices develop heads which are a function of flow rate. In the Kennison Nozzle, head is almost linear with flow above 10% of maximum flow rate. Accuracy is stated as 2% in this range. For the parabolic flume and the Leopold Lagco Flume, flow varies approximately as the three-halves power of head.

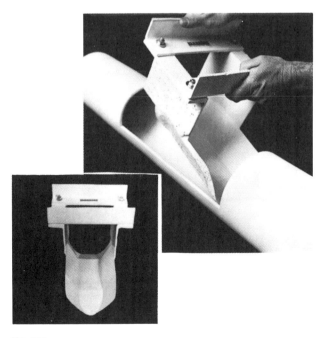

FIG. 29i
Flume insert elements. (Courtesy of Manning Environmental Corp.)

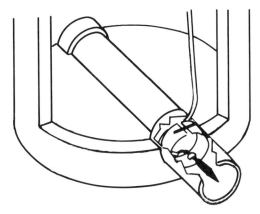

FIG. 29j
Volumetric flow computer measures depth and velocity in open channel and does not require a primary device. (Courtesy of Montedoro-Whitney Corp.)

These devices are available in medium to large sizes. Details as to structure, application, and characteristics are available from the manufacturers.

DETECTORS FOR OPEN-CHANNEL SENSORS

The level rise generated by flumes or weirs can be measured by just about any level detector including such simple devices as the air bubblers.

It is also possible to detect the flow in open channels without the use of flumes, weirs, or any other primary devices. One such design computes flow in round pipes or open channels by ultrasonically measuring the depth, calculating the flowing cross-sectional area on that basis, and multiplying the area by the velocity to obtain volumetric flow (Figure 29j).

Another open-channel flowmeter that does not need a primary element uses a robotically operated magnetic flowmeter probe to scan the velocity profile in the open channel (Figure 29k), In this design the computer algorithm separately calculates and adds up the flow segments through each slice of the velocity profile, as the velocity sensor moves down to the bottom of the channel.

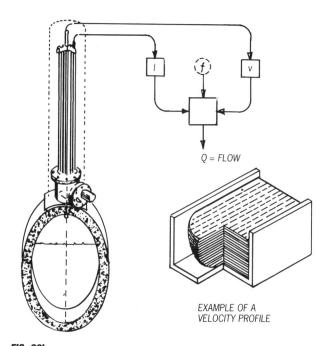

FIG. 29k
Robotically operated magmeter probe sensor is used to compute channel flow. (Courtesy of MSR Magmeter Mfg. Ltd.)

References

1. Streeter, V.L., "The Kinetic Energy and Momentum Corrections for Pipes and Open Channels of Great Width," *Civil Engineering,* Vol. 12, No. 4, April 1942, p. 212.
2. "Measuring Water in Irrigation Channels," *Farmers Bulletin 1682,* United States Dept. of Agriculture.
3. Paper Number 1948, *Proceedings of the Institute of Civil Engineers,* Vol. 101, 1936, p. 1195.
4. Wells, E.A., and Gotaas, H.B., "Design of Venturi Tubes in Circular Conduits," *Proceedings of the American Society of Civil Engineers, Journal of the Sanitary Engineering Division,* Vol. 82, New York: American Society of Civil Engineers, April 1956.

Bibliography

"Open Channel Flowmeters," *Measurements and Control,* December 1992.
Shinskey, G., "Characterizers for Flume and Weirs," *Instrument and Control Systems,* September 1974, p. 111.
Thorsen, T., and Oen, R., "How to Measure Industrial Wastewater Flow," *Chemical Engineering,* February 17, 1975, pp. 97–100.

A.1 International System of Units

The decimal system of units was conceived in the 16th century when there was a great confusion and jumble of units of weights and measures. It was not until 1790, however, that the French National Assembly requested the French Academy of Sciences to work out a system of units suitable for adoption by the entire world. This system, based on the metre as a unit of length and the gram as a unit of mass, was adopted as a practical measure to benefit industry and commerce. Physicists soon realized its advantages and it was adopted also in scientific and technical circles. The importance of the regulation of weights and measures was recognized in Article 1, Section 8, when the United States Constitution was written in 1787, but the metric system was not legalized in this country until 1866. In 1893, the international metre and kilogram became the fundamental standards of length and mass in the United States, both for metric and customary weights and measures.[1]

TABLE A.1
International System of Units

Quantity	Unit	SI Symbol	Formula	Quantity	Unit	SI Symbol	Formula
BASE UNITS							
length	metre	m	—	energy	joule	J	N · m
mass	kilogram	kg	—	entropy	joule per kelvin	—	J/K
time	second	s	—	force	newton	N	kg · m/s^2
electric current	ampere	A	—	frequency	hertz	Hz	(cycle)/s
thermodynamic temperature	kelvin	K	—	illuminance	lux	lx	lm/m^2
amount of substance	mole	mol	—	luminance	candela per square metre	—	cd/m^2
luminous intensity	candela	cd	—	luminous flux	lumen	lm	cd · sr
SUPPLEMENTARY UNITS:				magnetic field strength	ampere per metre	—	A/m
				magnetic flux	weber	Wb	V · s
plane angle	radian	rad	—	magnetic flux density	tesla	T	Wb/m^2
solid angle	steradian	sr	—	magnetomotive force	ampere	A	—
DERIVED UNITS:				**DERIVED UNITS:**			
acceleration	metre per second squared	—	m/s^2	power	watt	W	J/s
activity (of a radioactive source)	disintegration per second	—	(disintegration)/s	pressure	pascal	Pa	N/m^2
				quantity of electricity	coulomb	C	A · s
angular acceleration	radian per second squared	—	rad/s^2	quantity of heat	joule	J	N · m
				radiant intensity	watt per steradian	—	W/sr
angular velocity	radian per second	—	rad/s	specific heat	joule per kilogram-kelvin	—	J/kg · K
area	square metre	—	m^2				
density	kilogram per cubic metre	—	kg/m^3	stress	pascal	Pa	N/m^2
				thermal conductivity	watt per metre-kelvin	—	W/m · K
electric capacitance	farad	F	A · s/V	velocity	metre per second	—	m/s
electrical conductance	siemens	S	A/V	viscosity, dynamic	pascal-second	—	Pa · s
electric field strength	volt per metre	—	V/m	viscosity, kinematic	square metre per second	—	m^2/s
electric inductance	henry	H	V · s/A				
electric potential difference	volt	V	W/A	voltage	volt	V	W/A
				volume	cubic metre	—	m^3
electric resistance	ohm	Ω	V/A	wavenumber	reciprocal metre	—	(wave)/m
electromotive force	volt	V	W/A	work	joule	J	N · m

A.2 Engineering Conversion Factors

TABLE A.2
Engineering Conversion Factors

To Convert	Into	Multiply by	To Convert	Into	Multiply by
A			ares	sq. meters	100.0
abcoulomb	statcoulombs	2.998×10^{10}	Astronomical Unit	kilometers	1.495×10^8
acre	sq. chain		atmospheres	ton/sq. in.	.007348
	(Gunters)	10	atmospheres	cms of mercury	76.0
acre	rods	160	atmospheres	ft. of water (at	
acre	sq. links (Gunters)	1×10^5		4°C)	33.90
acre	hectare or		atmospheres	in. of mercury (at	
	sq. hectometer	.4047		0°C)	29.92
acres	sq. ft.	43,560.0	atmospheres	kgs/sq. cm	1.0333
acres	sq. meters	4,047.	atmospheres	kgs/sq. meter	10,332.
acres	sq. miles	1.562×10^{-3}	atmospheres	pounds/sq. in.	14.70
acres	sq. yards	4,840.	atmospheres	tons/sq. ft.	1.058
acre-feet	cu. ft.	43,560.0			
acre-feet	gallons	3.259×10^5		**B**	
amperes/sq. cm	amps/sq. in.	6.452	barrels (U.S., dry)	cu. in.	7056.
amperes/sq. cm	amps/sq. meter	10^4	barrels (U.S., dry)	quarts (dry)	105.0
amperes/sq. in.	amps/sq. cm	0.1550	barrels (U.S.,		
amperes/sq. in.	amps/sq. meter	1,550.0	liquid)	gallons	31.5
amperes/sq. meter	amps/sq. cm	10^{-4}	barrels (oil)	gallons (oil)	42.0
amperes/sq. meter	amps/sq. in.	6.452×10^{-4}	bars	atmospheres	0.9869
ampere-hours	coulombs	3,600.0	bars	dynes/sq. cm	10^6
ampere-hours	faradays	0.03731	bars	kgs/sq. meter	1.020×10^4
ampere-turns	gilberts	1.257	bars	pounds/sq. ft.	2,089.
ampere-turns/cm	amp-turns/in.	2.540	bars	pounds/sq. in.	14.50
ampere-turns/cm	amp-turns/meter	100.0	baryl	dyne/sq. cm	1.000
ampere-turns/cm	gilberts/cm	1.257	bolt (US cloth)	meters	36.576
ampere-turns/in.	amp-turns/cm	0.3937	Btu	liter–atmosphere	10.409
ampere-turns/in.	amp-turns/meter	39.37	Btu	ergs	1.0550×10^{10}
ampere-turns/in.	gilberts/cm	0.4950	Btu	foot-lbs	778.3
ampere-turns/meter	amp/turns/cm	0.01	Btu	gram-calories	252.0
ampere-turns/meter	amp-turns/in.	0.0254	Btu	horsepower-hrs	3.931×10^{-4}
ampere-turns/meter	gilberts/cm	0.01257	Btu	joules	1,054.8
Angstrom unit	in.	$3,937 \times 10^{-9}$	Btu	kilogram-calories	0.2520
Angstrom unit	meter	1×10^{-10}	Btu	kilogram-meters	107.5
Angstrom unit	micron	1×10^{-4}	Btu	kilowatt-hrs	2.928×10^{-4}
are	acre (US)	.02471	Btu/hr	foot-pounds/sec	0.2162
ares	sq. yards	119.60	Btu/hr	gram-cal/sec	0.0700
ares	acres	0.02471	Btu/hr	horsepower-hrs	3.929×10^{-4}

TABLE A.2
Engineering Conversion Factors Continued

To Convert	Into	Multiply by
Btu/hr	watts	0.2931
Btu/min	foot-lbs/sec	12.96
Btu/min	horsepower	0.02356
Btu/min	kilowatts	0.01757
Btu/min	watts	17.57
Btu/sq. ft/min	watts/sq. in.	0.1221
bucket (Br. dry)	cu. cm.	1.818×10^4
bushels	cu. ft.	1.2445
bushels	cu. in.	2,150.4
bushels	cu. meters	0.03524
bushels	liters	35.24
bushels	pecks	4.0
bushels	pints (dry)	64.0
bushels	quarts (dry)	32.0

C

To Convert	Into	Multiply by
calories, gram (mean)	Btu (mean)	3.9685×10^{-3}
candle/sq. cm	lamberts	3.142
candle/sq. in.	lamberts	.4870
centares (centiares)	sq. meters	1.0
Centigrade	Fahrenheit	(C° × 9/5) + 32
centigrams	grams	0.01
centiliter	ounce fluid (US)	.3382
centiliter	cu. in.	.6103
centiliter	drams	2.705
centiliters	liters	0.01
centimeters	feet	3.281×10^{-2}
centimeters	inches	0.3937
centimeters	kilometers	10^{-5}
centimeters	meters	0.01
centimeters	miles	6.214×10^{-6}
centimeters	millimeters	10.0
centimeters	mils	393.7
centimeters	yards	1.094×10^{-2}
centimeter-dynes	cm-grams	1.020×10^{-3}
centimeter-dynes	meter-kgs	1.020×10^{-8}
centimeter-dynes	pound-feet	7.376×10^{-8}
centimeter-grams	cm-dynes	980.7
centimeter-grams	meter-kgs	10^{-5}
centimeter-grams	pound-feet	7.233×10^{-5}
centimeters of mercury	atmospheres	0.01316
centimeters of mercury	feet of water	0.4461
centimeters of mercury	kgs/sq meter	136.0
centimeters of mercury	pounds/sq. ft.	27.85
centimeters of mercury	pounds/sq. in.	0.1934
centimeters/sec	feet/min	1.1969
centimeters/sec	feet/sec	0.03281
centimeters/sec	kilometers/hr	0.036
centimeters/sec	knots	0.1943
centimeters/sec	meters/min	0.6
centimeters/sec	miles/hr	0.02237
centimeters/sec	miles/min	3.728×10^{-4}
centimeters/sec/sec	feet/sec/sec	0.03281
centimeters/sec/sec	kms/hr/sec	0.036
centimeters/sec/sec	meters/sec/sec	0.01
centimeters/sec/sec	miles/hr/sec	0.02237
chain	inches	792.00
chain	meters	20.12
chains (surveyors' or Gunter's)	yards	22.00
circular mils	sq. cms	5.067×10^{-6}
circular mils	sq. mils	0.7854
circumference	radians	6.283
circular mils	sq. inches	7.854×10^{-7}
cords	cord feet	8
cord feet	cu. ft.	16
coulomb	statcoulombs	2.998×10^9
coulombs	faradays	1.036×10^{-5}
coulombs/sq. cm	coulombs/sq. in.	64.52
coulombs/sq. cm	coulombs/sq. meter	10^4
coulombs/sq. in.	coulombs/sq. cm	0.1550
coulombs/sq. in.	coulombs/sq. meter	1,550.
coulombs/sq. meter	coulombs/sq. cm	10^{-4}
coulombs/sq. meter	coulombs/sq. in.	6.452×10^{-4}
cubic centimeters	cu. ft.	3.531×10^{-5}
cubic centimeters	cu. in.	0.06102
cubic centimeters	cu. meters	10^{-6}
cubic centimeters	cu. yards	1.308×10^{-6}
cubic centimeters	gallons (U.S. liq.)	2.642×10^{-4}
cubic centimeters	liters	0.001
cubic centimeters	pints (U.S. liq.)	2.113×10^{-3}
cubic centimeters	quarts (U.S. liq.)	1.057×10^{-3}
cubic feet	bushels (dry)	0.8036
cubic feet	cu. cms	28,320.0
cubic feet	cu. in.	1,728.0
cubic feet	cu. meters	0.02832
cubic feet	cu. yards	0.03704
cubic feet	gallons (U.S. liq.)	7.48052
cubic feet	liters	28.32
cubic feet	pints (U.S. liq.)	59.84
cubic feet	quarts (U.S. liq.)	29.92
cubic feet/min	cu. cms/sec	472.0
cubic feet/min	gallons/sec	0.1247
cubic feet/min	liters/sec	0.4720
cubic feet/min	pounds of water/min	62.43
cubic feet/sec	million gals/day	0.646317

TABLE A.2

Engineering Conversion Factors *Continued*

To Convert	Into	Multiply by
cubic feet/sec	gallons/min	448.831
cubic inches	cu. cms	16.39
cubic inches	cu. feet	5.787×10^{-4}
cubic inches	cu. meters	1.639×10^{-5}
cubic inches	cu. yards	2.143×10^{-5}
cubic inches	gallons	4.329×10^{-3}
cubic inches	liters	0.01639
cubic inches	mil-feet	1.061×10^{5}
cubic inches	pints (U.S. liq.)	0.03463
cubic inches	quarts (U.S. liq.)	0.01732
cubic meters	bushels (dry)	28.38
cubic meters	cu. cms	10^{6}
cubic meters	cu. ft.	35.31
cubic meters	cu. in.	61,023.0
cubic meters	cu. yards	1.308
cubic meters	gallons (U.S. liq.)	264.2
cubic meters	liters	1,000.0
cubic meters	pints (U.S. liq.)	2,113.0
cubic meters	quarts (U.S. liq.)	1,057.
cubic yards	cu. cms	7.646×10^{5}
cubic yards	cu. ft.	27.0
cubic yards	cu. in.	46,656.0
cubic yards	cu. meters	0.7646
cubic yards	gallons (U.S. liq.)	202.0
cubic yards	liters	764.6
cubic yards	pints (U.S. liq.)	1,615.9
cubic yards	quarts (U.S. liq.)	807.9
cubic yards/min	cu. ft/sec	0.45
cubic yards/min	gallons/sec	3.367
cubic yards/min	liters/sec	12.74

D

To Convert	Into	Multiply by
dalton	gram	1.650×10^{-24}
days	seconds	86,400.0
decigrams	grams	0.1
deciliters	liters	0.1
decimeters	meters	0.1
degrees (angle)	quadrants	0.01111
degrees (angle)	radians	0.01745
degrees (angle)	seconds	3,600.0
degrees/sec	radians/sec	0.01745
degrees/sec	revolutions/min	0.1667
degrees/sec	revolutions/sec	2.778×10^{-3}
dekagrams	grams	10.0
dekaliters	liters	10.0
dekameters	meters	10.0
drams (apothecaries' or troy)	ounces (avoidupois)	0.1371429
drams (apothecaries' or troy)	ounces (troy)	0.125
drams (U.S., fluid or apothecaries)	cu. cm	3.6967

To Convert	Into	Multiply by
drams	grams	1.7718
drams	grains	27.3437
drams	ounces	0.0625
dyne/cm	erg/sq. millimeter	.01
dyne/sq. cm	atmospheres	9.869×10^{-7}
dyne/sq. cm	inch of mercury at 0°C	2.953×10^{-5}
dyne/sq. cm	inch of water at 4°C	4.015×10^{-4}
dynes	grams	1.020×10^{-3}
dynes	joules/cm	10^{-7}
dynes	joules/meter (newtons)	10^{-5}
dynes	kilograms	1.020×10^{-6}
dynes	poundals	7.233×10^{-5}
dynes	pounds	2.248×10^{-6}
dynes/sq. cm	bars	10^{-6}

E

To Convert	Into	Multiply by
ell	cm	114.30
ell	in.	45
em, pica	in.	.167
em, pica	cm	.4233
erg/sec	dyne–cm/sec	1.000
ergs	Btu	9.480×10^{-11}
ergs	dyne-centimeters	1.0
ergs	foot-pounds	7.367×10^{-8}
ergs	gram-calories	0.2389×10^{-7}
ergs	gram-cms	1.020×10^{-3}
ergs	horsepower-hrs	3.7250×10^{-14}
ergs	joules	10^{-7}
ergs	kg-calories	2.389×10^{-11}
ergs	kg-meters	1.020×10^{-8}
ergs	kilowatt-hrs	0.2778×10^{-13}
ergs	watt-hours	0.2778×10^{-10}
ergs/sec	Btu/min	$5,688 \times 10^{-9}$
ergs/sec	ft-lbs/min	4.427×10^{-6}
ergs/sec	ft-lbs/sec	7.3756×10^{-8}
ergs/sec	horsepower	1.341×10^{-10}
ergs/sec	kg-calories/min	1.433×10^{-9}
ergs/sec	kilowatts	10^{-10}

F

To Convert	Into	Multiply by
farads	microfarads	10^{6}
faraday/sec	ampere (absolute)	9.6500×10^{4}
faradays	ampere-hours	26.80
faradays	coulombs	9.649×10^{4}
fathom	meter	1.828804
fathoms	feet	6.0
feet	centimeters	30.48
feet	kilometers	3.048×10^{-4}
feet	meters	0.3048

TABLE A.2
Engineering Conversion Factors *Continued*

To Convert	Into	Multiply by	To Convert	Into	Multiply by
feet	miles (naut.)	1.645×10^{-4}	gallons	cu. in.	231.0
feet	miles (stat.)	1.894×10^{-4}	gallons	cu. meters	3.785×10^{-3}
feet	millimeters	304.8	gallons	cu. yards	4.951×10^{-3}
feet	mils	1.2×10^{4}	gallons	liters	3.785
feet of water	atmospheres	0.02950	gallons (liq. Br. Imp.)	gallons (U.S. liq.)	1.20095
feet of water	in. of mercury	0.8826	gallons (U.S.)	gallons (Imp.)	0.83267
feet of water	kgs/sq. cm	0.03048	gallons of water	pounds of water	8.3453
feet of water	kgs/sq. meter	304.8	gallons/min	cu. ft/sec	2.228×10^{-3}
feet of water	pounds/sq. ft.	62.43	gallons/min	liters/sec	0.06308
feet of water	pounds/sq. in.	0.4335	gallons/min	cu. ft/hr	8.0208
feet/min	cms/sec	0.5080	gausses	lines/sq. in.	6.452
feet/min	ft/sec	0.01667	gausses	webers/sq. cm.	10^{-8}
feet/min	kms/hr	0.01829	gausses	webers/sq. in.	6.452×10^{-8}
feet/min	meters/min	0.3048	gausses	webers/sq. meter	10^{-4}
feet/min	miles/hr	0.01136	gilberts	ampere-turns	0.7958
feet/sec	cms/sec	30.48	gilberts/cm	amp-turns/cm	0.7958
feet/sec	kms/hr	1.097	gilberts/cm	amp-turns/in.	2.021
feet/sec	knots	0.5921	gilberts/cm	amp-turns/meter	79.58
feet/sec	meters/min	18.29	gills (British)	cu. cm	142.07
feet/sec	miles/hr	0.6818	gills	liters	0.1183
feet/sec	miles/min	0.01136	gills	pints (liq.)	0.25
feet/sec/sec	cms/sec/sec	30.48	grade	radian	.01571
feet/sec/sec	kms/hr/sec	1.097	grains	drams (avoirdupois)	0.03657143
feet/sec/sec	meters/sec/sec	0.3048	grains (troy)	grains (avdp)	1.0
feet/sec/sec	miles/hr/sec	0.6818	grains (troy)	grams	0.06480
feet/100 feet	per cent grade	1.0	grains (troy)	ounces (avdp)	2.0833×10^{-3}
foot–candle	lumen/sq. meter	10.764	grains (troy)	pennyweight (troy)	0.04167
foot-pounds	Btu	1.286×10^{-3}	grains/U.S. gal	parts/million	17.118
foot-pounds	ergs	1.356×10^{7}	grains/U.S. gal	pouns/million gal	142.86
foot-pounds	gram-calories	0.3238	grains/Imp. gal	parts/million	14.286
foot-pounds	hp-hrs	5.050×10^{-7}	grains	dynes	980.7
foot-pounds	joules	1.356	grams	grains	15.43
foot-pounds	kg-calories	3.24×10^{-4}	grams	joules/cm	9.807×10^{-5}
foot-pounds	kg-meters	0.1383	grams	joules/meter (newtons)	9.807×10^{-3}
foot-pounds	kilowatt-hrs	3.766×10^{-7}	grams	kilograms	0.001
foot-pounds/min	Btu/min	1.286×10^{-3}	grams	milligrams	1,000.
foot-pounds/min	foot-pounds/sec	0.01667	grams	ounces (avdp)	0.03527
foot-pounds/min	horsepower	3.030×10^{-5}	grams	ounces (troy)	0.03215
foot-pounds/min	kg-calories/min	3.24×10^{-4}	grams	poundals	0.07093
foot-pounds/min	kilowatts	2.260×10^{-5}	grams	pounds	2.205×10^{-3}
foot-pounds/sec	Btu/hr	4.6263	grams/cm	pounds/inch	5.600×10^{-3}
foot-pounds/sec	Btu/min	0.07717	grams/cu. cm	pounds/cu. ft.	62.43
foot-pounds/sec	horsepower	1.818×10^{-3}	grams/cu. cm	pounds/cu. in.	0.03613
foot-pounds/sec	kg-calories/min	0.01945	grams/cu. cm	pounds/mil-foot	3.405×10^{-7}
foot-pounds/sec	kilowatts	1.356×10^{-3}	grams/liter	grains/gal	58.417
furlongs	miles (U.S.)	0.125	grams/liter	pounds/1,000 gal	8.345
furlongs	rods	40.0	grams/liter	pounds/cu. ft.	0.062427
furlongs	feet	660.0	grams/liter	parts/million	1,000.0

G

gallons	cu. cms	3,785.0
gallons	cu. ft.	0.1337

TABLE A.2
Engineering Conversion Factors *Continued*

To Convert	Into	Multiply by	To Convert	Into	Multiply by
grams/sq. cm	pounds/sq. ft.	2.0481	hundredweights (long)	tons (long)	0.05
gram-calories	Btu	3.9683×10^{-3}	hundredweights (short)	ounces (avoirdupois)	1,600
gram-calories	ergs	4.1868×10^{7}			
gram-calories	foot-pounds	3.0880	hundredweights (short)	pounds	100
gram-calories	horsepower-hrs	1.5596×10^{-6}			
gram-calories	kilowatt-hrs	1.1630×10^{-6}	hundredweights (short)	tons (metric)	0.0453592
gram-calories	watt-hrs	1.1630×10^{-3}	hundredweights (short)	tons (long)	0.0446429
gram-calories/sec	Btu/hr	14.286			
gram-centimeters	Btu	9.297×10^{-8}			
gram-centimeters	ergs	980.7		**I**	
gram-centimeters	joules	9.807×10^{-5}	inches	centimeters	2.540
gram-centimeters	kg-cal	2.343×10^{-8}	inches	meters	2.540×10^{-2}
gram-centimeters	kg-meters	10^{-5}	inches	miles	1.578×10^{-5}
			inches	millimeters	25.40
			inches	mils	1,000.0
	H		inches	yards	2.778×10^{-2}
hand	cm	10.16	inches of mercury	atmospheres	0.03342
hectares	acres	2.471	inches of mercury	feet of water	1.133
hectares	sq. feet	1.076×10^{5}	inches of mercury	kgs/sq. cm	0.03453
hectograms	grams	100.0	inches of mercury	kgs/sq. meter	345.3
hectoliters	liters	100.0	inches of mercury	pounds/sq. ft.	70.73
hectometers	meters	100.0	inches of mercury	pounds/sq. in.	0.4912
hectowatts	watts	100.0	inches of water (at 4°C)	atmospheres	2.458×10^{-3}
henries	millihenries	1,000.0			
hogsheads (British)	cu. ft.	10.114	inches of water (at 4°C)	inches of mercury	0.07355
hogsheads (U.S.)	cu. ft.	8.42184			
hogsheads (U.S.)	gallons (U.S.)	63	inches of water (at 4°C)	kgs/sq. cm	2.540×10^{-3}
horsepower	Btu/min	42.44			
horsepower	ft-lbs/min	33,000.	inches of water (at 4°C)	ounces/sq. in.	0.5781
horsepower	ft-lbs/sec	550.0			
horsepower (metric) (542.5 ft lb/sec)	horsepower (550 ft-lb/sec)	0.9863	inches of water (at 4°C)	pounds/sq. ft.	5.204
horsepower (550 ft lb/sec)	horsepower (metric) (542.5 ft-lb/sec)	1.014	inches of water (at 4°C)	pounds/sq. in.	0.03613
horsepower	kg-calories/min	10.68	International ampere	ampere (absolute)	.9998
horsepower	kilowatts	0.7457	International volt	volts (absolute)	1.0003
horsepower	watts	745.7			
horsepower (boiler)	Btu/hr	33.479			
horsepower (boiler)	kilowatts	9.803		**J**	
horsepower-hrs	Btu	2,547.	joules	Btu	9.480×10^{-4}
horsepower-hrs	ergs	2.6845×10^{13}	joules	ergs	10^{7}
horsepower-hrs	ft-lbs	1.98×10^{6}	joules	foot-pounds	0.7376
horsepower-hrs	gram-calories	641,190.	joules	kg-calories	2.389×10^{-4}
horsepower-hrs	joules	2.684×10^{6}	joules	kg-meters	0.1020
horsepower-hrs	kg-calories	641.1			
horsepower-hrs	kg-meters	2.737×10^{5}	joules	watt-hrs	2.778×10^{-4}
horsepower-hrs	kilowatt-hrs	0.7457	joules/cm	grams	1.020×10^{4}
hours	days	4.167×10^{-2}	joules/cm	dynes	10^{7}
hours	weeks	5.952×10^{-3}	joules/cm	joules/meter (newtons)	100.0
hundredweights (long)	pounds	112	joules/cm	poundals	723.3
			joules/cm	pounds	22.48

TABLE A.2
Engineering Conversion Factors *Continued*

To Convert	Into	Multiply by	To Convert	Into	Multiply by
K			kilometers/hr	ft/sec	0.9113
kilograms	dynes	980,665.	kilometers/hr	knots	0.5396
kilograms	grams	1,000.0	kilometers/hr	meters/min	16.67
kilograms	joules/cm	0.09807	kilometers/hr	miles/hr	0.6214
kilograms	joules/meter (newtons)	9.807	kilometers/hr/sec	cms/sec/sec	27.78
kilograms	poundals	70.93	kilometers/hr/sec	ft/sec/sec	0.9113
			kilometers/hr/sec	meters/sec/sec	0.2778
kilograms	pounds	2.205	kilometers/hr/sec	miles/hr/sec	0.6214
kilograms	tons (long)	9.842×10^{-4}	kilowatts	Btu/min	56.92
kilograms	tons (short)	1.102×10^{-3}	kilowatts	ft-lbs/min	4.426×10^4
kilograms/cu. meter	grams/cu. cm	0.001	kilowatts	ft-lbs/sec	737.6
kilograms/cu. meter	pounds/cu. ft.	0.06243	kilowatts	horsepower	1.341
kilograms/cu. meter	pounds/cu. in.	3.613×10^{-5}	kilowatts	kg-calories/min	14.34
kilograms/cu. meter	pounds/mil-foot	3.405×10^{-10}	kilowatts	watts	1,000.0
kilograms/meter	pounds/ft.	0.6720	kilowatt-hrs	Btu	3,413.
kilograms/sq. cm	dynes	980,665	kilowatt-hrs	ergs	3.600×10^{13}
kilograms/sq. cm	atmospheres	0.9678	kilowatt-hrs	ft-lbs	2.655×10^6
kilograms/sq. cm	feet of water	32.81	kilowatt-hrs	gram-calories	859,850.
kilograms/sq. cm	inches of mercury	28.96	kilowatt-hrs	horsepower-hrs	1,341
kilograms/sq. cm	pounds/sq. ft.	2,048.	kilowatt-hrs	joules	3.6×10^6
kilograms/sq. cm	pounds/sq. in.	14.22	kilowatt-hrs	kg-calories	860.5
kilograms/sq. meter	atmospheres	9.678×10^{-5}	kilowatt-hrs	kg-meters	3.671×10^5
kilograms/sq. meter	bars	98.07×10^{-6}	kilowatt-hrs	pounds of water evaporated from and at 212°F.	3.53
kilograms/sq. meter	feet of water	3.281×10^{-3}			
kilograms/sq. meter	inches of mercury	2.896×10^{-3}			
kilograms/sq. meter	pounds/sq. ft.	0.2048	kilowatt-hrs	pounds of water raised from 62° to 212°F	22.75
kilograms/sq. meter	pounds/sq. in.	1.422×10^{-3}			
kilograms/sq. mm	kgs/sq. meter	10^6			
kilogram-calories	Btu	3.968	knots	ft/hr	6,080.
kilogram-calories	foot-pounds	3,088.	knots	kilometers/hr	1.8532
kilogram-calories	hp-hrs	1.560×10^{-3}	knots	nautical miles/hr	1.0
kilogram-calories	joules	4,186.	knots	statute miles/hr	1.151
kilogram-calories	kg-meters	426.9	knots	yards/hr	2,027.
kilogram-calories	kilojoules	4.186	knots	ft/sec	1.689
kilogram-calories	kilowatt-hrs	1.163×10^{-3}			
kilogram meters	Btu	9.294×10^{-3}			
kilogram meters	ergs	9.804×10^7	**L**		
kilogram meters	foot-pounds	7.233	league	miles (approx.)	3.0
kilogram meters	joules	9.804	light year	miles	5.9×10^{12}
kilogram meters	kg-calories	2.342×10^{-3}	light year	kilometers	9.46091×10^{12}
kilogram meters	kilowatt-hrs	2.723×10^{-6}	lines/sq. cm	gausses	1.0
kilolines	maxwells	1,000.0	lines/sq. in.	gausses	0.1550
kiloliters	liters	1,000.0	lines/sq. in.	webers/sq. cm	1.550×10^{-9}
kilometers	centimeters	10^5	lines/sq. in.	webers/sq. in.	10^{-8}
kilometers	feet	3,281.	lines/sq. in.	webers/sq. meter	1.550×10^{-5}
kilometers	inches	3.937×10^4	links (engineer's)	inches	12.0
kilometers	meters	1,000.0	links (surveyor's)	inches	7.92
kilometers	miles	0.6214	liters	bushels (U.S. dry)	0.02833
kilometers	millimeters	10^6	liters	cu. cm	1,000.0
kilometers	yards	1,094.	liters	cu. ft.	0.03531
kilometers/hr	cms/sec	27.78	liters	cu. in.	61.02
kilometers/hr	ft/min	54.68	liters	cu. meters	0.001

TABLE A.2
Engineering Conversion Factors *Continued*

To Convert	Into	Multiply by
liters	cu. yards	1.308×10^{-3}
liters	gallons (U.S. liq.)	0.2642
liters	pints (U.S. liq.)	2.113
liters	quarts (U.S. liq.)	1.057
liters/min	cu. ft/sec	5.886×10^{-4}
liters/min	gals/sec	4.403×10^{-3}
lumens/sq. ft.	foot-candles	1.0
lumen	spherical candle power	.07958
lumen	watt	.001496
lumen/sq. ft.	lumen/sq. meter	10.76
lux	foot-candles	0.0929

M

To Convert	Into	Multiply by
maxwells	kilolines	0.001
maxwells	webers	10^{-8}
megalines	maxwells	10^6
megohms	microhms	10^{12}
megohms	ohms	10^6
meters	centimeters	100.0
meters	feet	3.281
meters	inches	39.37
meters	kilometers	0.001
meters	miles (naut.)	5.396×10^{-4}
meters	miles (stat.)	6.214×10^{-4}
meters	millimeters	1,000.0
meters	yards	1.094
meters	varas	1.179
meters/min	cms/sec	1.667
meters/min	ft/min	3.281
meters/min	ft/sec	0.05468
meters/min	kms/hr	0.06
meters/min	knots	0.03238
meters/min	miles/hr	0.03728
meters/sec	ft/min	196.8
meters/sec	ft/sec	3.281
meters/sec	kilometers/hr	3.6
meters/sec	kilometers/min	0.06
meters/sec	miles/hr	2.237
meters/sec	miles/min	0.03728
meters/sec/sec	cms/sec/sec	100.0
meters/sec/sec	ft/sec/sec	3.281
meters/sec/sec	kms/hr/sec	3.6
meters/sec/sec	miles/hr/sec	2.237
meter-kilograms	cm-dynes	9.807×10^7
meter-kilograms	cm-grams	10^5
meter-kilograms	pound-feet	7.233
microfarad	farads	10^{-6}
micrograms	grams	10^{-6}
microhms	megohms	10^{-12}
microhms	ohms	10^{-6}
microliters	liters	10^{-6}
microns	meters	1×10^{-6}
miles (naut.)	miles (statute)	1.1516
miles (naut.)	yards	2,027.
miles (statute)	centimeters	1.609×10^5
miles (statute)	feet	5,280.
miles (statute)	inches	6.336×10^4
miles (statute)	kilometers	1.609
miles (statute)	meters	1,609.
miles (statute)	miles (naut.)	0.8684
miles (statute)	yards	1,760.
miles/hr	cms/sec	44.70
miles/hr	ft/min	88.
miles/hr	ft/sec	1.467
miles/hr	kms/hr	1.609
miles/hr	kms/min	0.02682
miles/hr	knots	0.8684
miles/hr	meters/min	26.82
miles/hr	miles/min	0.1667
miles/hr/sec	cm/sec/sec	44.70
miles/hr/sec	ft/sec/sec	1.467
miles/hr/sec	kms/hr/sec	1.609
miles/hr/sec	meters/sec/sec	0.4470
miles/min	cms/sec	2,682.
miles/min	ft/sec	88.
miles/min	kms/min	1.609
miles/min	knots/min	0.8684
miles/min	miles/hr	60.0
mil-feet	cu. in.	9.425×10^{-6}
milliers	kilograms	1,000.
millimicrons	meters	1×10^{-9}
milligrams	grains	0.01543236
milligrams	grams	0.001
milligrams/liter	parts/million	1.0
millihenries	henries	0.001
milliliters	liters	0.001
millimeters	centimeters	0.1
millimeters	feet	3.281×10^{-3}
millimeters	inches	0.03937
millimeters	kilometers	10^{-6}
millimeters	meters	0.001
millimeters	miles	6.214×10^{-7}
millimeters	mils	39.37
millimeters	yards	1.094×10^{-3}
million gals/day	cu. ft/sec	1.54723
mils	centimeters	2.540×10^{-3}
mils	feet	8.333×10^{-5}
mils	inches	0.001
mils	kilometers	2.540×10^{-8}
mils	yards	2.778×10^{-5}
miner's inches	cu. ft/min	1.5
minims (British)	cu. cm	0.059192

TABLE A.2

Engineering Conversion Factors *Continued*

To Convert	Into	Multiply by	To Convert	Into	Multiply by
minims (U.S., fluid)	cu. cm	0.061612	pennyweights (troy)	ounces (troy)	0.05
minutes (angles)	degrees	0.01667	pennyweights (troy)	grams	1.55517
miles (naut.)	feet	6,080.27	pennyweights (troy)	pounds (troy)	4.1667×10^{-3}
miles (naut.)	kilometers	1.853	pints (dry)	cu. in.	33.60
miles (naut.)	meters	1,853.	pints (liq.)	cu. cms	473.2
minutes (angles)	quadrants	1.852×10^{-4}	pints (liq.)	cu. ft.	0.01671
minutes (angles)	radians	2.909×10^{-4}	pints (liq.)	cu. in.	28.87
minutes (angles)	seconds	60.0	pints (liq.)	cu. meters	4.732×10^{-4}
myriagrams	kilograms	10.0	pints (liq.)	cu. yards	6.189×10^{-4}
myriameters	kilometers	10.0	pints (liq.)	gallons	0.125
myriawatts	kilowatts	10.0	pints (liq.)	liters	0.4732
			pints (liq.)	quarts (liq.)	0.5
	N		Planck's quantum	erg/sec	6.624×10^{-27}
nepers	decibels	8.686	poise	gram/cm sec	1.00
Newton	dynes	1×10^5	pounds (avoirdupois)	ounces (troy)	14.5833
			poundals	dynes	13,826.
			poundals	grams	14.10
	O		poundals	joules/cm	1.383×10^{-3}
ohm (International)	ohm (absolute)	1.0005	poundals	joules/meter (newtons)	0.1383
ohms	megohms	10^{-6}			
ohms	microhms	10^6	poundals	kilograms	0.01410
ounces	drams	16.0	poundals	pounds	0.03108
ounces	grains	437.5	pounds	drams	256.
ounces	grams	28.349527	pounds	dynes	44.4823×10^4
ounces	pounds	0.0625	pounds	grains	7,000.
ounces	ounces (troy)	0.9115	pounds	grams	453.5924
ounces	tons (long)	2.790×10^{-5}	pounds	joules/cm	0.04448
ounces	tons (metric)	2.835×10^{-5}	pounds	joules/meter (newtons)	4.448
ounces (fluid)	cu. in.	1.805			
ounces (fluid)	liters	0.02957	pounds	kilograms	0.4536
ounces (troy)	grains	480.0	pounds	ounces	16.0
ounces (troy)	grams	31.103481	pounds	ounces (troy)	14.5833
ounces (troy)	ounces (avdp.)	1.09714	pounds	poundals	32.17
ounces (troy)	pennyweights (troy)	20.0	pounds	pounds (troy)	1.21528
			pounds	tons (short)	0.0005
ounces (troy)	pounds (troy)	0.08333	pounds (troy)	grains	5,760.
ounce/sq. in.	dynes/sq. cm	4309	pounds (troy)	grams	373.24177
ounces/sq. in.	pounds/sq. in.	0.0625	pounds (troy)	ounces (avdp.)	13.1657
			pounds (troy)	ounces (troy)	12.0
	P		pounds (troy)	pennyweights (troy)	240.0
parsec	miles	19×10^{12}			
parsec	kilometers	3.084×10^{13}	pounds (troy)	pounds (avdp.)	0.822857
parts/million	grains/U.S. gal	0.0584	pounds (troy)	tons (long)	3.6735×10^{-4}
parts/million	grains/Imp. gal	0.07016	pounds (troy)	tons (metric)	3.7324×10^{-4}
parts/million	pounds/million gal	8.345	pounds (troy)	tons (short)	4.1143×10^{-4}
pecks (British)	cu. in.	554.6	pounds of water	cu. ft.	0.01602
pecks (British)	liters	9.091901	pounds of water	cu. in.	27.68
pecks (U.S.)	bushels	0.25	pounds of water	gallons	0.1198
pecks (U.S.)	cu. in.	537.605	pounds of water/min	cu. ft/sec	2.670×10^{-4}
pecks (U.S.)	liters	8.809582	pound-feet	cm-dynes	1.356×10^7
pecks (U.S.)	quarts (dry)	8	pound-feet	cm-grams	13,825.
pennyweights (troy)	grains	24.0	pound-feet	meter-kgs	0.1383

TABLE A.2

Engineering Conversion Factors *Continued*

To Convert	Into	Multiply by	To Convert	Into	Multiply by
pounds/cu. ft.	grams/cu. cm	0.01602	revolutions/min/min	radians/sec/sec	1.745×10^{-3}
pounds/cu. ft.	kgs/cu. meter	16.02	revolutions/min/min	revs/min/sec	0.01667
pounds/cu. ft.	pounds/cu. in.	5.787×10^{-4}	revolutions/min/min	revs/sec/sec	2.778×10^{-4}
pounds/cu. ft.	pounds/mil-foot	5.456×10^{-9}	revolutions/sec	degrees/sec	360.0
pounds/cu. in.	gms/cu. cm	27.68	revolutions/sec	radians/sec	6.283
pounds/cu. in.	kgs/cu. meter	2.768×10^4	revolutions/sec	revs/min	60.0
pounds/cu. in.	pounds/cu. ft.	1,728.	revolutions/sec/sec	radians/sec/sec	6.283
pounds/cu. in.	pounds/mil-foot	9.425×10^{-6}	revolutions/sec/sec	revs/min/min	3,600.0
pounds/ft.	kgs/meter	1.488	revolutions/sec/sec	revs/min/sec	60.0
pounds/in.	gms/cm	178.6	rod	chain (Gunters)	.25
pounds/mil-foot	gms/cu. cm	2.306×10^6	rod	meters	5.029
pounds/sq. ft.	atmospheres	4.725×10^{-4}	rods (Surveyors' meas.)	yards	5.5
pounds/sq. ft.	feet of water	0.01602	rods	feet	16.5
pounds/sq. ft.	inches of mercury	0.01414			
pounds/sq. ft.	kgs/sq. meter	4.882		**S**	
pounds/sq. ft.	pounds/sq. in.	6.944×10^{-3}	scruples	grains	20
pounds/sq. in.	atmospheres	0.06804	seconds (angle)	degrees	2.778×10^{-4}
pounds/sq. in.	feet of water	2.307	seconds (angle)	minutes	0.01667
pounds/sq. in.	inches of mercury	2.036	seconds (angle)	quadrants	3.087×10^{-6}
pounds/sq. in.	kgs/sq. meter	703.1	seconds (angle)	radians	4.848×10^{-6}
pounds/sq. in.	pounds/sq. ft.	144.0	slug	kilogram	14.59
	Q		slug	pounds	32.17
quadrants (angle)	degrees	90.0	sphere	steradians	12.57
quadrants (angle)	minutes	5,400.0	square centimeters	circular mils	1.973×10^5
quadrants (angle)	radians	1.571	square centimeters	sq. ft.	1.076×10^{-3}
quadrants (angle)	seconds	3.24×10^5	square centimeters	sq. in.	0.1550
quarts (dry)	cu. in.	67.20	square centimeters	sq. meters	0.0001
quarts (liq.)	cu. cms	946.4	square centimeters	sq. miles	3.861×10^{-11}
quarts (liq.)	cu. ft.	0.03342	square centimeters	sq. millimeters	100.0
quarts (liq.)	cu. in.	57.75	square centimeters	sq. yards	1.196×10^{-4}
quarts (liq.)	cu. meters	9.464×10^{-4}	square feet	acres	2.296×10^{-5}
quarts (liq.)	cu. yards	1.238×10^{-3}	square feet	circular mils	1.833×10^8
quarts (liq.)	gallons	0.25	square feet	sq. cms	929.0
quarts (liq.)	liters	0.9463	square feet	sq. in.	144.0
	R		square feet	sq. meters	0.09290
radians	degrees	57.30	square feet	sq. miles	3.587×10^{-8}
radians	minutes	3,438.	square feet	sq. millimeters	9.290×10^4
radians	quadrants	0.6366	square feet	sq. yards	0.1111
radians	seconds	2.063×10^5	square inches	circular mils	1.273×10^6
radians/sec	degrees/sec	57.30	square inches	sq. cms	6.452
radians/sec	revolutions/min	9.549	square inches	sq. ft.	6.944×10^{-3}
radians/sec	revolutions/sec	0.1592	square inches	sq. millimeters	645.2
radians/sec/sec	revs/min/min	573.0	square inches	sq. mils	10^6
radians/sec/sec	revs/min/sec	9.549	square inches	sq. yards	7.716×10^{-4}
radians/sec/sec	revs/sec/sec	0.1592	square kilometers	acres	247.1
revolutions	degrees	360.0	square kilometers	sq. cms	10^{10}
revolutions	quadrants	4.0	square kilometers	sq. ft.	10.76×10^6
revolutions	radians	6.283	square kilometers	sq. in.	1.550×10^9
revolutions/min	degrees/sec	6.0	square kilometers	sq. meters	10^6
revolutions/min	radians/sec	0.1047	square kilometers	sq. miles	0.3861
revolutions/min	revs/sec	0.01667	square kilometers	sq. yards	1.196×10^6

TABLE A.2
Engineering Conversion Factors *Continued*

To Convert	Into	Multiply by	To Convert	Into	Multiply by
square meters	acres	2.471×10^{-4}	tons (short)/sq. ft.	kgs/sq. meter	9,765.
square meters	sq. cms	10^4	tons (short)/sq. ft.	pounds/sq. in.	2,000.
square meters	sq. ft.	10.76	tons of water/24 hrs	pounds of water/hr	83.333
square meters	sq. in.	1,550.	tons of water/24 hrs	gallons/min	0.16643
square meters	sq. miles	3.861×10^{-7}	tons of water/24 hrs	cu. ft/hr	1.3349
square meters	sq. millimeters	10^6			
square meters	sq. yards	1.196		**V**	
square miles	acres	640.0	volt/inch	volt/cm	.39370
square miles	sq. ft.	27.88×10^6	volt (absolute)	statvolts	.003336
square miles	sq. kms	2.590			
square miles	sq. meters	2.590×10^6		**W**	
square miles	sq. yards	3.098×10^6	watts	Btu/hr	3.4129
square millimeters	circular mils	1,973.	watts	Btu/min	0.05688
square millimeters	sq. cms	0.01	watts	ergs/sec	10^7.
square millimeters	sq. ft.	1.076×10^{-5}	watts	ft-lbs/min	44.27
square millimeters	sq. in.	1.550×10^{-3}	watts	ft-lbs/sec	0.7378
square mils	circular mils	1.273	watts	horsepower	1.341×10^{-3}
square mils	sq. cms	6.452×10^{-6}	watts	horsepower (metric)	1.360×10^{-3}
square mils	sq. in.	10^{-6}	watts	kg-calories/min	0.01433
square yards	acres	2.066×10^{-4}	watts	kilowatts	0.001
square yards	sq. cms	8,361.	watts (abs.)	Btu (mean)/min	0.056884
square yards	sq. ft.	9.0	watts (abs.)	joules/sec	1.
square yards	sq. in.	1,296.	watt-hours	Btu	3.413
square yards	sq. meters	0.8361	watt-hours	ergs	3.60×10^{10}
square yards	sq. miles	3.228×10^{-7}	watt-hours	foot-pounds	2,656.
square yards	sq. millimeters	8.361×10^5	watt-hours	gram-calories	859.85
			watt-hours	horsepower-hrs	1.341×10^{-3}
	T		watt-hours	kilogram-calories	0.8605
temperature (°C) + 273	absolute temperature (°K)	1.0	watt-hours	kilogram-meters	367.2
			watt-hours	kilowatt-hrs	0.001
temperature (°C) + 17.78	temperature (°F)	1.8	watt (International)	watt (absolute)	1.0002
temperature (°F) + 460	absolute temperature (°R)	1.0	webers	maxwells	10^8
			webers	kilolines	10^5
temperature (°F) − 32	temperature (°C)	5/9	webers/sq. in.	gausses	1.550×10^7
tons (long)	kilograms	1,016.	webers/sq. in.	lines/sq. in.	10^8
tons (long)	pounds	2,240.	webers/sq. in.	webers/sq. cm	0.1550
tons (long)	tons (short)	1.120	webers/sq. in.	webers/sq. meter	1,550.
tons (metric)	kilograms	1,000.	webers/sq. meter	gausses	10^4
tons (metric)	pounds	2,205	webers/sq. meter	lines/sq. in.	6.452×10^4
tons (short)	kilograms	907.1848	webers/sq. meter	webers/sq. cm	10^{-4}
tons (short)	ounces	32,000.	webers/sq. meter	webers/sq. in.	6.452×10^{-4}
tons (short)	ounces (troy)	29,166.66		**Y**	
tons (short)	pounds	2,000.	yards	centimeters	91.44
tons (short)	pounds (troy)	2,430.56	yards	kilometers	9.144×10^{-4}
tons (short)	tons (long)	0.89287	yards	meters	0.9144
tons (short)	tons (metric)	0.9078	yards	miles (naut.)	4.934×10^{-4}
			yards	miles (stat.)	5.682×10^{-4}
			yards	millimeters	914.4

INDEX

Accelerator-type flowmeter, 140–41
Accuracy, in selection of flowmeter, 14
AC-type magnetic flowmeters, 40
Aerodynamic probe, 99
Aeronautical Research and Test Institute, 105
Angular momentum, 65
Angular-momentum-type mass flowmeters, 65
   constant-torque-hysteresis clutch-type, 66
   Coriolis-type, 66–67
   gyroscopic-type, 67
   impeller-turbine-type, 65–66
   twin-turbine-type, 66
Annular orifice, 142, 143
Area-averaging pitot stations, 104–5
ASME-AGA Committee on Pulsation, 82
ASME flow nozzle, 176–77
ASME Fluid Meters Research Committee, 175, 176–77
Averaging pitot tubes, 100, 102
   advantage of, 102
   in calculating pressure differential, 103–4
   hot-tap installation of, 102–3
Axial flow gas turbine meter, 152

Bearing(s) turbine flowmeter, 149
Belt feeders, 132
   belt speed control, 134–35
   belt-type gravimetric meter, 133
   digital control, 135–36
   electromechanical gravimetric feeder, 132–33
Belt-type gravimetric feeders, feed rate of, 133–34
Bernoulli equation, 36
Bernoulli's Theorem, 80, 172
B.I.F. Universal Venturi, 176
BTU flowmeters
   for gaseous fuels, 19
      applications of, 20–21
      flow measuring loop, 19–20
      measuring heat flow, 19
   for heat exchangers
      electronic, 18
      energy audit, 17–18
      mechanical, 18
Bypass rotameters, 166–67
Bypass-type flow switch, 28
Bypass-type thermal mass flowmeters, 70–71

Calibrated orifices, 110
Capacitance-type flow/no flow switch, 28
Ceramic insert-type magnetic flowmeter, 43, 44
Chemical tracing, 22
Cippoletti weir, 192–93
Coanda Effect, 189
Coil curvature ratio versus critical Reynolds number, 37
Concentric, sharp-edged orifice, limitations of, 90
Concentric orifice plate, application of, 87
Conical edge orifice plate, application of, 87
Conical entrance orifice plate, 86–87
   Reynolds number for, 86

Constant-torque-hysteresis clutch-type flowmeter, 66
Coriolis flowmeters, 8, 50, 66–67
   advantages of, 62–63
   applications of, 62
   calibration of, 56
      inaccuracy in, 58, 59
      zeroing, 56, 59
      zero offset or zero stability, 59
   construction of, 53
      flow detectors, 56
      mechanical properties, 53
      sensor, 53, 54
      tube geometries, 53, 55
   electronics of
      outputs, 56
      signal processing, 56
   environmental conditions
      humidity, 61
      pipe stress, 61
      temperature, 61
      vibration, 61
   fluid properties, 59
      corrosion, 60–61
      density, 59
      erosion, 60
      plugging, 60
      pressure, 59–60
      pressure drop, 60
      temperature, 59
      viscosity, 60
   installation of, 61–62
   limitations of, 63
   operations of, 52
   theory for, 51–53
Corner taps, 85
Costs, in selection of flowmeter, 16
Critical flow nozzles, 110
Critical Reynolds number, 36
   versus coil curvature ratio, 37
Critical-velocity Venturi nozzles, 178
Cross-correlation flowmeter, 9, 22–23

Dall tube, 175–76
DC-type magnetic flowmeter, pulsed, 40–41
Detectors
   flow, 56
   jet deflection flow, 30–31, 32
   low flow rate, for gas metering, 9
   for open-channel sensors, 196
Deutsche Forschungs-und Versuchsanstalt für Luftund Raumfahrt, 105
Diaphragm meter, 107–8
Diaphragm pumps, 8, 75–76
   direct-driven, 76
   hydraulic-actuated metering pump, 76–77
   oil-driven, 76
   piston-head pump, 77
Differential pressure flowmeters
   advantages/disadvantages, 5
   elbow taps, 6–7
   energy costs of, 5–6
   orifice plates, 6
   pitot tubes, 6

   target/impact meters, 7
   Venturi tubes and nozzles, 6
Differential pressure measurement, 179
Differential pressure taps, locations for, 84–85
Doppler, Christian, 158
Doppler equation, 158
Doppler flowmeters, 158–59
   application and performance, 159
Doppler meters, 8
Double venturi, 99
   calibration of, 100
Dual-chamber gravimetric feeder, 139
Dual-frequency excitation, and magnetic flowmeters, 41
Dye tracing, 22
Dynamic solids flowmeters, 139
   accelerator-type, 140–41
   impulse-type, 139–40
   volumetric-type, 141

Eccentric orifice plate, 85–86
   application of, 87
Elbow taps, 6–7, 24
   installation of, 24–25
   selection of, 25
Electrical conductivities, 46
Electromagnetic flowmeters, 7
Electromechnanical gravimetric feeder, basic construction of, 133
Electronic BTU computers, 18
Energy audit, plant-wide, 17–18
Energy costs, of operating differential-pressure type flowmeters, 5–6

Fanning equation, 36, 37
Faraday's Law, 7, 39
Flange taps, 84, 85
Flow, conversion of units related to, 4
Flow measuring loop, for BTU flowmeters for gaseous fuels, 19–20
Flowmeters
   accuracy in, 14
   costs in, 16
   fluid duty selection table, 10
   installation requirements, 15–16
   Reynolds number in, 14–15
   and safety, 15
   selection of, 5, 9–14
Flow nozzles, 176–77
   applications, 178
   critical velocity venturi nozzles, 178
   in differential pressure measurement, 179
Flow provers, 115–16
Flow rate, relationship between, and stream velocity, 30–31
Flow sensor
   accuracy of, 180
   orientation table for, 2–3
Flow switches, 26
   design variations of, 27–28
   manual reset feature of, 26
   minimum settings for, 28
   solids, 28–29
Flow tubes, 175–75

in differential pressure measurement, 179
  pressure drop required for operation of, 5
Fluidic (Coanda effect) meter, 189
  characteristics of, 189–90
Fluidic oscillation, 7
Flume
  Leopold laggo, 195–96
  Palmer bowlus, 195
  parabolic, 195–96
  Parshall, 194–95
Folding paddle switch, 27
Four-chamber diaphragm meter, 107–8
Friction factor, in laminar flow, 37
Full circle loop with taps, 25
Full flow taps, 85

Gaseous fuels, BTU flowmeters for, 19–21
Gas metering, low flow rate detector for, 9
Gas turbine meters, 151–52
  axial flow, 152
  calibration of, 152
Glass tube rotameter, 166
Gravimetric feeder
  belt-type, 132
    electromechanical, 132–33
      feed rate of, 133–34
    digitally controlled, 136
    dual-chamber, 139
    electromechanical, 132–33
    vertical, 136–37
    vertical gate, 129–30
Gravimetric feeding system
  utilizing rotary valve volumetric feeder, 130
  utilizing variable-speed screw feeder, 130–31
  utilizing vibratory feeder, 131
Gravimetric meter, belt-type, 130
  feed rate control of, 133–34
Gravity-type variable-area flowmeters, 162–63
Gyroscopic mass flowmeter, 67

Hagen-Poiseuille equation, 35, 36
Hagen-Poiseuille Law, 33, 34
Head meters, 79–80
Head-type flowmeters, 79–80
  characteristics of
    beta ratio, 81
      density of flowing fluid, 81
      square root relationship, 81
    choice of differential pressure range, 82
    compressible fluid flow, 82
    flow noise, 83
    orifice meter, 83–85
    pulsating flow, 82–83
    Reynolds number for, 81–82
  theory of, 80
Head-type flow sensor, variable-area flowmeter as, 162–69
Heat exchangers, BTU flowmeters for, 17–18
Heat flow rate of gaseous fuel, 19
Heat-transfer flowmeters
  bypass-type designs, 70–71
  operation principle of, 68–69
Herschel, Clemens, 79
High beta flow nozzle, 177
High-precision displacement flowmeter, 115
High-precision gas flowmeter, 109
Hot-tapping procedure, 31
Hot-wire-type mass flowmeters, 71–72

Impact meters, 7
Impeller-turbine-type mass flowmeter, 65–66
Impeller-type flowmeters, 152–53
Impulse-type solids flowmeter, 139–40

Inductance pickup coil system, 145, 146
Inline ballistic flow prover, 15
Insertion-type flowmeters, 153–54
  optical photoflow sensor as, 154
  paddlewheel flowmeters as, 154
Installation requirements, in selection of flowmeter, 15–16
Integral orifice flowmeters, 87–88
Intelligent flowmeters, 159–60

Jet deflection flow detectors, 30–31, 32
  hot-tapping, 31

Kármán, Tódor von, 182
Kennison nozzle, 195–96
K factor, 146–47

Laminar flow, 9, 34
Laminar flowmeters, 33, 38
  commercially available units, 38
  design calculations for liquid service, 35
  design parameters for, 34–35
  errors in, 35–36
  Hagen-Poiseuille Law, 33, 34
  range extension techniques, 36–37
  theory of, 33–34
Leopold laggo flume, 195–96
Linear mass flowmeter, 67
Liquid PD meters, 8
Liquid sealed drum meter, 107
Liquid turbine meters, 145
  bearing turbine flowmeter, 149
  calibration curve for, 147
  electrical installation, 151
  electronic display units, 145–46
  flow capacity for, 148
  linearity and repeatability, 146–47
  mechanical installation of, 150–51
    recommended pipework, 148
    strainer recommendations for, 151
  meter characteristics and features, 149–50
  meter sizing, 147–49
  Pelton wheel meters, 149
  pressure distribution through, 148, 149
  viscosity and density effects, 147
Lobed impeller meter, 108–9
Loss-in-weight flowmeters, 137–38
  continuous, 137
  equipment, 138
    system sizing, 138–39
  weight-sensing section, 137–38
Low beta flow nozzle, 177
Low flow rate detector for gas metering, 9

Magnetic flowmeters, 39
  AC and DC excitation, 40–41
  advantages of, 48–49
  applications, 45–47
  capacity and range, 45
  construction of, 41–43
    ceramic liners, 43, 44
    magmeter electronics and intelligence, 45
    probe-type units, 43–45
  dual-frequency excitation, 41
  installation of, 47–48
  limitations of, 49
  theory for, 39–40
Manufactured gases, combustion constants and composition of, 20
Mass flow equation, 94
Mass flowmeters, 8
  angular-momentum-type, 65
  Coriolis, 8, 50–63
  linear, 67

  radiation-type, 64, 65
  thermal, 68–72
  Wheatstone-type, 8
Mechanical BTU meters, 18
Membrane pumps. See Diaphragm pumps
Metallic tube rotameter, 166
Metering pumps, 8, 73–74, 78
  diaphragm pumps, 75–76
  hydraulic-actuated, 76–77
  peristaltic pumps, 74
  piston pumps, 74–75
  plunger pump, 8
  proportioning pumps, 77–78
Meter prover, construction of, 109, 110
Microwave flow switches, 28–29
Multiple-opening pitot tubes, 100, 102
Multi-pulse shift reflection method, 156

National Bureau of Standards, 100
Natural gases, combustion constants and composition of, 20
Newton's second law of angular motion, 65
Nuclear belt scales, 137
Nutating disk meter, 111–12

Open-channel sensors, detectors for, 196
Operating cost, questions on, 16
Optical flow sensors, 154
Optical photoflow sensor, 154
Orifice(s), 79
  bore calculations for, 91, 94–95
  eccentric and segmental orifice plates, 85–86
  head meters, 79–80
    orifice meter, 83–85
  installation of, 88–90
  integral, 87–88
  limitations of, 90
  metering accuracy of, 95
  quadrant edge and conical entrance orifice plates, 86–87
  straight-run requirements, 89
Orifice fitting, 88
Orifice flowmeter, capacity table for, 92–94
Orifice meter, 83
  flow through orifice plate, 83–84
  location of pressure taps, 84–85
Orifice plates, 6
  pressure drop required for operation of, 5
Oscillating piston meter, 112–13
Oscillating vane in orifice bypass, 190–91
Oval-gear flowmeter, 114
  servo-version of, 114

Paddle-type flow switches, 27
  minimum settings for, 28
Paddlewheel flowmeters, 154
Palmer bowlus flume, 195
Parabolic flume, 195–96
Parshall, R. L., 194
Parshall flume, 194–95
Particle tracking, 22
Peristaltic pumps, 74, 77–78
Pickup coil system
  inductance, 145, 146
  reluctance, 145, 146
Pilot rotameters, 166–67
Pipe taps, 85
Piston flowmeter, 167
Piston pumps, 74–75
Pitot, Henri de, 97

Pitot tube, 6
  advantages/disadvantages of, 97
  area-averaging pitot stations, 104–5
  averaging, 100, 102
    advantage of, 102
    calculating pressure differential produced by, 103–4
    hot-tap installation of, 102–3
  calibration of, 100
  multiple-opening, 100, 102
  performance data for, 101
  pressure differential produced, 103–4
  for pulsating flow, 105
  single-ported, 99–100
  static pressure measurement, 98–99, 100
  theory of, 97–98
Pitot-type magnetic flowmeter, 42–43
Pitot Venturi, 99
  calibration of, 100
Plunger metering pump, 8
Positive displacement gas flowmeters, 107
  application notes, 109–10
  diaphragm meter, 107–8
  high-precision gas flowmeter, 109
  liquid sealed drum meter, 107
  lobed impeller meter, 108–9
  testing and calibration of, 110
Positive displacement liquid meters and provers, 111, 116–17
  accessories and intelligent electronics, 115
  flow-provers, 115–16
  high-precision displacement flowmeter, 115
  nutating disk meter, 111–12
  oscillating piston meter, 112–13
  reciprocating piston meter, 113
  rotating impeller flowmeter, 114
  rotating lobe flowmeter, 113–14
    oval-gear flowmeter, 114
  rotating vane flowmeter, 112
  viscous helix flowmeter, 114–15
Positive displacement meter, 8
  effect of viscosity variation on, 14
  pressure drop required for operation, 5
Pressure head, 80
Probe-type magnetic flowmeter, 43–45
Probe-type solids flow switches, 28
Propeller-type flowmeters, 152–53
Pulsating flow, pitot tubes for, 105
Pulsator-head pumps, 77
Pulse tracking, 22
Purge flow regulators, 118
Purge rotameters, 118
  with differential pressure regulator, 119
  with needle control valves, 118–19

Quadrant edge orifice plate, 86–87
  Reynolds number for, 86
Quadrant orifice plate, application of, 87

Radiation-type mass flowmeters, 64, 65
Radius taps, 85
Reciprocating piston meter, 113
Rectangular weirs, 193
Reluctance pickup coil system, 145, 146
Revolving plate feeder, 132
Reynolds, Osborne, 81
Reynolds number, 33, 38
  for conical entrance orifice plate, 86
  critical, 36
  for head-type flowmeters, 81–82
  for liquid flow, 34
  for quadrant edge orifice plate, 86
  in selection of flowmeter, 14–15

Robotically operated magmeter probe sensor, 196
Roll feeder, 131–32
Rotameters, 162–63
  characteristics of, 164–65
  sizing, 164
  types of, 165–66
Rotary displacement-type flowmeter, metering cycle of, 112
Rotary vane feeder, 130
Rotating impeller flowmeter, 114
Rotating lobe flowmeter, 113–14
  oval-gear flowmeter, 114
Rotating vane flowmeter, 112
Rotating vane meter, 108, 109

Safety, in selection of flowmeter, 15
Sampling-type flowmeter, 42–43
Screw feeder, 130–31
Segmental orifice plate, 85–86
Segmental square edge orifice plate, application of, 87
Segmental wedge flowmeter, 121–22
  capacities in GPM units, 122
  for clean fluid service, 121
  for corrosive/slurry service, 121, 122
Shaker feeder, 131
Short-form magnetic flowmeter, 43
Shunt flowmeter, 152, 153
Sight flow indicators, 126
  cross-section of, 124, 126
  drip-tube design, 124
  flapper design, 124
  paddle/propeller design, 124
  pressure drops on water applications, 125
Sing-around flowmeters, 156
Single-ported pitot tube, 99–100
Snell's Law, 158
Solids flowmeters and feeders, 127–28
  belt feeders, 132–33
    belt speed control, 134–35
    digital control, 135–36
    feed rate control of, 133–34
  dual-chamber gravimeter feeder, 139
  dynamic solids flowmeters, 139
    accelerator-type, 140–41
    impulse-type, 139–40
    volumetric-type, 141
  hopper and their accessories, 128–29
  loss-in-weight flowmeters, 137–38
    continuous, 137
    equipment, 138
    system sizing, 138–39
    weight-sensing section, 137–38
  nuclear belt scales, 137
  throttling flow to feeders, 129
    revolving plate feeder, 132
    roll feeder, 131–32
    rotary vane feeder, 130
    screw feeder, 130–31
    shaker feeder, 131
    vertical gate gravimetric feeder, 129–30
    vibratory feeder, 131
    vertical gravimetric feeder, 136–37
Solids flow switches, 28–29
Sonic flow nozzle, 9
Sonic Venturi flowmeter, 9
Spool piece, 156
Square edge orifice plate, application of, 87
Static pressure, measurement of, 98–99, 100
Streamline flow, 34
Stream velocity, relationship between, and flow rate, 30–31

Strouhal number, 182
Surge hopper, 128–29
Swinging vane flow switch, 27

Tagging techniques for flow measurement, 22
Tapered plug flowmeter, 167
Target meters, 7, 142
  annular orifice, 142, 143
  drag-body flowmeter, 142–43
Thermal flowmeters, 9
  categories of, 68
  heat transfer flowmeters, 68–71
  hot wire probes, 71–72
Thermal flow switch, 27
  minimum settings for, 28
Thermally isolating waveguides, 160
Transitional flow, 34
Transit-time flowmeters, 156
  application and performance, 158
  construction of, 157–58
  frequency difference type, 156
  single/multi-beam, 8
Trapezoidal weir, 192–93
Triboflow, 28
Turbine and other rotary element flowmeters, 144–45
  gas turbine meters, 151–53
  impeller and shunt flowmeters, 152–53
  insertion-type flowmeters, 153–54
  liquid turbine meters, 145
Turbine meters, 7
  effect of viscosity variation on, 14
  pressure drop required for operation of, 5
Twin-turbine mass flowmeter, 66

Ultrasonic flowmeters, 155–56
  displays, receivers, and intelligent units, 159–60
  doppler flowmeters, 158–59
    application and performance, 159
  early problems with, 156
  transit-time flowmeters, 156
    application and performance, 158
    construction of, 157–58
    frequency difference type, 156
Ultrasonic flow switch, minimum settings for, 28
Ultrasonic meters
  Doppler meters, 8
  transit-time flowmeters, 8

Valve body type flow switches, 27
Vane-type variable-area flowmeters, 168–69
Variable area flowmeter, 9, 162–63
  gates and vanes, 167–69
  gravity-type, 162–63
  piston, 167
  rotameters, 162–67
    bypass, 166–67
    characteristics of, 164–65
    pilot, 166–67
    sizing, 164
    types of, 165–66
    tapered-plug, 167
Variable-area meters, 7–8
Variable-area type flow switches, 27
  minimum settings for, 28
Variable-gate flowmeter, 167–68
Velocity head, 80
Vena contracta taps, 84–85
Venturi-Cone flowmeter, 170–71
Venturi nozzles, 6
Venturi tubes, 6, 79, 172–73

accuracy of, 178–79
advantages and disadvantages of, 179–80
classic, 173
in differential pressure measurement, 179
short-form, 173–74
    flow calculations, 175
    installation of, 174–75
Venturi-type thermal mass flowmeter, 71
Vertical gate gravimetric feeder, 129–30
Vertical gravimetric feeder, 136–37

Vibratory feeder, 131
Viscous flow, 34
Viscous helix flowmeter, 114–15
Volumetric flowmeters, 141
Vortex meters, 7, 188–89
Vortex procession, 7
Vortex shedding phenomenon, 5, 182
V-votch weirs, 193

Wall tap, for static pressure measurement, 100

Weigh hoppers, 138
Weir box, 193
Weirs, 192–94
Weymouth, 80
Wheatstone-type mass flowmeter, 8
Winter-Kennedy taps, 25
Wobble Index, 19

Zeroing of Coriolis flowmeter, 56, 59
Zero stability/offset, 59